SCHAUM'S OUTLINE OF

THEORY AND PROBLEMS

OF

INTRODUCTORY

GEOLOGY

RICHARD W. OJAKANGAS, Ph.D.

Professor of Geology
University of Minnesota

SCHAUM'S OUTLINE SERIES

McGRAW-HILL, INC.

New York St. Louis San Francisco Auckland Bogotá Caracas
Hamburg Lisbon London Madrid Mexico Milan Montreal
New Delhi Paris San Juan São Paulo Singapore
Sydney Tokyo Toronto

RICHARD W. OJAKANGAS, currently Professor of Geology, University of Minnesota—Duluth, received his Ph.D., in Geology from Stanford University, his M.A. from the University of Missouri—Columbia, and his B.A. from the University of Minnesota—Duluth. He has published about 75 papers, most dealing with the origin of sedimentary rocks. His research has taken him to Antarctica, India, Australia, USSR, Finland, Canada, as well as to several other areas in North America. He has co-authored *The Earth—Past and Present* for college use, *Challenges in Earth Science* for junior high schools, and *Minnesota's Geology*.

Schaum's Outline of Theory and Problems of
GEOLOGY

1 2 3 4 5 6 7 8 9 0 10 11 12 13 14 15 16 17 18 19 20 SHP SHP 9 2 1 0

ISBN 0-07-047704-3

Sponsoring Editor, John Aliano
Production Supervisor, Fred Schulte
Editing Supervisors, Meg Tobin, Maureen Walker
Cover design by Amy E. Becker.

Library of Congress Cataloging-in-Publication Data

Ojakangas, Richard
Schaum's outline of theory and problems of introductory geology/
Richard Ojakangas.
p. cm.—— (Schaum's outline series)
ISBN 0-07-047704-3
1. Geology——Outlines, syllabi, etc. 2. Geology——Problems, exercises, etc. I. Title. II. Title: Theory and problems of introductory geology.
QE41.038 1991
550' .2'02—dc20

89-77819
CIP

Preface

This solved-problems Outline has been written to cover the field of geology in a broad way, including both physical and historical geology. Thus it is suitable as a supplement for a general introductory geology course, a physical geology course, and a historical geology course. Although astronomy, except for our solar system, and weather/climate are not covered in this book, it should nevertheless be useful in most earth science courses as well.

Chapters 1 through 16 cover the standard topics of a physical geology course. Plate tectonic theory, the grand unifying theory of the geological sciences, is treated quite thoroughly in chapter 17 and is referred to numerous times in most other chapters. The historical geology chapters, chapters 18 through 24, provide a detailed look at the geological history of North America and numerous references to geology elsewhere in the world. Chapter 25, People and the Environment, deals with the use and misuse of earth resources by its inhabitants. The final chapter provides an introduction to the nature of the other bodies in our solar system and compares their features to earth's features and processes.

The introductory material of each chapter is by design quite brief. Not all of the chapter problems can be answered directly from the introductory material. If that were the case the introductory materials would by themselves constitute another general geology text. The solved problems themselves constitute the main teaching tool and give the book a stand-alone quality. However, its value is enhanced when used in conjunction with a standard geology text.

The problems of each chapter cover the main points of that chapter. The supplementary problems cover other important but perhaps not mainline aspects of geology. Some more detailed and complicated problems are also included in the supplementary problems. While any study guide should include some emphasis on commonly used terms, the thrust of this book is to emphasize basic geologic concepts. Concepts are always more important than definitions. As Albert Einstein said, "Ideas are much better than information." A conceptual approach will result in real learning and synthesis of processes and data, whereas the regurgitation of definitions results in the learning of isolated facts of little value, except perhaps in a geological trivia game.

During the preparation of this volume, I consulted numerous geology texts. Especially useful were *The Earth—Past and Present* by R. W. Ojakangas and D. G. Darby (1976, McGraw-Hill); *Evolution of the Earth* by R. H. Doff, Jr. and R. L. Batten (McGraw-Hill); and *Physical Geology* by L. D. Leet, S. Judson, and M. E. Kauffman (1978, Prentice-Hall). Additional useful texts were *The Earth's Dynamic Systems* by W. K. Hamblin (Burgess), *Essentials of Geology* by F. K. Lutgens and E. J. Tarbuck (Merrill), and *Plate Tectonics for Introductory Geology* by J. R. Carpenter and P. M. Astwood (Kendall-Hunt). This book will be a complement to these texts as well as to numerous other good texts that are available today.

RICHARD W. OJAKANGAS

Contents

Chapter 1

Earth: The Big Picture

CONTINENTS

The earth is a unique planet, with 71 percent of its surface covered by water. The seven continents extend under the oceans, encompassing the continental shelves and the continental slopes. Thus, the shelves and the slopes are parts of the landmasses rather than of the ocean basins.

On the continents, the mountain ranges form the most spectacular topographic features. Plateaus, generally of medium elevation, and the plains, generally of the lowest elevation and the lowest relief, are the other two prominent features on the land surface. The highest mountain ranges are generally located around the Pacific Ocean or lie along an east-west line between Africa, Europe, and Asia.

OCEAN BASINS

The ocean basins, until the 1940s, were thought to be deep and rather featureless. Oceanographic studies have revealed a vast network of midoceanic ridges such as the Mid-Atlantic Ridge and off-center oceanic rises such as the East Pacific Rise, both of which, if on land, would be prominent mountain ranges. These are zones of active volcanism and faulting (breaking and moving) of the earth's crust or exterior layer.

Around the rim of the Pacific Ocean basin are numerous *island arcs* such as the Aleutian Islands, Japan, and the Philippine Islands. These are zones of active volcanism, and similar zones lie adjacent to the Pacific along the west coast of South America (the Andes) and western North America (the Cascades). Seaward of the arcs are *oceanic trenches,* long and narrow troughs that contain the deepest points in the oceans. The ridges, island arcs, and trenches make up the major features of the ocean basins. Other features are *active volcanoes* (such as Mauna Loa on Hawaii), *seamounts* (extinct volcanoes), *guyots* (seamounts with flat, eroded tops), *escarpments* (clifflike features perpendicular to the ridges and rises), and *submarine fans* comprising the thick sediment accumulations called *continental rises* at the bases of the continental slopes. Cutting the slopes are numerous *submarine canyons.*

EARTH'S INTERIOR

The nearly spherical earth consists of a very thin crust (8 to 35 km thick), a thick mantle (about 2900 km thick), a fluid outer core (2300 km thick), and a solid inner core (radius of about 1200 km). The crust and the mantle are made of rock material, and both parts of the core are largely made of iron.

The earth's overall density is 5.52; since crustal rocks have densities of about 2.6 (the granitic rocks of the continents) to 3.0 (the basalts of the ocean basins), the earth must have a dense interior.

Solved Problems

1.1 What are the three "spheres" at or near the earth's surface, excluding the biosphere?

Atmosphere, hydrosphere, and lithosphere.

1.2 How thick is the atmosphere?

About 10,000 km. At that distance above the earth's surface, hydrogen density is about equal to that in interplanetary space. But 90 percent of the atmosphere is in the lowest 18 km and 99.99 percent is within 50 km of the surface (Fig. 1.1).

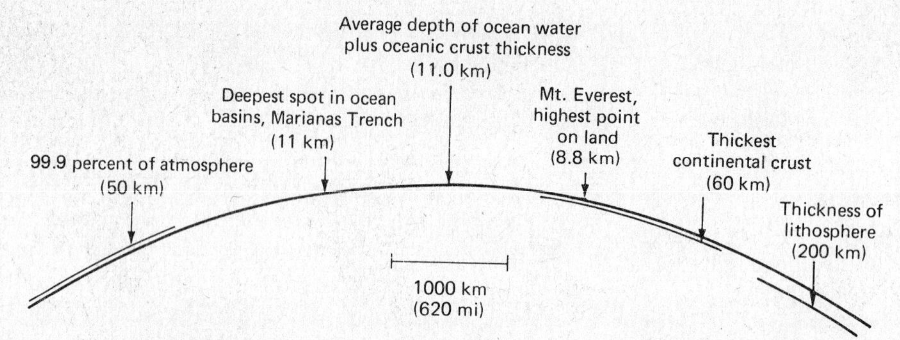

Fig. 1.1 Earth's surface, crust, and lithosphere, drawn approximately to scale. Most features are encompassed within the thickness of a drafted line. (*From Ojakangas and Darby, 1976.*)

1.3 Where is most of the water of earth's hydrosphere found?

97 percent is in the ocean basins, which have an average depth of about 4000 m.

1.4 What percent of the earth's surface is covered by oceans and seas?

71 percent.

1.5 What is the *lithosphere?*

The lithosphere is the "rock" sphere at the earth's surface. It extends to a depth of about 200 km. At the surface, it consists of many rock types, but basalt and granite are dominant. The lithosphere is defined more precisely in Chapters 15 and 17. See Fig. 1.1.

1.6 What are the major large-scale topographic features on the continents?

Mountain ranges (see Fig. 1.2), plateaus, and plains.

1.7 When the earth's surface is drawn to scale, are the mountains prominent features?

No. See Fig. 1.1.

1.8 What are the two major large-scale, high-relief topographic features on the ocean floors?

Ridges (or oceanic rises) and trenches. (See Fig. 1.2.)

1.9 What relationships do earthquakes and volcanoes have to ridges and trenches?

Most are closely related. This will be explained in more detail in Chapters 14 and 17.

1.10 What is the common spatial relationship of volcanic island arcs (such as the Aleutian Islands and the Philippine Islands), volcanic arcs on land (such as the Andes of South America and the Cascades of North America), and oceanic trenches?

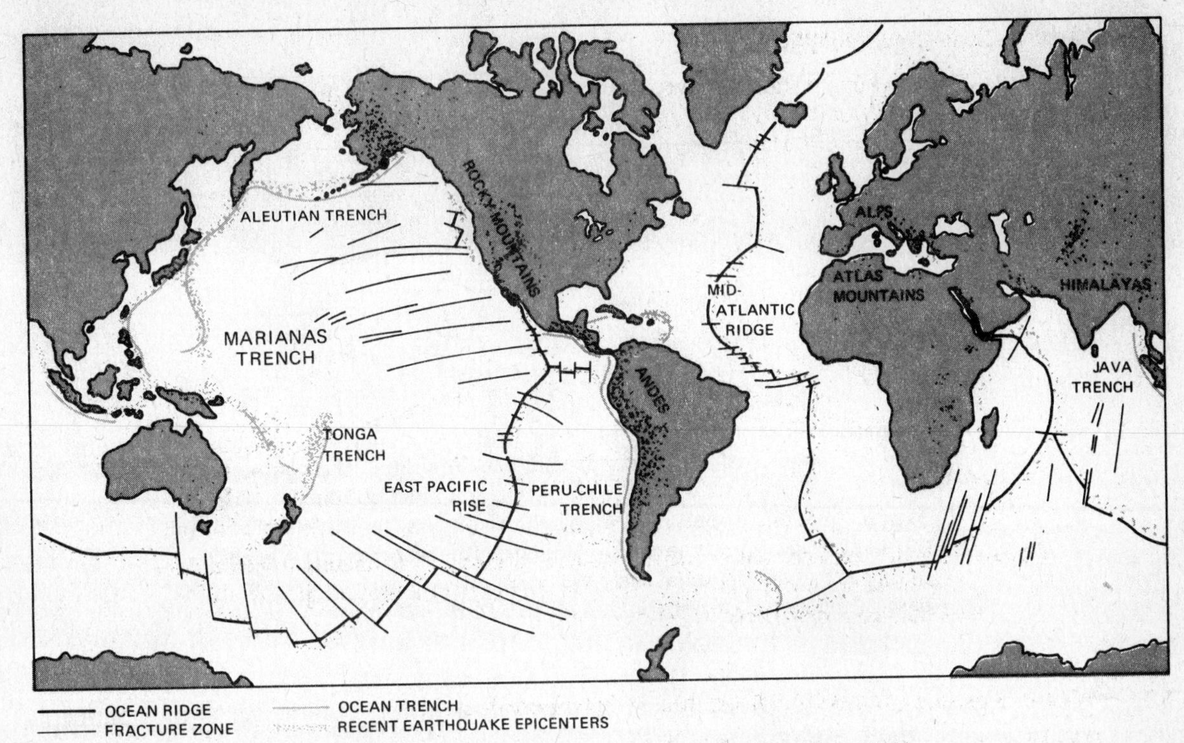

Fig. 1.2 Major surface features of the world. (*From Ojakangas and Darby, 1976.*)

Island arcs, as well as the arcs on the continents, are landward of and parallel to the trenches. (See Fig. 1.2.)

1.11 What are the relationships of the continental shelves, continental slopes, and the deep oceanic basins? (See Fig. 1.3.)

The shelves border the exposed continents and are actually the submerged portions of the continents. The slopes are located between the shelves and the deep ocean basins.

1.12 How wide are the continental shelves?

They vary considerably. On the West Coast of the United States, for example, they are only a few kilometers wide, whereas off the East Coast, they are as wide as 350 km.

1.13 Are the shorelines of the continents the edges of the continents?

No, the true limits of the continents are at the bases of the continental slopes.

1.14 What are the angles of slope, compared to a horizontal surface, of the continental shelves and the continental slopes?

The shelves usually dip oceanward at angles of much less than 1°, and the continental slopes usually dip oceanward at angles of less than 5°.

1.15 Why are continental slopes always depicted in diagrams as quite steep, about 45 to 60°, as in Fig. 1.3?

Because vertical exaggeration in scale is necessary in order to show the gradual slopes as topographic features. Note the differences in horizontal and vertical scales of Fig. 1.3.

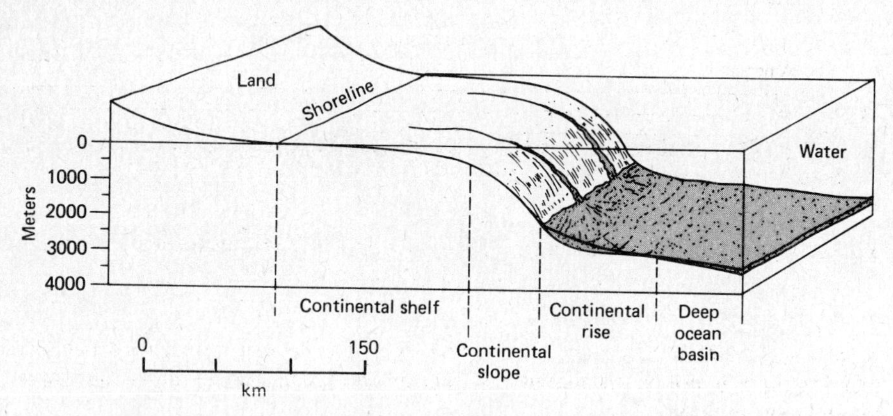

Fig. 1.3 Diagrammatic representation of major features beneath the ocean at the edge of a continent. The vertical scale is exaggerated; the continental shelves are very gently inclined surfaces, and most continental slopes have inclinations of only 3 to 6°. The horizontal scale is generalized, for the shelf varies greatly in width. Water depths also vary, but general figures are about 180 m at the outer edge of the shelf and 1800 m at the base of the slope. (*From Ojakangas and Darby, 1976.*)

1.16 What are general figures for the depths of ocean water at (*a*) the outer edges of the continental shelves and (*b*) the bases of the slopes?

(*a*) 180 m (600 ft). (*b*) 1800 m (6000 ft).

1.17 What are *continental rises?*

Thick wedges of sediment deposited at the bases of the continental slopes (Fig. 1.3). Many appear to be coalescing submarine fans.

1.18 Where is sediment found in the oceans, and how thick is it?

All over the ocean floor, but it is very thin on the deep ocean floor (a few hundred meters), very thick at the bases of the slopes (as thick as 10,000 m), and again thinner on the shelves (as thick as a few thousand meters).

1.19 What major features are found as deep "cuts" on the continental slopes? Give an example.

Submarine canyons (Fig. 1.3). A well-studied example is the Hudson Submarine Canyon, on the slope opposite the Hudson River and New York City. It has a slope of 4 to 7°, is 600 to 1200 m (2000 to 4000 ft) deep, and is more than 320 km (200 mi) long.

1.20 How did submarine canyons originate?

There are two general theories. One is that they are related to lower stands of sea level during the ice age, and the other is that they were cut by turbidity currents (sediment-laden density currents) beneath sea level.

1.21 What is the Mid-Atlantic Ridge?

A broad ridge that is a zone of volcanism and faults (cracks along which there has been movement). It approximately bisects the Atlantic Ocean (see Fig. 1.2). It will be described in more detail in Chapter 17.

1.22 What is the relief on the Mid-Atlantic Ridge? See Fig. 1.4.

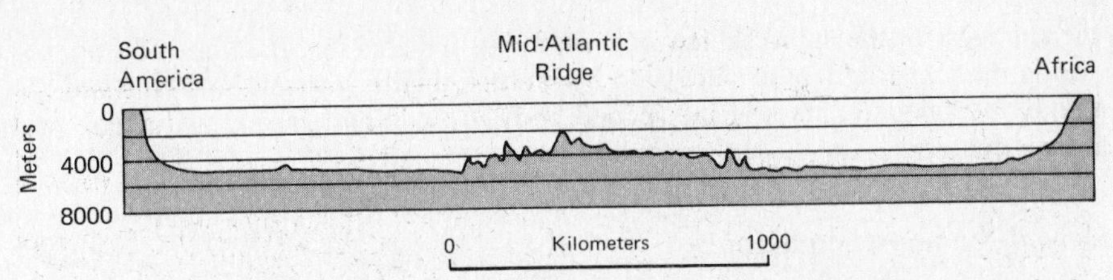

Fig. 1.4 Topographic profile of the Mid-Atlantic Ridge. Note the great vertical exaggeration. (*From Ojakangas and Darby, 1976.*)

It rises 3600 m (12,000 ft) above the adjacent abyssal plain (ocean deeps). However, its width is more than 1000 km, so it is not as prominent a feature as one might imagine based on diagrams that have a high vertical exaggeration.

1.23 Is there any expression of the Mid-Atlantic Ridge above sea level?

Yes. Iceland and the Azore Islands are examples of volcanoes that have reached the surface.

1.24 Where are the deepest points in the oceans?

In the trenches. The greatest depth is thought to be 11,040 m (36,198 ft) below sea level, in the Marianas Trench (Fig. 1.2). For contrast, the highest point on land is the tip of Mt. Everest, 8853 m (29,028 ft) above sea level.

1.25 What are *seamounts* and *guyots?*

Seamounts are extinct volcanoes on the ocean floor. Guyots are flat-topped seamounts, with the tops of the volcanoes removed by erosion when they were situated at or near sea level.

1.26 Are the rock types that make up the continents and the ocean basins the same?

No. In general, the continents are made of granitic rocks with a density of about 2.6 to 2.7, and the ocean basins are floored by basalt with a density of 3.0.

1.27 Was Christopher Columbus the first person to say that the earth was "round" rather than flat?

No. As early as 500 B.C., the Greeks had determined that the earth is round.

1.28 How did the Greeks determine that the earth was round?

By observing the earth's shadow on the moon during eclipses of the moon, ships coming into sight over the horizon (with the tips of their sails visible first), and mountain tops becoming visible before the bases when moving toward them.

1.29 Is the earth spherical?

Not really, although it is nearly so. It has an equatorial bulge due to the earth's rotation, with the equatorial diameter of 12,757 km (7927 mi), 43 km longer than the polar diameter of 12,714 km (7900 mi), making it slightly "pear-shaped." It is generally described as an oblate ellipsoid. Yet it is still more spherical than a bowling ball. (See Fig. 1.1.)

1.30 What is the significance of the following equation?

$$F = \frac{Gm_1m_2}{d^2}$$

This is the equation for the universal law of gravitation. It states that the force of attraction between any two bodies in the universe can be calculated by multiplying the gravitational constant G by the mass of the first body m_1 times the mass of the second body m_2, and dividing by the square of the distance d^2 between their centers.

1.31 Who derived the above equation?

Sir Isaac Newton, in 1664.

1.32 Where would the force of gravitational attraction between the earth and a basketball be the greatest—on the equator or at the North Pole?

At the pole. This can be determined by studying the equation of Problem 1.30. The basketball would be m_1, and the earth would be m_2. G is a constant. The distance d between the center of the basketball and the center of the earth would be 22 km shorter when the ball is at the pole than when it is at the equator. As d is a term in the denominator, the force of gravitational attraction F will become larger as d becomes smaller.

1.33 If the earth's volume is 1.087×10^{27} cm^3 and its mass is 6×10^{27} grams, what is the density of the earth?

Density is defined as the mass per unit volume, or $d = m/v$. Therefore, $d = 6 \times 10^{27}$ g/1.087×10 cm^3, or 5.52 g/cm^3. This is 5.52 times as dense as an equal volume of water which has a density of 1.0.

1.34 How do scientists know that the earth is denser at depth than at the surface?

Because rocks at the surface have densities of only 2.6 (granite) to 3.0 (basalt) g/cm^3, whereas the density of the earth as a whole is 5.52. Therefore, the earth must have a denser interior.

1.35 What is the general nature of the earth's interior?

It consists of a thin rocky crust, a thick rocky mantle, and a thick metallic iron (plus nickel and silicon) core (Fig. 1.5). See Chapter 15 for more details. Note the thickness of the crust on Fig. 1.1.

1.36 Meteorites provide scientists with clues to the nature of the earth's interior. How?

Meteorites apparently come from our own solar system and consist of three main types—stony, stony-iron, and iron. They are interpreted to represent the materials of the terrestrial planets such as earth.

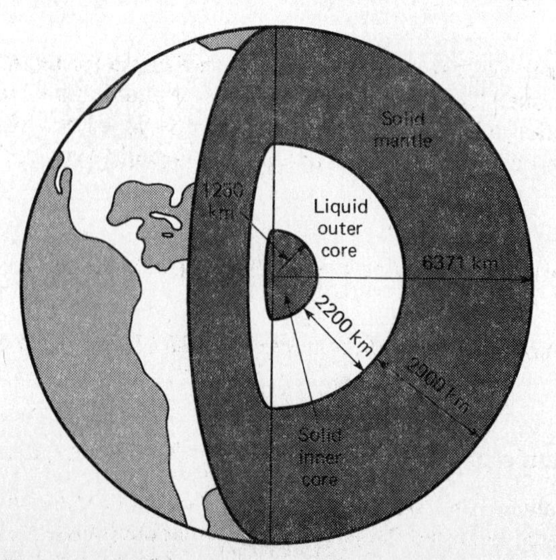

Fig. 1.5 Model of the interior of the earth.

Supplementary Problems

1.37 Is the magnetic field of the earth (Fig. 1.6) generated internally or externally?

It is known that only about 5 percent of the field is external; this part is the result of changes in the electric currents in the upper atmosphere and is related to the sun's activity. The large part, 95 percent, is internal.

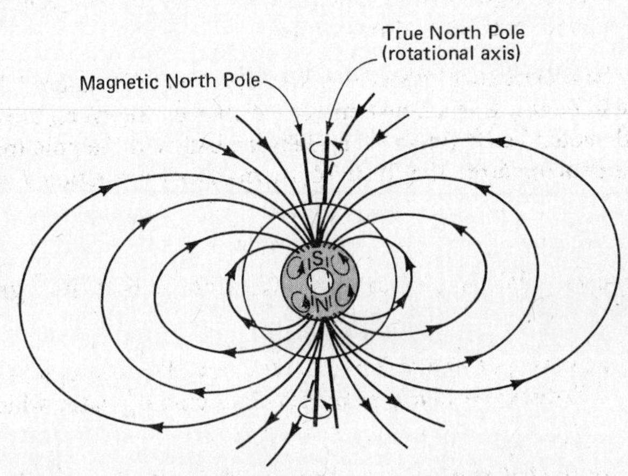

Fig. 1.6 Earth's magnetic field. (*From Ojakangas and Darby, 1976.*)

1.38 What is the evidence that the earth's magnetic field is not emanating from permanently magnetized material, even though it might seem that the earth has a strong bar magnet at its center?

A strong magnetic field, such as the earth's, must be due to magnetized metal and therefore is generated in the earth's core (Fig. 1.6). However, it cannot be coming from a solid metallic bar magnet at the center of the earth, for the following reasons. The interior of the earth is estimated to be much hotter than the temperature at which iron and nickel lose their magnetism (760°C or 1400°F and 349°C and 660°F, respectively). Furthermore, the melting point of iron is about 2000°C, and the outer core of the earth is thought to be molten.

1.39 If earth's magnetism is not derived from a permanently magnetized solid metallic core, what causes the magnetism?

Various lines of evidence suggest a deep-seated internal origin for the magnetic field. It probably emanates from the earth's outer core, which apparently is a liquid and which may be generating the magnetic field (in conjunction with the rotation of the earth), as eddies of molten material are moved by thermal convection. In effect, the earth is acting as a gigantic dynamo, converting mechanical energy into electric energy.

1.40 Drill holes into the earth's crust indicate a relationship between temperature and depth. What is this relationship?

The earth's crust gets hotter with depth, at a rate of about 1°C per 30 m of depth. This is the *geothermal gradient*.

1.41 At the rate of temperature increase given in Problem 1.40, how hot is the center of the earth?

The earth's radius is about 6363 km (see Problem 1.29), or 6,363,000 m. At 1°C per 30 m, the temperature at the center would be about 212,000°C! Other evidence suggests that it is probably on the order of 2800°C (5000°F). Therefore, the rate of temperature increase is not uniform to the center of the earth.

1.42 Calculate the vertical exaggeration of Fig. 1.3.

The vertical exaggeration is calculated by dividing the horizontal scale by the vertical scale. (Measure the same distance on each scale and divide.) On this figure, 1 cm is 2000 m (that is, 2 km) on the vertical scale and 1 cm is 50 km on the horizontal scale. Therefore, 50/2 = 25X. (The exaggeration can also be calculated if verbal scales are given, such as 1:50,000 or 1:2,000.)

1.43 Calculate the vertical exaggeration of Fig. 1.4.

On the horizontal scale, 1 cm = about 300 km. On the vertical scale, 1 cm = about 7700 m (that is, 7.7 km). Therefore the approximate vertical exaggeration is 300/7.7 or about 39X.

1.44 The size of the earth was probably first calculated by Erastosthenes, a Greek living in Egypt, in 250 B.C. Figure 1.7 shows the elements he used in his calculation. How could he have arrived at a circumference of about 40,000 km, not far off from today's accepted value of 40,077 km, for the equatorial circumference?

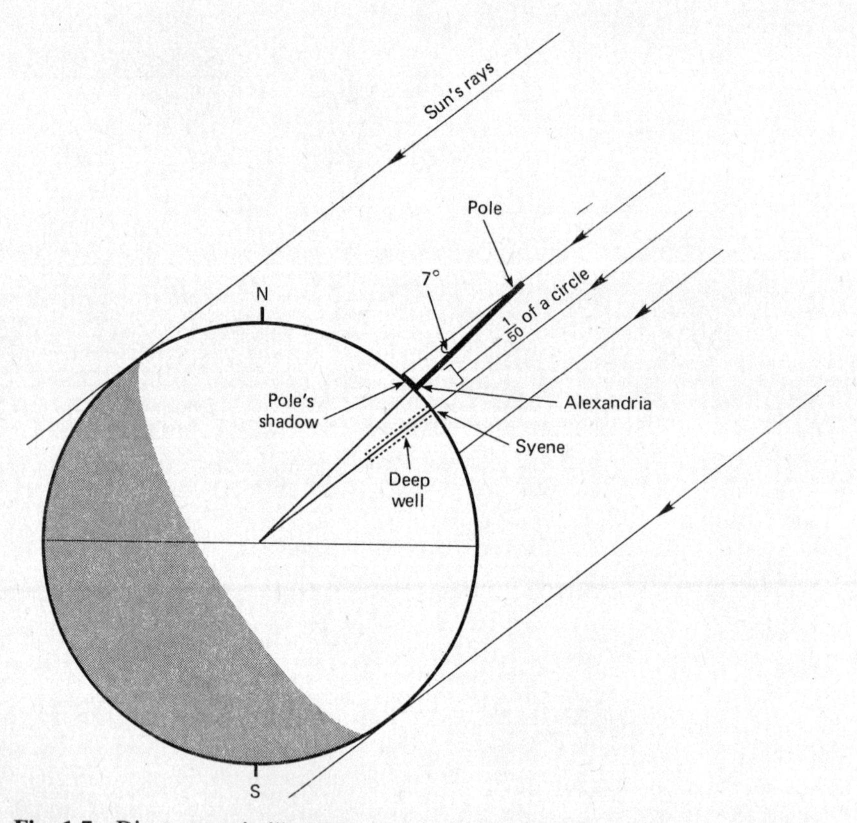

Fig. 1.7 Diagrammatic illustration of the elements used by Eratosthenes in his calculations of the size of the earth in 250 B.C.

He assumed that the earth was spherical. He reasoned that because the sun always appears to be the same size, regardless of where on earth the observer is located, the sun is very distant from the earth and its rays are therefore essentially parallel lines of light when they reach earth. He reasoned further that as the sun's reflection was seen in the bottom of a deep well at noon on a certain day in Syene (now Aswan), the sun must be directly overhead. On the same day to the north in Alexandria, the sun's rays made an angle of about 7.2° with a vertical column, as determined by the length of the column's shadow. Knowing the distance between Syene and Alexandria (based on the rate of travel of camel caravans), he calculated that the distance between the two points was an arc equivalent to about 1/50 of a full circle. (The distance is equivalent to about 800 km.) Finally, 50 × 800 = 40,000 km.

1.45 How was it first determined that the earth is not perfectly spherical?

In 1671, the French astronomer Jean Richter traveled from Paris (49°N latitude) to French Guiana on the northeast coast of South America (15°N latitude) and noted that his very accurate pendulum clock was losing 2½ min a day (based on astronomical observations). As a pendulum is affected by the gravitational attraction of the earth, he concluded that the attraction must be less near the equator. Newton in 1686, using his laws of gravitation (see Problem 1.30) and motion, showed that the earth is an oblate ellipsoid, as described in Problem 1.29. The centrifugal force of rotation causes the equatorial bulge.

Chapter 2

Atoms and Minerals: The Small Picture

ATOMS

All physical things consist of matter in the form of a solid, liquid, or gas. Matter occupies space, has mass (quantity), and consists of small particles called *atoms* (from the Greek word for "indivisible"). There are 92 naturally occurring elements (13 others have been created in nuclear reactors), ranging from hydrogen (the lightest) to uranium (the heaviest). Atoms are small—there can be 100 million side by side in a linear inch. In spite of the meaning of the word *atom,* atoms are divisible and consist of 3 main smaller particles—positively charged *protons,* negatively charged *electrons,* and uncharged *neutrons* (probably consisting of a proton and an electron), and about 30 other still smaller particles such as neutrinos and antineutrinos.

In the most straightforward atomic model (and it *is a model* that may well change in the future), protons and neutrons form the nucleus and electrons orbit in three-dimensional energy-level shells and subshells around the nucleus at nearly the speed of light. Some scientists speak of an "electron mist," for it is not possible to determine where a given electron is at a given instant of time. About 99.95 percent of the mass but only 1/1 billionth of the volume of an atom is in the nucleus. In each atom, the number of protons is equal to the number of electrons; therefore, an atom is electrically neutral. The number of protons defines the *atomic number* of an element (1 for hydrogen and 92 for uranium), and the number of protons plus neutrons defines the *mass number* or the approximate *atomic weight* of an element.

ISOTOPES

Whereas a given element always has a specified number of protons, the number of neutrons can vary, creating atoms which have identical physical and chemical properties but which have different atomic weights; these atoms are *isotopes* of that element, and most elements have two or more isotopes.

Some isotopes are radioactive; spontaneous changes occur in the nucleus of the atom and new elements are the result. (In normal chemical reactions, changes occur only in the electron shells.) These nuclear reactions are of two types, fusion and fission. The amount of energy given off by a nuclear reaction is very large and is given by Einstein's equation, $E = mc^2$. The logical explanation for the origin of the elements in the first place is by nuclear reactions, probably in dense stars where the temperature is several million degrees Celsius and where neutrons and protons are moving so fast that, when they hit each other, they fuse together. In such a process, the stars give off immense amounts of energy.

IONS

The outer energy-level shells of atoms can contain a maximum of eight electrons. Atoms "strive toward stability," most attempting to attain eight electrons in their outer shells. This is accomplished, e.g., by adding an electron if the atom has seven electrons in its outer shell, or, e.g., by giving up one electron if it has only one in its outer shell. If electrons are gained or lost, an atom is no longer electrically neutral, and it becomes a charged atom or *ion*. Positively charged ions are *cations,* and negatively charged ions are *anions*. Because opposite charges attract, oppositely charged ions can be attracted to each other, combine, and form minerals. This bonding of oppositely charged ions is called *ionic bonding*. The other important type of bonding is *covalent bonding,* in which atoms attain eight electrons in their outer shells by sharing electrons with adjacent atoms, rather than by giving up or gaining electrons. Ionic and covalent bonds are the most important types of bonding in minerals. There are also much weaker bonds between ions or atoms in solids; these are called *van der Waals bonds,* and they are weak electrical attractions. *Metallic bonding* is present in metals and is related to electrons that are moving through the metal, thereby allowing the metal to have an electrical balance.

MINERALS

A *mineral* is a naturally occurring, homogeneous, inorganic crystalline substance with definitive chemical and physical properties. *Crystalline* means that the ions are combined in a definite geometric pattern or crystal structure. Assuming that many different kinds of ions are present, ionic size and charge determine which ions go into making up a given crystal structure. Table 2.1 is a list of the common ions with their charges and ionic radii (in angstroms; 1 Å = 1×10^{-8} cm, or 0.00000001 cm).

Table 2.1 Radii and Charges of Common Ions

Ion	Radius	Element
K^{1+}	= 1.33 Å	(potassium)
Ca^{2+}	= 0.99 Å	(calcium)
Na^{1+}	= 0.97 Å	(sodium)
Si^{4+}	= 0.42 Å	(silicon)
Al^{3+}	= 0.51 Å	(aluminum)
Mg^{2+}	= 0.66 Å	(magnesium)
Fe^{2+}	= 0.74 Å	(iron)
O^{2-}	= 1.32 Å	(oxygen)

Only 8 of the 92 naturally occurring elements are abundant in the earth's crust (Table 2.2). The elements combine to make the minerals; more than 2000 are known but only a few are common. The 8 minerals in Table 2.3 make up more than 95 percent of the rocks in the crust. Note that these main minerals are all *silicates,* minerals made of silicon, oxygen, usually aluminum, and other positively charged ions (cations). The basic "building block" of the earth's crust is the *silicon-oxygen tetrahedron* of four oxygens with one silicon between the four oxygens. Other main mineral groups beside the silicates are the oxides, carbonates, sulfides, sulfates, and chlorides.

Table 2.2 The Common Elements in Earth's Crust

Element	Approximate Abundance by Weight (Percent)
Oxygen (O)	46.6
Silicon (Si)	27.7
Aluminum (Al)	8.1
Iron (Fe)	5.0
Calcium (Ca)	3.6
Sodium (Na)	2.8
Potassium (K)	2.6
Magnesium (Mg)	2.1
	98.5

Table 2.3 The Common Crust-Forming Minerals

Mineral	Formula
Muscovite mica	HKAlSiO
Biotite mica	HKFeMgAlSiO
Hornblende	HKNaCaMgFeAlSiO
Pyroxene	CaMgFeAlSiO
Orthoclase feldspar	KAlSiO
Plagioclase feldspar	CaNaAlSiO
Olivine	(Fe, Mg)SiO
Quartz	SiO_2

*For simplicity, formuli are not actual quantitative formuli.

A few minerals, such as quartz, form under various conditions of temperature and pressure, and thus have a wide stability range. However, most minerals form under more specific conditions and therefore have narrower stability ranges. When the conditions change, the minerals change too, to minerals that are stable under the new conditions.

ROCKS

Aggregates of minerals (and coal or glass) are *rocks*. There are three generic classes of rocks—igneous, sedimentary, and metamorphic. *Igneous rocks* crystallize from magmas; *sedimentary rocks* are formed from sediment (pieces of other rocks) or are precipitated out of water; and *metamorphic rocks* are formed from other rocks by changes in temperature, pressure, and solutions. Because minerals change under different conditions, so do the rocks.

Solved Problems

2.1 Describe the makeup of the two simplest atoms, hydrogen (atomic number 1) and helium (atomic number 2).

See Fig. 2.1. Hydrogen has a nucleus of one proton and has one electron orbiting in a three-dimensional "shell." Helium has a nucleus of two protons and two neutrons and has two electrons orbiting in a single three-dimensional "shell."

2.2 How much of the mass and how much of the volume of an atom are in the nucleus?

99.95 percent of the mass but only 1/1 billionth of the volume is in the nucleus of an atom. Therefore, an atom is mostly space. An atom has been loosely compared to our solar system, with a nucleus (the sun) and electrons (the planets) in orbit around it. Obviously, the comparison is not a perfectly good one, for electrons move in three-dimensional shell-like orbits rather than two-dimensional planar orbits as do the planets.

2.3 What is the ratio of protons to electrons in an atom, and what is the net electric charge of an atom?

The number of protons is equal to the number of electrons. Since each proton has a positive charge of 1 and each electron has a negative charge of 1, an atom is electrically neutral.

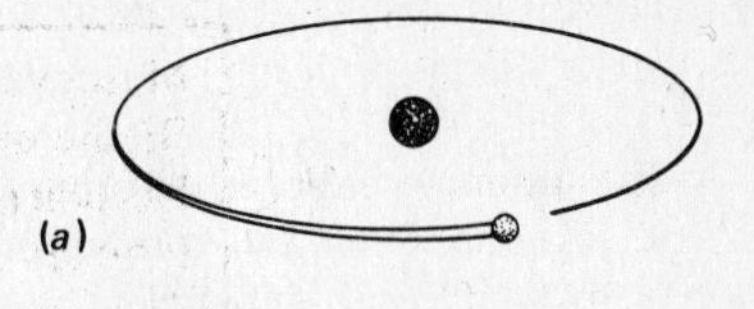

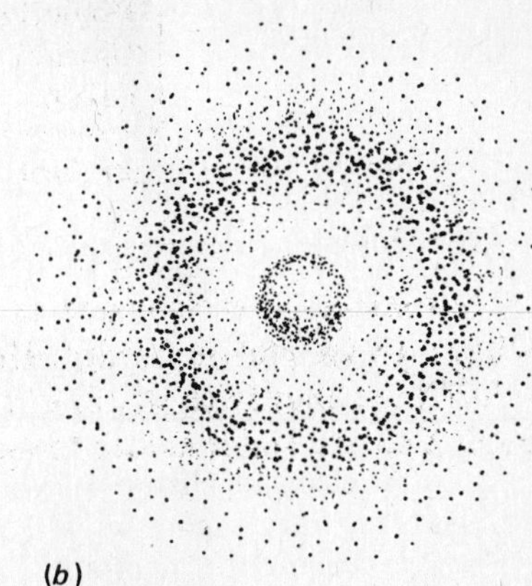

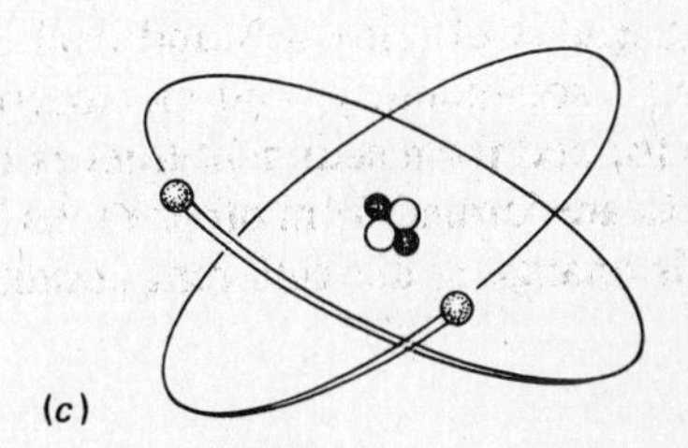

Fig. 2.1 Models of simple atoms. (*a*) Hydrogen, the simplest atom, with a nucleus of only one proton and with one electron in a three-dimensional spherical orbit around the nucleus. (*b*) Hydrogen with neither the nucleus nor the electron shown as a distinct particle. (*c*) Helium, the next simplest atom, with a nucleus of two protons and two neutrons; two electrons orbit in a single three-dimensional shell. (*From Ojakangas and Darby, 1976.*)

2.4 How is the atomic number of an atom determined?

It is equal to the number of protons.

2.5 How is the atomic weight (mass number) of an atom determined?

It is equal to the number of protons plus the number of neutrons.

2.6 What is an isotope?

Atoms of a given element which differ in atomic weight but still have the same number of protons and the same physical and chemical properties are isotopes of that element.

2.7 There are three types of oxygen—^{16}O, ^{17}O, and ^{18}O. What are these varieties called?

Isotopes. ^{16}O is the common isotope, with an atomic weight of 16.

2.8 Hydrogen has three isotopes—^{1}H, ^{2}H, and ^{3}H. How do they differ?

The nucleus of ^{1}H consists of one proton, the nucleus of ^{2}H consists of one proton and one neutron, and ^{3}H has one proton and two neutrons. Yet all three are hydrogen atoms. ^{1}H is the common isotope. ^{2}H and ^{3}H are deuterium and tritium, respectively, and are also known as "heavy hydrogen."

2.9 Uranium 238 (^{238}U), the heaviest naturally occurring element, has an atomic number of 92. How many protons, neutrons, and electrons does it have?

The atomic number is determined by the number of protons, and since the atomic number is 92, it has 92 protons. Since the number of protons is equal to the number of electrons in an atom, it has 92 electrons (in several shells). The mass number, 238, is determined by the number of protons and the number of neutrons. Therefore, 238 minus 92 gives the answer of 146 neutrons.

2.10 Uranium has another isotope with an atomic weight of 235 (^{235}U). How many protons, neutrons, and electrons does it have?

By the same reasoning as in Problem 2.9, the answer is 92, 143, and 92, respectively.

2.11 Some isotopes undergo radioactive decay and are called radioactive isotopes. For example, ^{238}U (uranium) decays to ^{206}Pb (lead). How does this happen?

Heavier elements change to lighter elements by emitting from the nucleus protons, neutrons, electrons, and short-wavelength energy. Since the atomic number of uranium is 92 and the atomic number of lead is 82, uranium loses 10 protons. As the difference in atomic weight is 32, the number of neutrons lost by uranium is 32 minus 10, or 22. This occurs in a number of steps. The number of electrons lost is not important here. The energy will be discussed below.

2.12 The process described in Problem 2.11 is a nuclear reaction because it involves changes in the nucleus. Is it nuclear fusion or nuclear fission?

Nuclear fission, for it involves "splitting" the uranium atom to make a lighter element.

2.13 What are the differences in the two nuclear processes, fission and fusion?

Fission is the "splitting" of an atomic nucleus, making lighter elements out of heavier ones, as in Problem 2.11. *Fusion* is the joining of lighter atomic nuclei to make a heavier nucleus, as when four hydrogen nuclei (each consisting of one proton) join to make up one helium nucleus which consists of two protons and two neutrons. (See the following problem.) Scientists actually use two deuterium (heavy hydrogen) isotopes, each with a nucleus consisting of one proton and one neutron.

2.14 What is a neutron?

In our model of the atom, a neutron can be thought of as consisting of one proton with a charge of +1 and one electron with a charge of −1, and thus a neutron is electrically neutral. To make a proton from a neutron, an electron must be ejected from the nucleus—this is a beta particle, not to be confused with the electrons in the shells of atoms.

2.15 What is the significance of Einstein's equation, $E = mc^2$?

This is the equation for atomic energy. It shows that a great amount of energy E is produced by the transformation of a mass m into energy. The amount of energy is equal to the "transformed" mass times the square of the velocity of light. Although the transformed mass is a very small quantity, the velocity of light is a very large number (300,000 km/s or 3×10^{10} cm/s or 186,000 mi/s).

2.16 Explain the statement, "Energy and mass are two aspects of the same thing."

Energy and mass can be equated by $E = mc^2$. Mass cannot be destroyed, but it can be converted to energy. And energy can be converted into mass.

2.17 When four hydrogen nuclei, each composed of one proton, join to make one helium nucleus of two protons and two neutrons (as in Problem 2.13 above), the mass of the helium nucleus is calculated to be less than the mass of the original four protons, by the amount of 0.0029 mass units. What is the significance of such a mass "loss"?

This is the amount of mass transformed to energy in the equation $E = mc^2$. It shows how energy is generated by nuclear fusion, as in the hydrogen bomb. It is hoped that the process can be perfected and contained so as to be a source of nuclear power.

2.18 How are elements probably formed?

By nuclear fusion reactions (stellar nucleosynthesis in stars), as in the previous problem. In dense stars, the temperature is several million degrees Celsius. Protons and neutrons are moving very fast, colliding, fusing together, and giving off vast amounts of energy in the process, in much the same manner as in Problem 2.17.

2.19 What is the source of the sun's energy?

Nuclear fusion. Although the temperature of the sun's surface is estimated at only 5500°C, it is estimated at 14 million degrees centigrade at its center. Hydrogen (more than 80 percent of the sun's mass) is being converted to helium, which makes up most of the remainder of the mass. (See Problem 2.17.)

2.20 What is an *ion?*

An atom with an electric charge, either + or −, called a *cation* or an *anion,* respectively. (Recall that an atom per se is electrically neutral, for the number of negatively charged electrons is equal to the number of positively charged protons.)

2.21 What is a mineral?

A *mineral* can be defined as a naturally occurring, homogeneous, inorganic, crystalline substance with relatively definite chemical and physical properties.

2.22 Why are ions important geologically?

It is the charged atoms, or ions, that take part in chemical reactions and form the minerals.

2.23 The electrons of atoms are arranged in three-dimensional energy-level "shells" or spheres. What is the stable maximum number of electrons in the outer energy-level shell of an atom?

Eight.

2.24 Atoms "strive toward stability" by trying to fill the outer energy-level shell to the maximum stable number of 8. This is commonly illustrated by the bonding of sodium and chlorine atoms (Fig. 2.2). Explain how this happens.

When a sodium atom, with one electron in its outer shell, gives its outer lone electron to a chlorine atom that has seven electrons in its outer shell, the result is that both have eight electrons in their outer shells. Each is "satisfied."

2.25 Refer to Fig. 2.2. After sodium gives an electron to chlorine, are they still electrically neutral atoms?

No. They are now charged atoms or ions, Na^{+1} and Cl^{-1}, because sodium now has 11 protons and only 10 electrons, for a net charge of +1, and chlorine now has 17 protons and 18 electrons, for a net charge

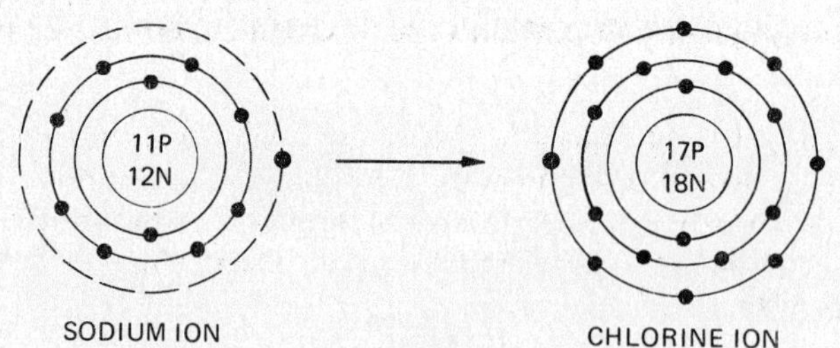

Fig. 2.2 Bonding of sodium and chlorine atoms. When a sodium atom gives its outer electron to a chlorine atom, both end up with eight electrons in their outer shells. As neither atom now has as many electrons as protons, both become charged (sodium now has a positive charge and chlorine a negative charge) and are called *ions*. (*From Ojakangas and Darby, 1976.*)

of −1. Oppositely charged ions are attracted to each other, and the chlorine and sodium ions thus combine into the electrically neutral compound NaCl. The ions have lost their identities.

2.26 What is the compound that results from the chemical reaction illustrated by Fig. 2.2? See Fig. 2.3.

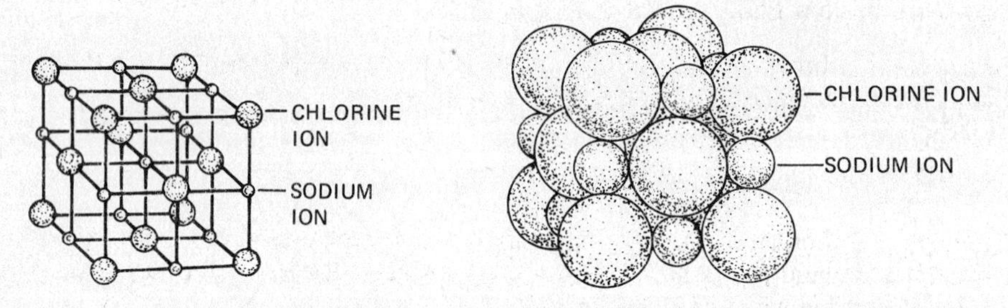

Fig. 2.3 Crystal structure of the mineral halite (sodium chloride, or common table salt). The left diagram is an "exploded" model, permitting us to see into the crystal structure. X-ray studies of halite (the first mineral to be studied by x-rays) reveal that the ions are actually touching each other, as in the model to the right. (*From Ojakangas and Darby, 1976.*)

It is NaCl, or common table salt, which is the mineral halite. The reaction can be expressed as $Na^{+1}+Cl^{-1} = NaCl$.

2.27 Why are the inert gases—neon, krypton, radon, xenon, and argon—inert, thereby seldom participating in chemical reactions to make compounds?

Because they already have eight electrons in their outer energy-level shells and do not tend to add or give off electrons as in Problem 2.24 above.

2.28 What type of chemical bonding is illustrated by Fig. 2.2?

Ionic bonding, the forming of a compound by the bonding of oppositely charged ions. The oppositely charged ions are attracted to each other by electrostatic forces. "Opposite charges attract."

2.29 Explain the other geologically important type of chemical bonding, covalent bonding; give an example.

Covalent bonding is a type of bonding in which atoms attain the stable configuration of eight electrons in their outer shells by sharing rather than by losing or gaining electrons. A common example is the water molecule, H_2O. The one electron in the shell of each hydrogen atom and the six electrons in the outer shell of the oxygen atom are shared by all three atoms. In effect, each of the three atoms has eight electrons in its outer shell.

2.30 Diamond and graphite are both pure carbon. Yet diamond is the hardest natural substance and graphite is one of the softest. Why?

The strength of bonding between ions varies with the configuration of the ions relative to each other. In diamond, formed at a very high temperature of about 3500°C and a pressure 250,000 times that at the earth's surface, each carbon atom is equidistant from four other carbons in a three-dimensional array and the bonds are equally strong. They are covalent bonds—each C atom shares two electrons with each of the four adjacent atoms. Thus, each C atom has eight electrons in its outer shell. In graphite, which is formed under directional pressure and a lower temperature, the carbon atoms are arranged in hexagonal arrays in layers, with covalent bonds. The layers are bonded to each other by weak Van der Waals forces (a weak short-range electrostatic force).

2.31 What is *cleavage?*

The property of a mineral to break along flat, smooth surfaces. (Such surfaces are shiny, for they reflect light.)

2.32 Does graphite have a good cleavage?

Yes, because it will break along the planes of weak bonds (see Problem 2.30). The sheets easily slide over each other, thus making graphite an excellent lubricant.

2.33 Are petroleum and water minerals?

Petroleum is made of organic molecules and does not have a crystalline structure, so it is not a mineral. Water when it is a liquid or gas does not have a crystalline structure, but it does when it is ice. So, ice is a mineral, albeit one with a low melting point.

2.34 The mineral halite is shown in Fig. 2.3. How many cleavages does it have, and what are their relationships to each other?

Three good cleavages, at 90° to each other. (Look at salt from a salt shaker with a magnifying glass.)

2.35 In Fig. 2.3, two models of halite are shown. One is an "exploded" model which enables one to see the location of the ions in the crystal structure. The model on the right shows the actual positioning of the ions next to each other. How do mineralogists know that the ions touch?

X-rays beamed into a crystal structure reveal the positions of ions. Halite was the first mineral to be studied by x-rays.

2.36 What are the two main factors that determine which ions of the many types present in a watery solution or in a magma will be bonded together to form minerals?

The sizes and charges of the ions. Because ions touch each other, as in Fig. 2.3, the size of the ion is critical. If it is too large, it will not fit into the crystalline structure. Similarly, a mineral must be electrically neutral, so the charge of the ion is important.

2.37 Note the ionic radii and ionic charges of the eight most common ions in the earth's crust in Table 2.1. Which pairs of ions have very similar sizes and charges and can substitute easily for each other in a crystal structure?

Only Fe^{+2} and Mg^{+2} readily substitute for each other without disturbing the crystal structure. The mineral olivine shows that this is true, for olivine ranges in composition from Fe_2SiO_4 to Mg_2SiO_4, with all ratios of iron and magnesium in between. This is known as a solid solution series. Most olivines contain both iron and magnesium, so its general chemical formula is $(Fe,Mg)_2SiO_4$.

2.38 Which of the eight common elements is most abundant in the earth's crust, hydrosphere, and atmosphere? (See Table 2.2.)

Oxygen. It constitutes 47 percent of the crust by weight but 94 percent by volume because it is so large. It makes up 89 percent of water by weight and 21 percent of the atmosphere by weight.

2.39 Note from Table 2.3 that the most common minerals in the earth's crust are all silicates. The basic building block is the silicon-oxygen tetrahedron (see Fig. 2.4). What is the net electric charge of the tetrahedron and why is this geologically important?

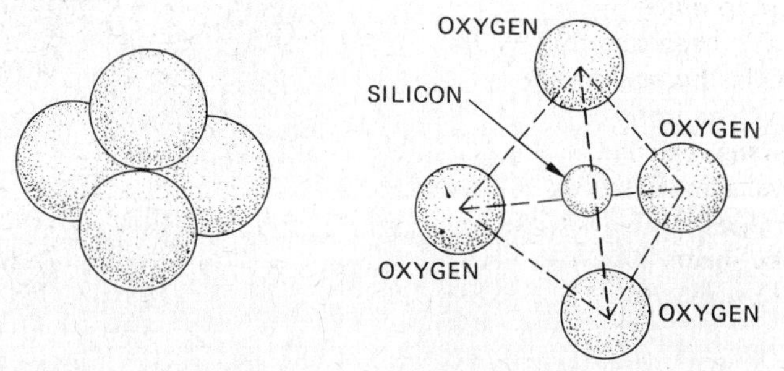

Fig. 2.4 The silicon-oxygen tetrahedron. Four oxygen ions surround a smaller silicon ion, as shown in the exploded model. (*From Ojakangas and Darby, 1976.*)

The silicon ion has a charge of +4 and the four oxygens each have a charge of −2. The net charge is −4, and the tetrahedron acts as a charged "complex ion." It will combine with cations to form the silicate minerals. Olivine illustrates this. Fe^{+2} and Mg^{+2} provide the four positive charges needed to balance the four negative charges of the tetrahedron.

2.40 What is illustrated by Fig. 2.5?

That oxygens are shared by adjacent silicon-oxygen tetrahedra. Single chains will be joined by cations to other single chains to form minerals such as pyroxene, double chains will be joined by cations to form minerals such as hornblende, and sheets will be joined by cations to form minerals such as muscovite and biotite. The crystal structures of the feldspars and quartz consist of tetrahedra joined in three dimensions (not illustrated), with each oxygen shared by an adjacent tetrahedron.

2.41 Why are quartz and the feldspars (plagioclase and orthoclase) such hard minerals?

Because of the three-dimensional bonding described above.

2.42 Name the main mineral groups, based on composition.

The silicates, oxides, carbonates, sulfides, sulfates, and chlorides.

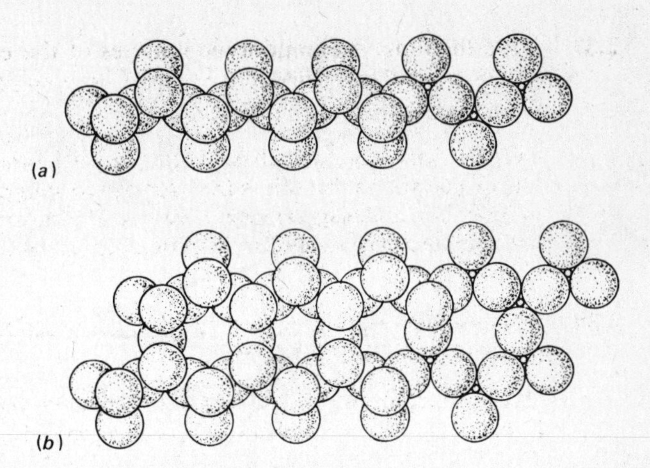

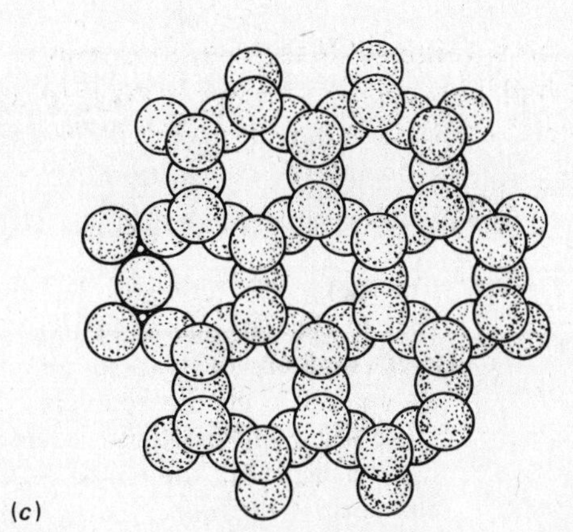

Fig. 2.5 Single chains (*a*), double chains (*b*), and sheets (*c*), each illustrating a sharing of oxygens by adjacent tetrahedra. In (*a*), each tetrahedron shares two oxygens; in (*b*), some share two and some share three; and in (*c*), each shares three oxygens. Single chains are joined to other single chains, and double chains are joined to other double chains by positively charged ions (such as iron, magnesium, sodium, calcium, and potassium). Similarly, sheets are joined to other sheets by such ions, making stacks of sheets. (*From Ojakangas and Darby, 1976.*)

2.43 The 8 common rock-forming minerals are listed in Table 2.3. If the list were of the 10 most common minerals, which two should be included?

Calcite ($CaCO_3$) and clay (HAlSiO). Clay is really a complex group of minerals with sheet structures as in Fig. 2.5. Both calcite and clay are formed at the earth's surface and are the main ingredients of the common sedimentary rocks limestone and shale, respectively.

2.44 A major principle is illustrated by the "rock cycle" in Fig. 2.6, although it may not be readily apparent. Think about it, and then see if the answer does indeed make sense.

That minerals, and also rocks which are aggregates of minerals, are stable in the environment in which they formed. Each mineral forms under a specific range of temperature and pressure and other conditions. For example, olivine crystallizes out of magma at depth, and if brought to the earth's surface, it will break down because the iron in it will combine with oxygen (it will oxidize) to form iron oxide minerals such as hematite (Fe_2O_3) or limonite ($Fe_2O_3 \cdot H_2O$).

2.45 The three main groups of rocks, based on origin, are shown on Fig. 2.6. Describe each.

Igneous rocks: crystallize from magmas.

Sedimentary rocks: are formed from pieces (sediment) of other rock material or are precipitated from water.

Metamorphic rocks: are formed from other rocks by changes in temperature and/or pressure, commonly in the presence of solutions.

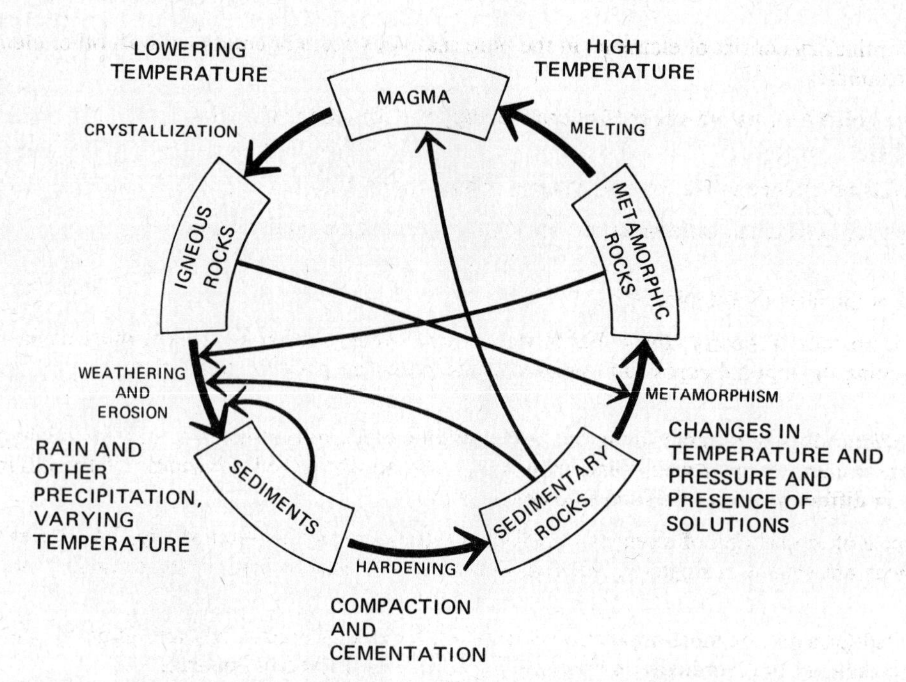

Fig. 2.6 The rock cycle illustrates that changes in its surroundings will cause a rock to change so as to be more stable under the new conditions. Thus, rocks (and obviously, also the minerals which make up the rocks) are stable only in the environments in which they form. Note that the rock cycle has many shortcuts; rarely does a rock take the circle route. And if it did, wouldn't the only clue to its travels be its chemical composition? (*From Ojakangas and Darby, 1976.*)

Supplementary Problems

2.46 What elements form the oxide mineral group? Give an example.

Oxygen, combined with cations. Hematite (Fe_2O_3), the common iron ore, is an example. Other iron ore minerals are limonite ($Fe_2O_3 \cdot H_2O$) and magnetite (Fe_3O_4).

2.47 What elements form the carbonate mineral group? Give an example.

The carbonate ion—CO_3—with a net charge of -2, combined with cations. Calcite ($CaCO_3$), one of the most common nonsilicate minerals, is an example.

2.48 What elements form the sulfide mineral group? Give an example.

Sulfur—S—combined with cations. Pyrite (FeS_2) is the most common example. The sulfide group includes most of the ores of copper, lead, zinc, and several other metals.

2.49 What elements form the sulfate mineral group? Give an example.

The sulfate complex ion—SO_4—with a net charge of -2, combined with cations. Gypsum ($CaSO_4$) is the most common example.

2.50 What elements form the chloride mineral group? Give an example.

Chlorine—Cl—combined with cations. Halite (NaCl) is the most common example.

2.51 A few minerals consist of elements in the pure (native) state, not combined with other elements. Name four examples.

Native gold (Au), native silver (Ag), native copper (Cu), and sulfur (S).

2.52 Minerals crystallize in six crystal systems. Name them.

Isometric, hexagonal, tetragonal, orthorhombic, monoclinic, and triclinic.

2.53 What is the streak of a mineral?

It is a physical property, the color of the mineral when powdered. This is most easily observed by scratching the mineral across an unglazed white porcelain plate (a streak plate).

2.54 The silicon-oxygen tetrahedron is the basic building block of the silicates, and one might guess that all silicate mineral formuli would end with SiO_4. Instead, the specific formulas of the common silicates contain different ratios of silicon to oxygen. Why?

Because of the sharing of oxygens by adjacent tetrahedra as illustrated in Fig. 2.5. For example, fewer oxygens are shared in single chains than in quartz or feldspar where all the oxygens are shared.

2.55 The feldspars are the most important minerals in the earth's crust. They are similar to quartz in that all the oxygens of the tetrahedra are shared. How do they differ from quartz?

In the feldspars, some of the silicon ions are replaced by aluminum ions. Because their charges differ (see Table 2.1), cations enter the crystal structure to keep the mineral electrically neutral. In orthoclase, it is the potassium ion with a charge of +1. In plagioclase, it is sodium (Na^{+1}) and calcium (Ca^{+2}).

2.56 One of the common rock-forming minerals is plagioclase feldspar which contains variable amounts of Na and Ca in addition to Al, Si, and O. Pure Na-plagioclase has the formula $NaAlSi_3O_8$. Because the sizes of Na^{+1} and Ca^{+2} ions are similar (see Table 2.1), Ca^{+2} can substitute for Na^{+1}. Would the formula of Ca-plagioclase be $CaAlSi_3O_8$? Explain.

No. Note that Na-plagioclase is electrically neutral—there are 16 positive charges and 16 negative charges. If a Ca ion substitutes for an Na ion, there will be 17 positive charges and 16 negative charges, and a mineral structure must be electrically neutral. (See the next problem.)

2.57 Most plagioclase contains both Na and Ca. How is the apparent charge imbalance accommodated? (*Hint:* The solution involves more ionic substitutions.)

Note from Table 2.1 that the sizes of Si^{+4} and Al^{+3} are not too different. The total positive charge of the plagioclase can be reduced by 1 if an Al^{+3} substitutes for a Si^{+4}. The formula of Ca-plagioclase would therefore be $CaAl_2Si_2O_8$. As most plagioclase contains both Na^{+1} and Ca^{+2}, some of the silicons must be replaced by aluminums. This process can be called coupled ionic substitution.

2.58 See Fig. 2.3. Which sodium ion belongs to which chlorine ion in a halite "molecule" of NaCl?

One cannot speak in terms of molecules when one is describing a crystalline material because each ion is bonded to several other ions. Here, each sodium is equidistant from six chlorines, and vice versa. Each ion is a part of the total crystalline structure, and a given ion is not part of a "NaCl molecule." Therefore, mineralogists do not generally refer to minerals in terms of molecules. However, when halite is dissolved in water, it is proper to refer to molecules of NaCl in the water. Chemists use "molecules" when referring to gases and liquids.

2.59 How are the hardnesses of minerals determined?

By the ability of one mineral to scratch another, in terms of the Mohs scale of hardness. The scale is based on a sequence of minerals, each harder than the mineral below it on the scale, hardness 10 being the hardest. Thus, calcite (hardness 3) will scratch gypsum (hardness 2). Some additional hardness "tools" are your fingernail (hardness 2.5 if you drink your milk), a penny (hardness 3), and a knife blade (hardness 5.5). The scale is as follows:

1. Talc (soft)
2. Gypsum
3. Calcite
4. Fluorite
5. Apatite
6. Orthoclase
7. Quartz
8. Topaz
9. Corundum
10. Diamond (hard)

However, this is a relative scale only. Diamond is much harder than the other minerals.

2.60 What is metallic bonding?

Metallic bonding is a type of chemical bond that is neither ionic nor covalent. Instead the electrons are free to move anywhere in the metal. This type of bond makes some metals soft so that they can be bent without breaking.

2.61 A 100-g nugget of gold that you found panning in a stream in the mountains seems to be pure, but you wonder whether there might be some quartz in the center. You drop it (carefully) into a graduated cylinder and note that it displaces 5.2 cm^3 (cubic centimeters) of water. Calculate its density.

The density is 19.2 g/cm^3, a good value for pure gold. To arrive at this number, you will use the equation: Density = mass/volume. The density is thus 100 divided by 5.2, or about 19.2 g/cm^3.

2.62 At a glance, quartz and calcite may look alike. What physical properties distinguish them?

Quartz has a hardness of 7, calcite 3. Quartz has no cleavage, and calcite has three good cleavages at angles of about 75°. Calcite reacts with (fizzes in) hydrochloric acid, and quartz does not.

2.63 The alchemists of past centuries were always looking for a magical way to change lead into gold. With nuclear reactions, how would a modern alchemist try to accomplish this transition? (One isotope of lead has 82 protons and 125 neutrons, whereas gold has 79 protons and 118 neutrons.)

By removing 3 protons and 7 neutrons from the nucleus of lead. This, of course, involves splitting the lead nucleus, no small feat.

2.64 Name the main physical properties of minerals.

Hardness, cleavage, color, streak, specific gravity, luster, and crystal form.

2.65 A quartz crystal is illustrated in Fig. 2.7. Which crystal system does it represent? (See Problem 2.52.)

The crystal is hexagonal, with three crystal axes of equal length (see the cross-section) and one axis of a different length.

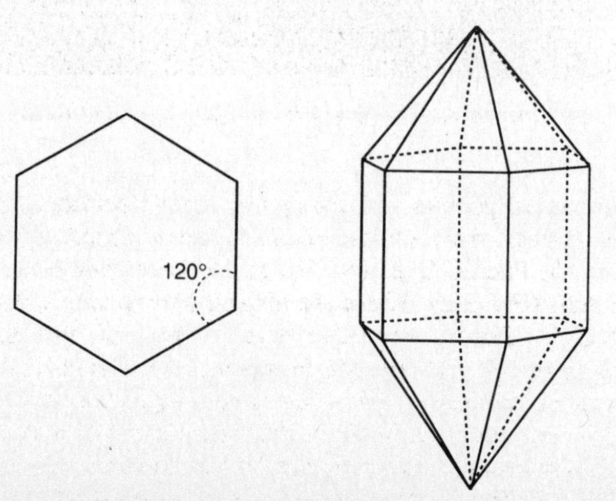

Fig. 2.7 Sketch of a quartz crystal and a cross-section of a quartz crystal.

Chapter 3

Igneous Rocks and Volcanism

IGNEOUS ROCK BODIES

Igneous rocks are rocks that have cooled from magmas. *Magmas* are very hot solutions (500 to 1700°C) of molten rock, made up of ions of many elements, primarily oxygen, silicon, aluminum, iron, calcium, sodium, potassium, magnesium, and water. Where magmas have intruded into preexisting rocks beneath the earth's surface and cooled, *intrusive (plutonic) rocks* are formed. If the magmas reach the surface, *extrusive (volcanic) rocks* are formed. Intrusive rocks crystallize as very large *batholiths,* smaller *stocks, dikes* which cut across existing rock layers, and *sills* which were intruded between existing rock layers (Fig. 3.1). Extrusive magmas (lavas) form either *lava flows* or *pyroclastic* (''fire particle'') deposits, the latter the result of volcanic explosions such as occurred at Mt. St. Helens in 1980.

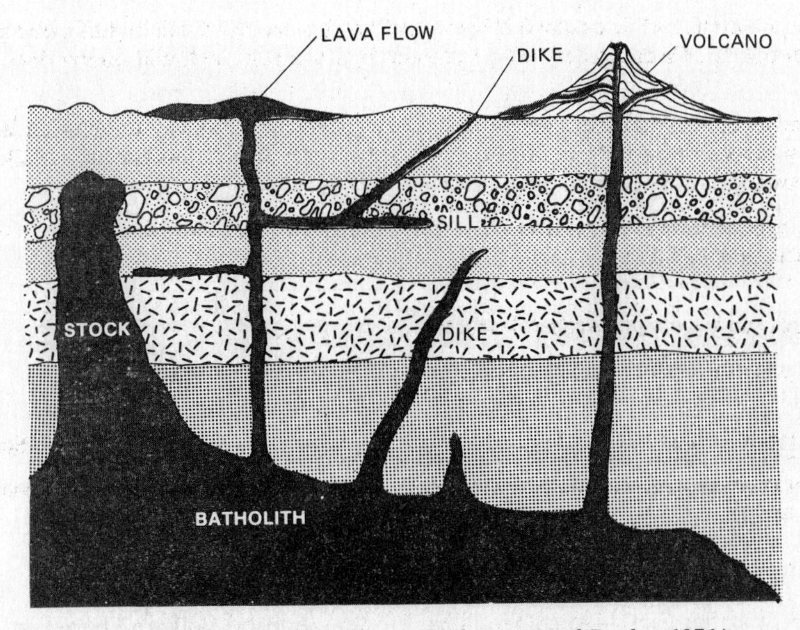

Fig. 3.1 Igneous rock bodies. (*From Ojakangas and Darby, 1976.*)

Volcanism encompasses igneous activity on the earth's surface. *Volcanoes* are built up by volcanic activity at central vents. Ancient volcanic rocks are widespread, but most active volcanoes are found in a belt around the Pacific Ocean and in an east-west belt extending through the Mediterranean Sea region and Asia (Fig. 3.2). There are also many volcanoes along midocean ridges, such as the Mid-Atlantic Ridge. There are three main types of volcanoes. *Composite cones* or *stratovolcanoes,* such as Mt. St. Helens, Mt. Hood, and Mt. Ranier in the Pacific Northwest, are the most impressive, for most are at least a few kilometers high; they consist of both pyroclastic material and lava flows. *Cinder cones* are much smaller and consist of pyroclastics. *Shield volcanoes* are volcanoes which are much broader than they are high and mainly consist of lava flows. The subdued shape of shield volcanoes is the result of low-viscosity lavas which readily flow away from the volcanic vents, whereas cinder cones and composite volcanoes are formed by more viscous (stickier) lavas from which gases cannot readily escape and which therefore cause volcanic explosions. In certain areas, lavas are extruded from long cracks (fissures) rather than from vents. Those lavas are usually basaltic in composition and have low viscosities. Therefore they form widespread lava flows, as on the Columbia Plateau of northwestern United States or the Deccan Plateau of India, and are termed "flood basalts."

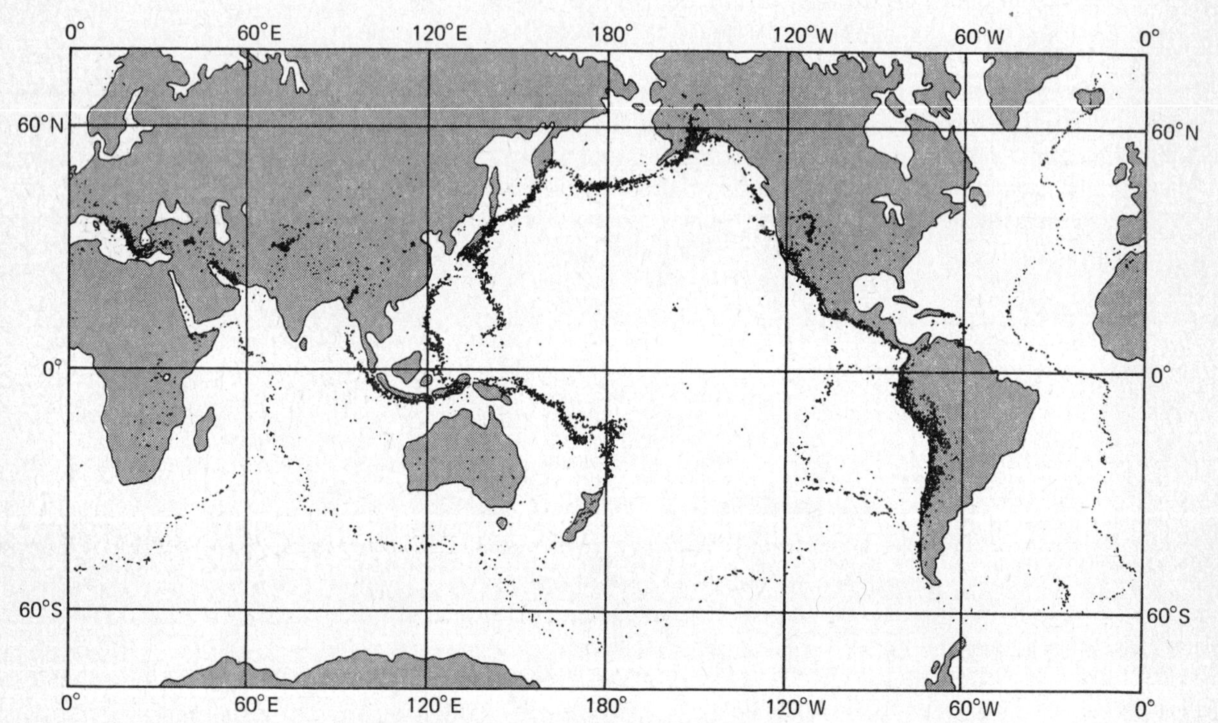

Fig. 3.2 Map showing zones of major volcanoes and earthquakes. They are not differentiated on this map, but earthquakes are far more numerous than volcanoes. There are about 500 active volcanoes in the world. (*From Ojakangas and Darby, 1976.*)

IGNEOUS ROCK TEXTURES

Lavas cool quickly at the earth's surface where the temperature is low. Therefore, the ions in the lava do not have time to arrange themselves into large crystals before the lava solidifies. The result is numerous small crystals and the formation of a *fine-grained (aphanitic) rock*. If a lava cools very rapidly, a *natural glass* without crystals will result. Magmas cooling at depth are insulated by rock, a poor conductor of heat, and cool slowly; crystals have time to grow large and a *coarse-grained (phaneritic) rock* will be formed. Magmas with much water and other gases, usually the residual fluids of larger bodies of crystallizing magma, allow for easy movement of ions. Hence large crystals and very coarse-grained rocks (*pegmatites*) can be formed in this manner, too. Igneous rocks which have cooled at two different rates have two sizes of crystals; such rocks with larger *phenocrysts* distributed among much smaller sized crystals or groundmass are *porphyritic igneous rocks*. Gases in lava flows tend to rise to the tops of flows, forming gas bubbles or cavities called *vesicles*. These may later be filled by minerals precipitating out of solutions moving through the flows, forming *amygdules*.

IGNEOUS MAGMA AND ROCK COMPOSITION

Different magmas have different compositions and thus the rocks that form when they solidify will also differ in composition. Magmas and rocks containing abundant iron, magnesium, and calcium, in addition to silicon, aluminum, and oxygen, are *mafic* in composition. Those containing more abundant silicon, potassium, and sodium are *felsic*. There is a gradation between these two types, with many magmas and rocks (such as diorite and andesite) of intermediate composition. *Ultramafic* (very mafic) magmas and rocks also occur.

How do magmas of differing composition form? Some mafic magma may be generated at depth in the lower crust or the upper mantle by *partial melting* of preexisting ultramafic rock, whereby the minerals which melt at the lowest temperatures melt and leave the higher-temperature unmelted minerals behind. Higher in the crust, any preexisting rock may melt quite completely, thereby forming a magma of a similar composition. As a magma cools, specific minerals crystallize at specific temperatures; thus an order of crystallization (Fig. 3.3) of the common igneous minerals is the result. This order of crystallization can lead to a *fractional crystallization* or *differentiation* of a magma, as shown earlier in this century by N. L. Bowen of the U.S. Geophysical Laboratory.

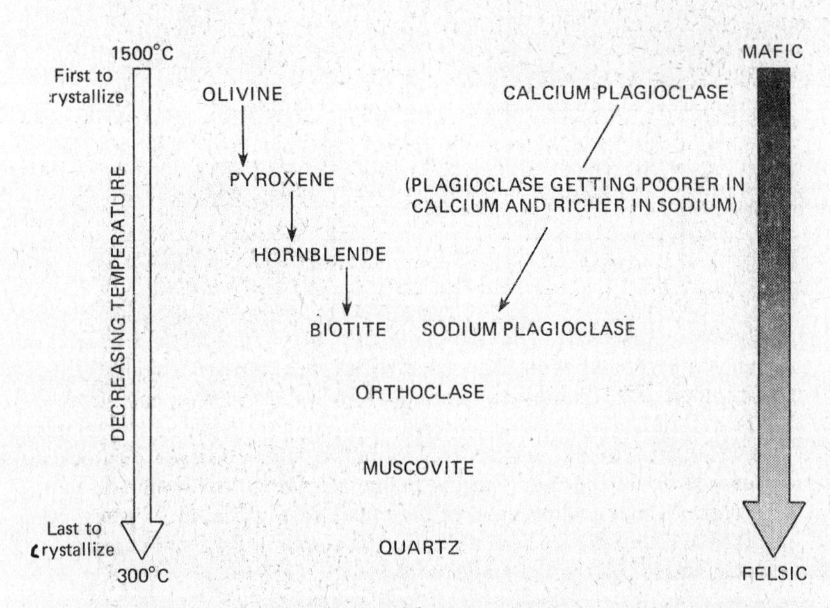

Fig. 3.3 Bowen's reaction series, which gives the order of crystallization of the igneous minerals from a mafic magma. In detail, it shows how some minerals react with the magma to form other minerals (e.g., olivine reacts to form pyroxene) and shows that the plagioclase that crystallizes at lower temperatures is richer in sodium than is the plagioclase that crystallizes earlier.

As a mafic magma cools, the first minerals to crystallize are olivine, pyroxene, and calcium plagioclase. Because these minerals are high in iron, magnesium, and calcium, the remaining magma is somewhat depleted in these elements; silicon, aluminum, potassium, and sodium thus make up greater percentages of the magma than originally. If the first-formed minerals stay in contact with the magma, they may react with the magma and be remelted. However, if they settle or are in some other way separated from the magma, the remaining magma will have an intermediate composition as the temperature drops further; hornblende and sodium plagioclase will crystallize. If these crystals are removed from the magma, any remaining magma will be felsic, and finally potassium feldspar, the micas, and quartz will crystallize.

IGNEOUS ROCK CLASSIFICATION

Igneous rocks are usually classified and named on the basis of texture (grain size) and mineral composition, although a chemical classification is necessary for much detailed work. The most common igneous rock types are shown on Fig. 3.4.

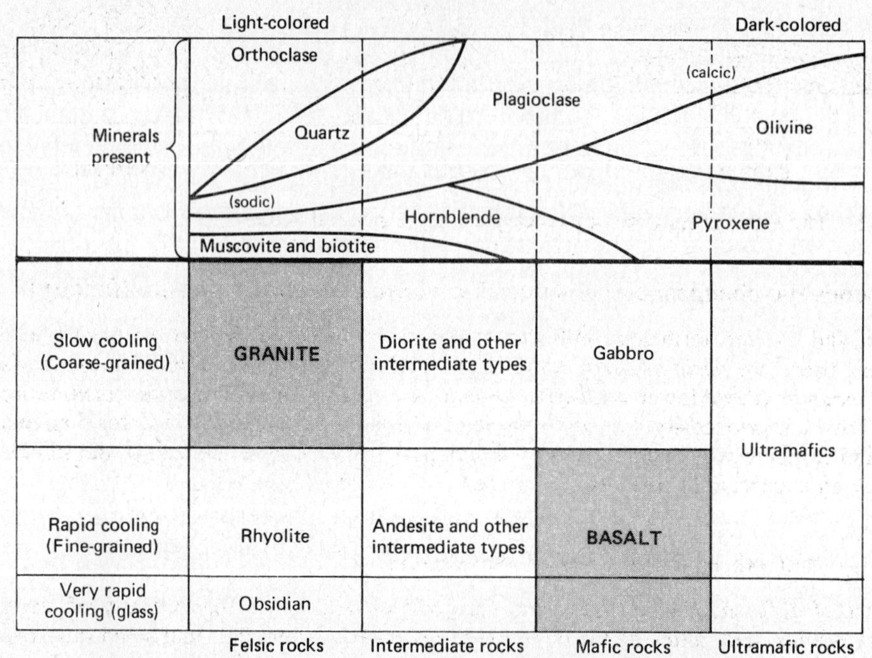

Fig. 3.4 Igneous rock chart. The two major rock types, granite and basalt, are shown by bolder print. The other rock types (hundreds could be included) are volumetrically minor. Minerals present in each rock can be determined by reading upward from the rock name. For example, granite can contain muscovite, biotite, hornblende, sodic plagioclase, quartz, and orthoclase, but the essential minerals are orthoclase and quartz, which occupy most of the upper part of the box. On the other extreme, ultramafics contain only pyroxene, olivine, and calcic plagioclase. (*After Ojakangas and Darby, 1976.*)

Solved Problems

3.1 The three main types of volcanoes are illustrated in cross-section in Fig. 3.5. Note scales. Name each type.

(*a*) Composite cone or stratovolcano. (*b*) Cinder cone. (*c*) Shield volcano.

3.2 What are the differences in composition of the two major types of magma?

Mafic (basaltic) magmas contain relatively high amounts of iron, magnesium, and calcium and low amounts of silicon, potassium, and sodium relative to felsic (granite) magmas; conversely, felsic magmas are relatively low in iron, magnesium, and calcium and high in silicon, potassium, and sodium.

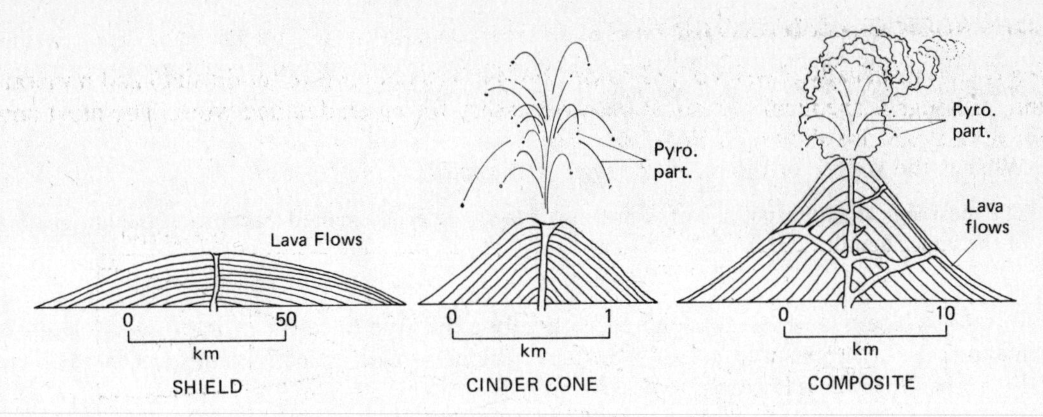

Fig. 3.5 The three main types of volcanoes. Note general scales.

3.3 How does the chemical composition of a magma affect its viscosity (fluidity)?

Silicon and oxygen form long polymer chains which become entangled, making a viscous (sticky) magma. Therefore *felsic magmas* (those high in silicon, sodium, and potassium) are viscous, whereas *mafic magmas* (those lower in silicon, sodium, and potassium and higher in calcium, iron, and magnesium) have a low viscosity, i.e., are very fluid. Felsic lava contains more water than mafic magma, but this fluid component is overpowered by the viscosity-producing tendencies of the silicon. (See Problem 3.15 for an exception to this last statement.)

3.4 Why are there three main types of volcanoes?

Because of differences in the chemical composition, and hence the viscosity of magmas. A basaltic magma is very fluid, allowing for rapid flow (tens of kilometers per hour!) and easy escape of the volcanic gases, thereby minimizing pressure buildup and the resultant explosions and pyroclastic debris. The fluid basaltic magmas thus tend to produce relatively low-lying and broad shield volcanoes. Lava with an intermediate chemical composition has a moderate viscosity, flows relatively slowly, and thereby solidifies more readily even on relatively steep slopes. The moderate viscosity hinders gas escape and the ensuing violent explosions break up already formed volcanic rock, and hurl rock fragments as well as blobs of magma into the air to cool quickly and form glass. Thus the composite cones are composed of both flows and pyroclastic debris. Cinder cones generally range in composition from andesite to rhyolite; thus pressure buildup is common and pyroclastic debris is the result. Some composite and cinder cones may have plugs of rhyolite in their craters, the result of very viscous rhyolitic magmas.

3.5 Bowen's idea of fractional crystallization of basaltic magma as a means of producing the broad variety of igneous rocks, while certainly a natural and laboratory process, does not completely answer the big picture. Why not?

Granitic rocks are restricted to the continents and basaltic rocks are most abundant in the ocean basins—why are they not together if both formed from the same magmas?

The total differentiation of a basalt would lead to only about 10 percent of granite—there is far too much granite to be accounted for in this way.

One might expect a total spectrum of igneous rock types in roughly equal proportions; however, intermediate rock types (between basalt and granite) are volumetrically much less important.

3.6 What are some other theories which explain why there are different kinds of magma?

Because magmas are generated at depth, direct observation of magmas in the process of formation is not possible. Laboratory and theoretical studies indicate that while some magmas are the result of the differentiation (fractional crystallization) of mafic magmas (as Bowen suggested), some may be the result of the *complete* melting of original rocks of different compositions (e.g., a melted mudstone would have a different composition than a melted granite or a melted quartz sandstone), and some may form by the *partial* melting of other rocks (especially in the upper mantle).

3.7 What is the source of the basaltic lavas of the midoceanic ridges?

The most logical source is the partial melting of the most easily melted fractions of the ultramafic rocks of the upper mantle.

3.8 Basaltic volcanism does not tend to be highly explosive because of the low viscosity of the magma. Yet, the basaltic volcano Surtsey which formed off of Iceland on the Mid-Atlantic Ridge in 1963 was extremely explosive in its early stages. Why?

Because seawater had access to the hot magma in the subsea volcanic vent, thereby generating steam which caused the violent explosions. Blocks of rock were thrown several hundred meters into the air. Once the debris pile was higher than sea level and lava flows were able to rise to the surface, the water no longer had access to the vent, and subsequent volcanic activity was of a quieter nature.

3.9 Is an igneous rock likely to be composed only of quartz and olivine? Why or why not? (Refer to Fig. 3.3.)

No, because olivine crystallizes early at a high temperature and quartz is the last mineral to crystallize at a low temperature.

3.10 If, as a mafic (basaltic) magma crystallizes (Figs. 3.3 and 3.4) underground, the crystals are removed from contact with the magma, and the remaining magma crystallizes separately, which kind of a rock will form first and which will form last?

First will be gabbro (composed of olivine, pyroxene, and calcic plagioclase), then will be one of the intermediate rocks such as diorite (composed of hornblende and intermediate plagioclase), and last will be granite (composed of quartz and orthoclase plus one or more of muscovite, biotite, and hornblende).

3.11 If a magma crystallizes at a temperature of about 1500°C, what is its likely composition—felsic or mafic? (See Fig. 3.3.)

Mafic. Felsic magmas crystallize at lower temperatures.

3.12 How does the rate of cooling of a magma affect grain size of the resultant rock, and why? (See Fig. 3.4.)

If a magma cools slowly, there is time available for the ions in the magma to move to a few sites of crystallization, resulting in larger crystals (e.g., as in a granite or a gabbro). On the other hand, if a magma cools rapidly, there will not be time for crystals to grow large (e.g., as in a basalt or a rhyolite). If a magma cools very rapidly, no crystals can form, and a supercooled liquid or glass (e.g., obsidian) is the result.

3.13 See the microscopic view of igneous rocks in Fig. 3.6. Which rock cooled the slowest and which cooled the fastest?

The rock which cooled the slowest is shown in Fig. 3.6(*a*); the fastest-cooling rocks are in Figs. 3.6(*b*) and 3.6(*c*).

3.14 How does a porphyritic rock, as in Fig. 3.6(*b*), form?

Two sizes of crystals, large phenocrysts in a fine-grained groundmass, are the result of two different rates of cooling.

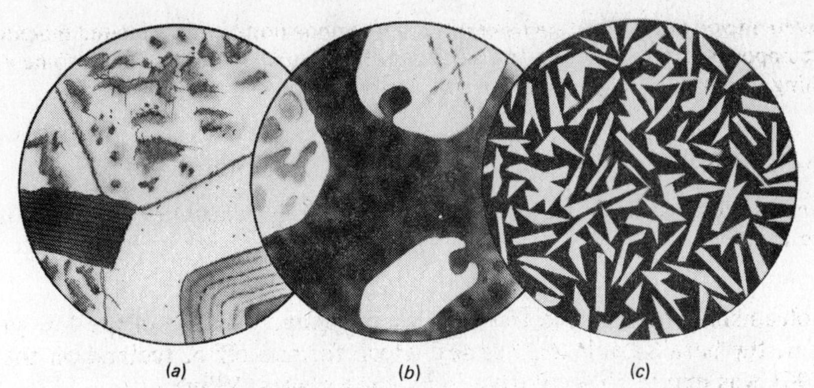

Fig. 3.6 Microscopic views of igneous rocks: (*a*) granite, (*b*) porphyritic rhyolite, (*c*) basalt. The fields of view are about 2 mm in diameter.

3.15 Name another factor, other than the rate of cooling, that can affect crystal size. Very coarse-grained intrusions of igneous rocks (pegmatites) are an example of this factor at work.

The presence of abundant volatiles (gases) in a magma will reduce viscosity and allow ions to move readily to a few sites of crystallization. Pegmatites form from the last watery solutions of a magma. Some rare crystals in pegmatites are several meters long, and crystals tens of centimeters long are common.

3.16 Why do many of the late-forming pegmatites of the previous question contain rare elements such as lithium, tin, and beryllium?

Because these ions have sizes and charges which do not allow them to fit into the crystal structures of the common igneous minerals of Fig. 3.3. Therefore, they tend to be excluded from earlier crystallization and concentrated in the pegmatites.

3.17 What kind of igneous rock bodies are the light-colored rocks shown in Fig. 3.7? (Refer to Fig. 3.1.)

Dikes, because they cross-cut preexisting rock structures.

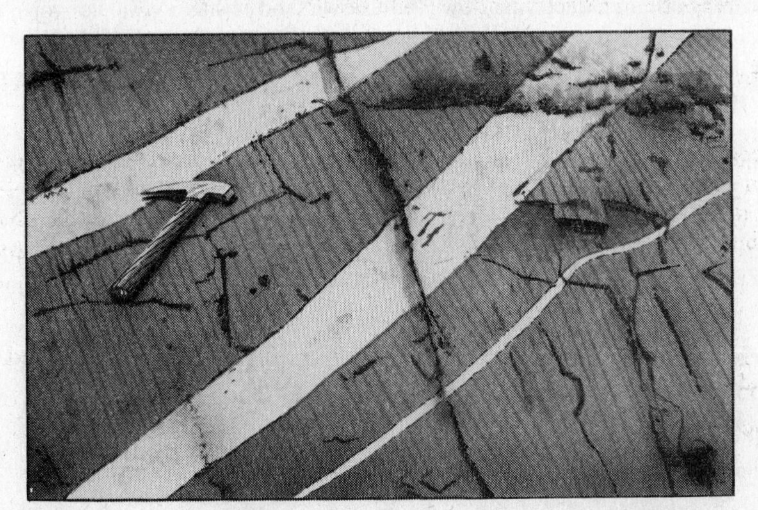

Fig. 3.7 Bodies of light-colored granite cutting across darker metamorphic rock. The dikes cooled below the surface but have been exposed by erosion. Note geology hammer for scale.

3.18 A layer of granite 1 m thick is present in a sequence of sedimentary rocks, parallel to the bedding in the sedimentary rocks (Fig. 3.8). Is it a dike, a sill, or a stock? (Refer to Fig. 3.1.)

Fig. 3.8 Intrusions of light-colored granite intruded between and parallel to layers of metamorphosed sedimentary rock.

It is a sill, which by definition is concordant with (parallel to) the original structures in the country rock.

3.19 Refer to the previous problem. How can one tell whether it is a sill which intruded between rock layers or whether it is a lava flow which was extruded on top of sedimentary rock and then was covered with other sediments?

Rocks immediately adjacent to a sill, on both top and bottom, will show effects (baking) of the hot magma whereas a lava flow can obviously only bake the rock upon which it was laid down. Also, rubbly and irregular tops often characterize flows, and flows tend to have vesicular tops.

3.20 What is the most obvious evidence that lava extruded under water?

The presence of pillow structures (Fig. 3.9) which form as lava repeatedly breaks out of rapidly chilled bulbous forms or pillows to form new ones. As pillows commonly form in piles, with later pillows upon earlier formed pillows, they have pointed bases and rounded tops, making them useful as indicators of the original tops in tilted sequences of lava flows. They can range in size from a few tens of centimeters to several meters. Submersibles along the Mid-Atlantic Ridge and elsewhere and divers off Hawaii where lavas reach the sea have photographed pillows on the sea floor, thus proving their origin.

3.21 Figure 3.9(*b*) depicts a pile of pillows tilted to a vertical position. In which direction are the pillows topping (that is, which is the top of the sequence?)

The top of the sequence is to the left. Note the points are to the right and the rounded tops are to the left. Stretching of pillows during deformation can obscure this relationship.

3.22 Figure 3.2 shows the locations of the world's active volcanoes. Is a volcano likely to suddenly appear in, say, Iowa or Minnesota?

(a) (b)

Fig. 3.9 (*a*) Pillowed lava flow of basaltic composition. The tops point toward the top of the sketch. (*b*) Vertical pillows. Deformation has tilted the pile of pillows into a vertical orientation in both (*a*) and (*b*).

No. Note that most of the world's 400-plus active volcanoes are found in a zone around the Pacific (the "Ring of Fire") and in a zone from the Mediterranean eastward across Asia. A new volcano would most likely appear in those zones. (The reason for volcanoes occurring in these zones will be given in Chapter 17.)

3.23 Igneous rocks are classified on the basis of texture (grain size) and composition (mineralogy). A general indicator of igneous rock composition is the color. Name the following igneous rocks by referring to Fig. 3.4:
(*a*) A coarse-grained, light-colored rock
(*b*) A coarse-grained, dark-colored rock
(*c*) A fine-grained, light-colored rock
(*d*) A fine-grained, dark-colored rock

(*a*) Granite. (*b*) Gabbro. (*c*) Rhyolite. (*d*) Basalt.

3.24 Why was obsidian a much sought after rock which was traded far and wide by the Native Americans? (For example, obsidian from Yellowstone Park was traded as far east as Ohio.)

Its lack of minerals, and hence lack of crystalline structure (the result of extremely rapid cooling), cause it to break with a smooth fracture like thick bottle glass. This allows it to be shaped by chipping or pressure into arrowheads, spear points, and other tools.

3.25 What coarse-grained (plutonic) rock consists of abundant calcic plagioclase plus olivine and pyroxene?

Gabbro.

3.26 What is the fine-grained equivalent of gabbro?

Basalt.

3.27 Name the coarse-grained (plutonic) rock consisting primarily of hornblende and sodic plagioclase.

Diorite, one of the more common intermediate rocks.

3.28 What is the volcanic rock which has the same composition as a granite?

Rhyolite.

3.29 What are the two most abundant igneous rocks?

Granite, which cores the continents, and basalt, which floors the ocean basins. (Note that they are written in capital letters on Fig. 3.4.)

3.30 What are the rounded structures shown in Fig. 3.10?

Fig. 3.10 Two lava flows with gas cavities (vesicles) which have been filled later by mineralizing solutions. The vesicles are now amygdules. Gas cavities generally form in the top parts of lava flows because the gases rise through the liquid lava.

They are amygdules, originally gas cavities (vesicles) in lava flows which were filled by solutions moving through the crystallized rock.

3.31 How would one distinguish between an amygdaloidal basalt and a porphyritic basalt?

Amygdules are round and commonly layered or radiating. The phenocrysts of a porphyritic rock [as in Fig. 3.6(*b*)] are commonly elongated single crystals. Crystal faces may be visible because the early-formed phenocrysts grew in a magma without interference from adjacent crystals.

Supplementary Problems

3.32 Volcanoes are classified as active, dormant, or extinct. If they have erupted during the last 50 years, they are clearly classified as active. On what bases are dormant or extinct volcanoes classified, and how can geologists be sure an extinct or dormant volcano may not suddenly become active?

If a composite volcano has a fairly undissected cone shape, implying geologically recent activity, it may be classed as dormant. If the cone is deeply eroded, it may be called extinct. However, any large volcano, especially if it is located in the two major belts of Fig. 3.2, may come to life with little or no warning. Increased frequency of minor earth tremors and bulging on the volcano surface are indicators of impending major activity, but activity could theoretically start so suddenly as to be totally unpredicted.

3.33 Mt. Somma (ancestral Mt. Vesuvius), which towered above the Roman seaside resort cities of Pompeii and Herculaneum, hadn't erupted for 800 years, when in A.D. 79, it erupted. What happened to the two cities?

Pompeii, along with 2000 inhabitants, was buried beneath 8 to 10 m of pyroclastic debris and Herculaneum was buried beneath muddy volcanic ash flows which were hot enough to char logs.

3.34 What were the major geologic effects of the May 18, 1980, eruption of Mt. St. Helens in southwestern Washington (Fig. 3.11)?

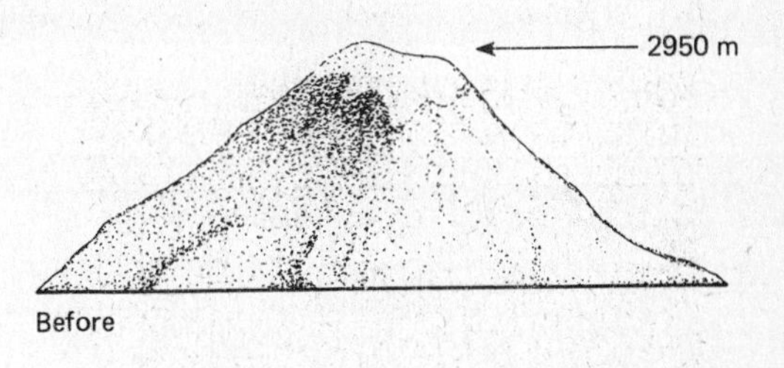

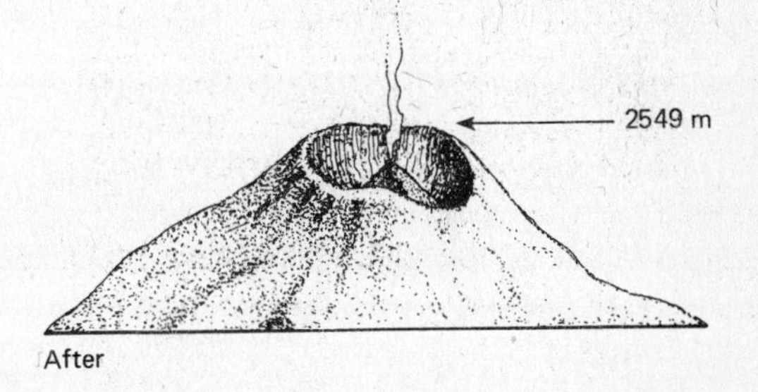

Fig. 3.11 Before and after views of Mt. St. Helens. It erupted with great violence in 1980. Summit elevations are shown. A felsic dome is now forming in the crater.

A 500-km^2 area to the north of the volcano was devastated as 400 m of the summit was blown off, a column of ash and gas rose to a height of 19 km, ash was carried as far as 900 km northeastward by strong prevailing winds (a total of about 2 km^3), minor ash was carried across the United States, a mudflow surged 30 km down the North Fork of the Toutle River, pyroclastic flows moved as much as 25 km northward along the valley of the above-mentioned river, an estimated total of 800 million board feet of timber was flattened, and 68 people were killed.

3.35 Note the similarities in the sketches of Mt. St. Helens (Fig. 3.11) and Crater Lake, Oregon (Fig. 3.12). Surmise how Crater Lake may have formed.

It may appear that an eruption similar to that of Mt. St. Helens may have occurred 6000 to 6600 years ago on the basis of radiocarbon dating of trees killed in that volcanic event. However, calculations of the volume of debris spread outward from Crater Lake indicate that much of the original volume of the volcano Mt. Mazama must have collapsed into the underlying magma chamber.

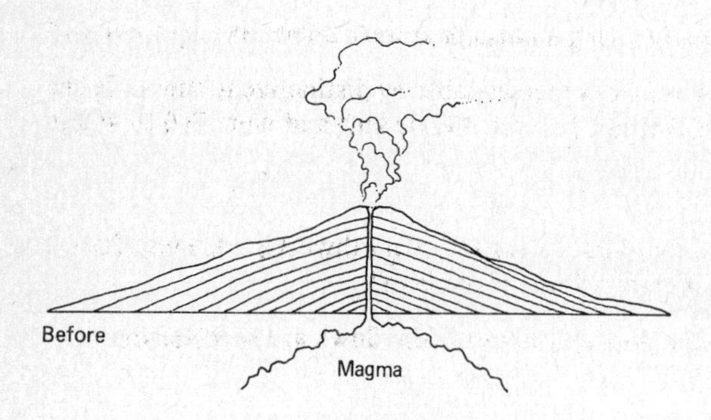

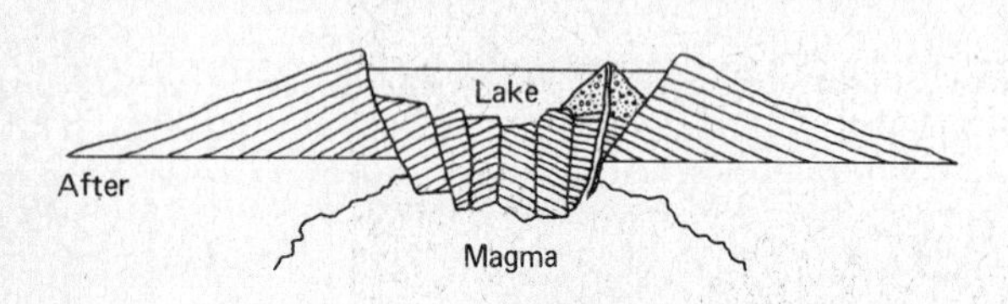

Fig. 3.12 Crater Lake, Oregon. Initially, Mt. Mazama was an impressive composite cone. After the magma chamber emptied about 6600 years ago, the cone collapsed into the chamber, forming a large caldera. A new cone formed in the caldera.

3.36 What could happen to the cities shown in Fig. 3.13 if any of the nearby volcanoes erupted?

Many could be affected by either ash falls or by mudflows. (See Problems 3.33 and 3.34.)

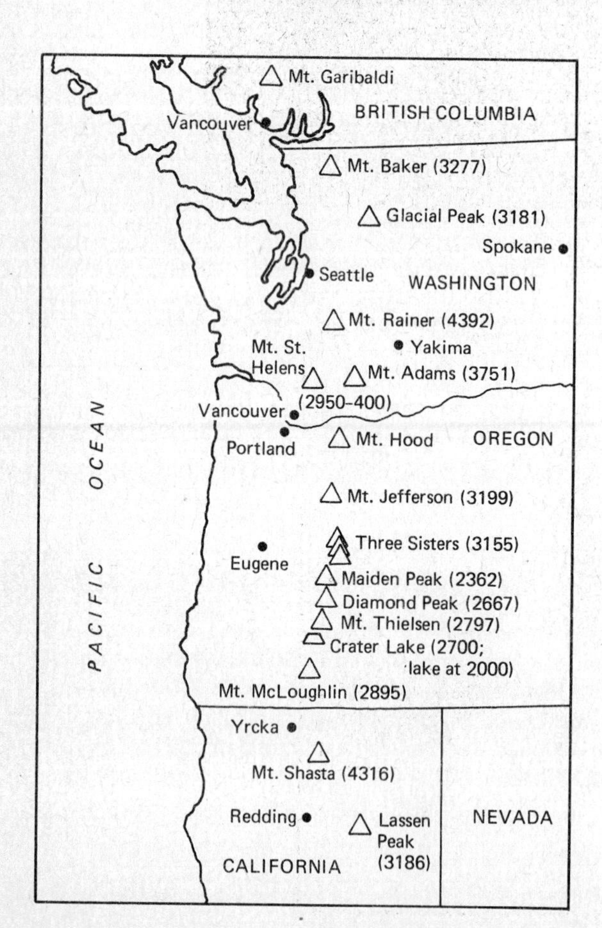

Fig. 3.13 The Cascade Range. Elevations are given in meters. Mt. St. Helens became active in 1980 after more than a hundred years of dormancy. Mt. Lassen erupted rather recently (minor activity in 1914 to 1917). Mt. Shasta and Mt. Baker have acted up in the last 2000 years. Mt. Hood and Crater Lake figure in Indian legends; Crater Lake formed when Mt. Mazama erupted and collapsed into its magma chamber about 6600 years ago. Mt. Ranier lets off steam.

3.37 How do geologists determine whether hot ash flows (glowing avalanches) were minor or major events?

By studying the extent of individual ash flows, now rock but recognizable by distinctive textures. It has been determined that some individual ash flows in Nevada covered 18,000 km^2 and were 300 to 450 m thick.

3.38 Figure 3.14 is a sketch of lava flows in the Columbia Plateau region of northwestern United States. What do the flows indicate about the viscosity and composition of the lava?

The lava had a low viscosity and was basaltic in composition, because the flows are so extensive.

(*a*)

Fig. 3.14 (*a*) Lava flows of the Columbia Plateau.

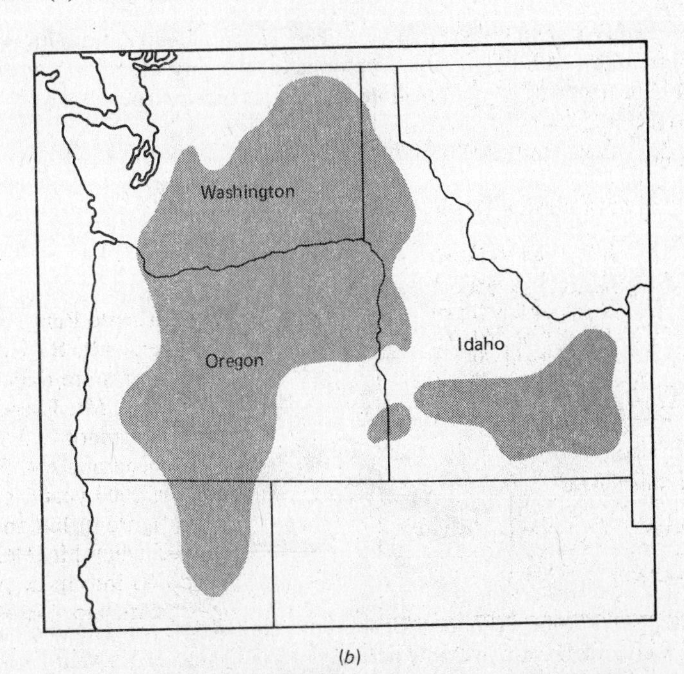

(*b*)

Fig. 3.14 (*b*) Distribution of lava flows of the Columbia and Snake River Plateaus.

3.39 Note the map of major igneous rock bodies along western North America (Fig. 3.15). Name the type of igneous bodies shown, and name the likely dominant rock type present.

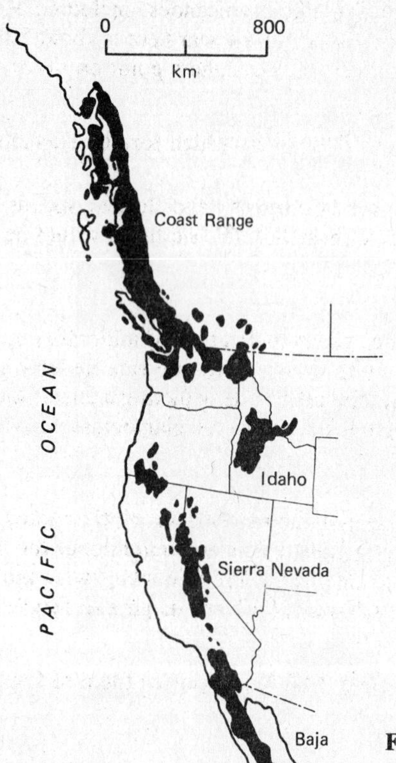

Fig. 3.15 Map of batholiths of western North America. The plutonic rock has been uplifted and exposed by erosion.

They are batholiths because of their large size. The Sierra Nevada batholith covers about 70,000 km^2 (27,000 mi^2). Like many batholiths, it is a composite of many individual intrusions which may have cooled at depth beneath thick piles of sedimentary and volcanic rocks. Most have a granitic composition.

3.40 Figure 3.16 includes three sketches of the Hawaiian Islands. (*a*) What is the origin of the islands?

They are volcanoes.

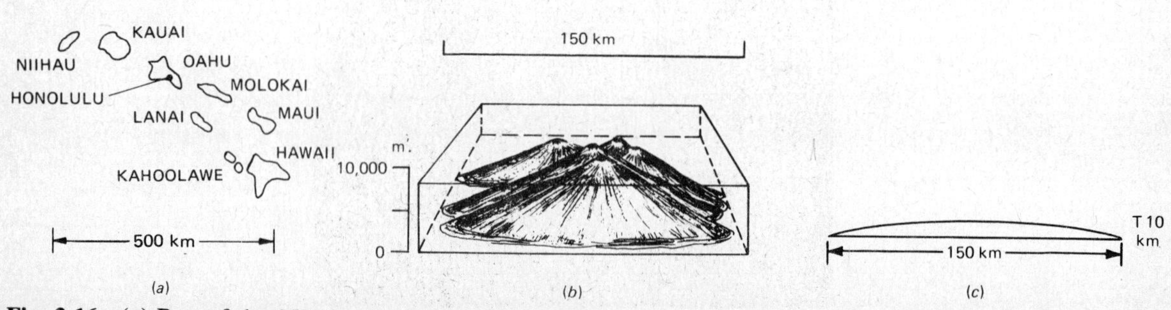

Fig. 3.16 (*a*) Part of the 2500-km-long Emperor Seamount chain (Hawaiian Island chain) of volcanoes. Only those on the island of Hawaii are now active. (*b*) The island of Hawaii. The vertical scale is greatly exaggerated. (*From Ojakangas and Darby, 1976.*) (*c*) In this cross-section, the vertical and horizontal scales are the same.

(*b*) Figure 3.16(*b*) is the island of Hawaii. What types of volcanoes make up the island?

Shield volcanoes. In the sketch, they may at a glance appear to be composite cones, but note the scale. In order to depict it, the diagram has a large vertical exaggeration. A truer scale is shown in the cross-section of Fig. 3.16(*c*). The island of Hawaii consists of five shield volcanoes, including Mauna Loa, which is 150 km across and stands 4200 m above sea level. As the ocean near here is about 5000 m deep, Mauna Loa is higher than Mt. Everest (8848 m above sea level), the highest point on the continents.

3.41 Refer to Fig. 3.16. What is the viscosity and composition of the lavas which formed Hawaii?

Modern flows in Hawaii (eruptions are common) have a low viscosity and flow at speeds of tens of kilometers per hour. The composition of the flows is mafic (basaltic). (Felsic flows would be too sticky or viscous to flow rapidly.)

3.42 Refer to Fig. 3.3 which has been introduced as the order of crystallization of minerals out of a mafic magma. More specifically, it is known as Bowen's Reaction Series, with one side depicting discontinuous reactions of minerals with the remaining magma and the other side depicting a continuous reaction of minerals with the remaining magma. What does it tell us about (*a*) plagioclase and (*b*) olivine, pyroxene, and hornblende?

The right side shows that plagioclase crystals will become more and more sodic as they grow in magma. They are, in fact, commonly zoned with the inner zones the most calcic and outer zones the most sodic. The left side or discontinuous reaction side shows that olivine, if left in contact with the remaining magma, will react to form pyroxene and that pyroxene will react with the magma to form hornblende.

3.43 Note the columnar joints in igneous rocks in Figs. 3.14 and 3.17. What causes them to form, and how are they oriented relative to the shape of the body?

Fig. 3.17 Columnar joints in Devils Tower, Wyoming, which has been interpreted as a volcanic rock or as an erosional remnant of a large sill.

They are cracks that may form when intrusive magma or extrusive lava cools, because the solid rock occupies less volume than the liquid. The columns are commonly formed perpendicular to the broad flat cooling surfaces of the flows, sills, and dikes. Thus the columns are vertical in a horizontal flow or sill and horizontal in a vertical dike.

3.44 What minerals commonly fill vesicles to form amygdules as in Fig. 3.10?

Several minerals are common, including quartz, calcite, various radiating hydrous aluminum-silicates of the zeolite family, and the green minerals chlorite and epidote. World-famous Lake Superior agates are simply microcrystalline quartz amygdules with reddish color bands caused by minor amounts of iron oxide.

3.45 If volcanoes are dangerous, as illustrated when Mt. St. Helens erupted in May 1980, why are some volcanic regions such as Sumatra in the East Indies so densely populated?

Because volcanic ash is quickly converted, in less than 50 years, to a rich soil, especially in a subtropical climate. People like to eat.

Chapter 4

Weathering and Soils

MECHANICAL WEATHERING

Rocks are stable in the environment in which they formed (see Fig. 2.6). Plutonic rocks and metamorphic rocks form beneath the earth's surface and are therefore generally unstable at the surface where air and water can attack them.

Mechanical weathering is the breakdown of rocks into smaller pieces. This is accomplished by many agents, but frost action is the most important in temperate and polar climates. Water, which expands in volume by about 9 percent when it freezes, can penetrate the tiniest of cracks. It freezes at the top of the crack first, thus sealing the remainder of the crack. As the remaining water turns to ice and expands, pressures as great as 30,000 lb/in^2 can be generated. This breaks rock into smaller pieces. This process is very important, as more surface area thus becomes available for air and water to attack (Fig. 4.1).

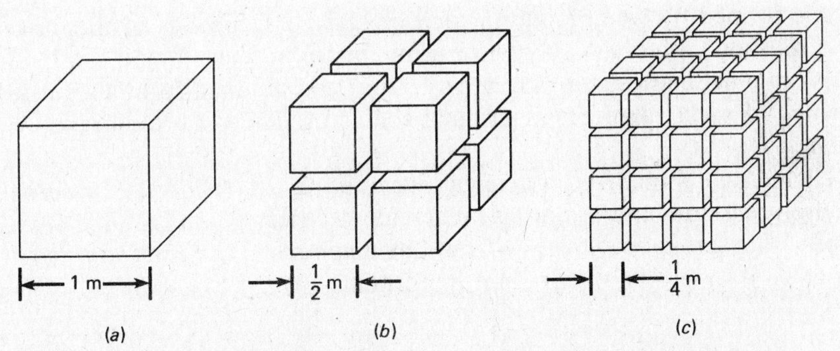

Fig. 4.1 As a cube of rock breaks into smaller pieces, the surface area increases. The cube in (*a*) is one meter on a side and has a surface area of 6 m^2. If it is broken into 8 cubes each 0.5 m on a side, the total surface area of all the cubes together will be 12 m^2. If broken into cubes 0.25 m on a side, the area will again double. Finally, if broken into small cubes 1/6000 m on a side, the total surface area will be 216,000 m^2, which is equal to the surface area of 50 standard football fields! (*From Ojakangas and Darby, 1976.*)

CHEMICAL WEATHERING

Chemical weathering is the chemical breakdown of rock, with new products formed in the process. Air (especially the oxygen fraction) and water (especially if it contains carbon dioxide—CO_2—which combines with water to form carbonic acid—H_2CO_3) react chemically with the rock. The oxygen combines with the iron in minerals, and iron oxides are the result. As the iron is removed from the crystal structure during this *oxidation*, the structure disintegrates further.

A small amount of rock can be dissolved by even distilled water, because part of the water dissociates into hydrogen ions (H^+) and hydroxyl ions (OH^-) in the process called *hydrolysis*. The H^+ attacks the silicate minerals and can be thought of as replacing the cations of the minerals (the K^+, Na^+, Ca^{2+}, and Mg^{2+}), thus sending them into solution and forming clay minerals that can be represented by the general qualitative formula HAlSiO, which does not show the relative quantities of each element. (In addition, most clays contain some of the just named cations, and others as well.) Acidic waters are more effective, for the acids also dissociate and result in more hydrogen ions available for attacking the minerals. Carbonic acid is the most common acid in natural waters; the CO_2 comes from the air and from plant decay in the soil. Chemical weathering involving carbonic acid is called *carbonation*. Carbonation attacks the silicate minerals to produce clay and also dissolves calcite, the mineral in the common sedimentary rock limestone.

The order of weathering of the common igneous minerals (the Goldich Mineral Stability Series) is the same as the order in which they crystallize from a magma (see Fig. 3.3). Olivine, pyroxene, and calcium plagioclase are the least stable, and quartz is the most stable.

CLIMATE AND SOILS

The chemical weathering processes discussed above are most effective in a hot and wet climate, for the rate of chemical reactions is higher than in a cold and dry climate. The cations released during weathering may be leached out and carried to the sea if the rainfall is high. In a more temperate climate, some cations remain to be utilized in the new clay minerals formed during chemical weathering, thus resulting in clays rich in the nutrients required by plants (for example, K^+, Na^+, Ca^{2+}, Mg^{2+}, P^{3+}, and many others). Therefore, climate is the major control on the type of clay minerals formed and on the quality of the soil.

Soils contain clay minerals, iron oxides (which are relatively insoluble), quartz, and organic matter. A soil profile formed by weathering on rock or sediment consists of three horizons—the dark uppermost or A horizon with much clay and organic material plus silt and sand; a lighter-colored B horizon with less organic matter and rich in clay, iron, and alumina from above; and the C horizon with an abundance of rock fragments or original sediment. The C horizon rests on the parent bedrock or sediment.

W. D. Keller has summarized chemical weathering as follows: "Mountain building and volcanism require great amounts of energy, whereas weathering has plodded along on a few thousand calories per mole throughout the ages, but has made soils, concentrated minerals, and carried organic evolution as a by-product." So, is it important?

Solved Problems

4.1 What are the two main types of weathering?

Mechanical (disintegration), the breaking of rock into smaller pieces, and chemical (decomposition), the chemical breakdown of rock.

4.2 What are the agents of mechanical weathering?

Frost action is most important in temperate climates. Other factors are burrowing by animals, the growth of plant roots, human activity, the expansion of rock due to "unloading" by erosion of overlying rock, the swelling of rock due to the formation during chemical weathering of minerals that have greater volumes than the original minerals, and possibly the daily thermal expansion and contraction of minerals in desert environments.

4.3 What causes chemical weathering?

Air (especially the oxygen in it) and water (especially if it contains acids, as does most rainwater and groundwater).

4.4 In which type of climate is chemical weathering most effective?

Subtropical to tropical climates, where it is warm and where moisture is abundant. The rate of chemical reactions approximately doubles for every 10°F increase in temperature. Water helps to bring the reactants together and carries away the products in solution.

4.5 Do mechanical and chemical weathering operate independently of each other?

Rarely. However, in polar climates the temperature is so low that chemical reactions are very slow, and in deserts the humidity is low and this slows reactions.

4.6 Why is mechanical weathering so important to chemical weathering?

Because as mechanical weathering proceeds, the amount of surface area available for attack by air and water increases, and thus more chemical weathering occurs. See Fig. 4.1.

4.7 Joints, or cracks in rocks, are natural passageways for air and water. What should happen to joints with time? See Fig. 4.2.

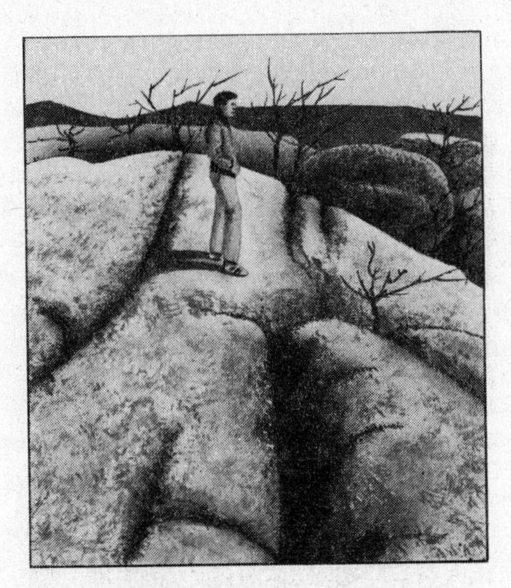

Fig. 4.2 Weathered joints in granite in Missouri.

They should become wider, largely because of chemical weathering.

4.8 How might the big rounded granite "boulders" in Fig. 4.3 have formed? (*Clue:* They have been formed at this site and have not been carried here by ice, water, or humans.)

By weathering along joints. Imagine three sets of joints at right angles to each other. Cubelike shapes would be bounded by the joints. As chemical weathering proceeds, the corners of the cubes are attacked from three sides, and thus more material is removed there compared to a side of the cube which is attacked on only one surface. The result can be rounded rocks like these.

Fig. 4.3 "Elephant rocks" in granite in Missouri. The location is the same as Fig. 4.2, showing that the widening of joints is a step along the way to big rock forms such as these.

4.9 How are *talus* piles below cliffs, as in Fig. 4.4, related to weathering?

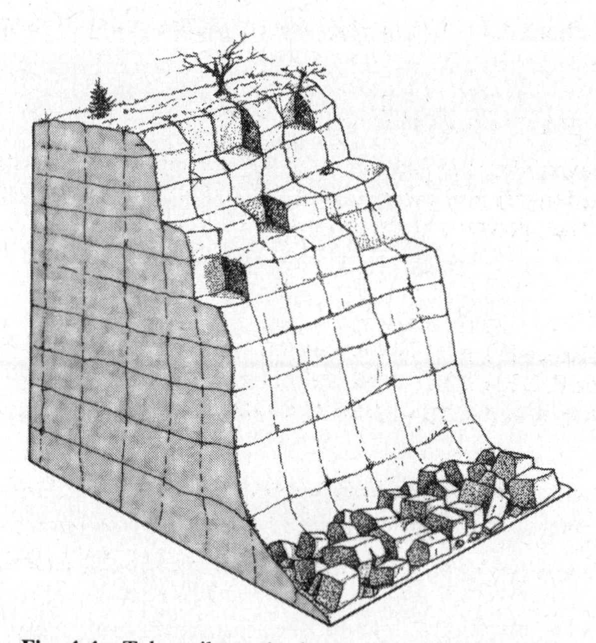

Fig. 4.4 Talus pile at the base of a cliff in Utah.

Generally they are the result of frost action that pushes blocks off jointed cliffs, and they fall under the influence of gravity. Upon falling, they may break into smaller pieces.

4.10 Which minerals does the chemical weathering process called *oxidation* affect most readily?

Those minerals that contain iron. The common iron-bearing minerals are olivine, pyroxene, hornblende, and biotite.

4.11 What are the common products of oxidation?

In everyday language, rust. In a mineralogical sense, the products are the iron oxide minerals, hematite (Fe_2O_3) and limonite ($Fe_2O_3 \cdot H_2O$). The process of hydration (the addition of water to the crystal structure) is also involved in the production of limonite.

4.12 Which chemical weathering processes are most important in the production of clay minerals?

Hydrolysis and carbonation.

4.13 Explain carbonation.

Carbonation is the process whereby carbonic acid (H_2CO_3), a common constituent of groundwater, attacks rocks and causes a chemical reaction. The main product is clay, with a general qualitative formula of HAlSiO. (Recall that most clays contain the "minerals"—really the cations—that end up in foods.)

4.14 Give a generalized chemical equation that illustrates the effect of carbonation on limestone, the common sedimentary rock made of the mineral calcite ($CaCO_3$).

$$H_2CO_3 + CaCO_3 = Ca(HCO_3)_2$$

In words, this states that carbonic acid plus calcite forms calcium bicarbonate, the soluble form of calcium carbonate found in water.

4.15 Give a generalized chemical equation that illustrates the effect of carbonation on silicate minerals.

$$MAlSiO + H_2CO_3 = HAlSiO + MCO_3 + HSiO$$

In words, this can be expressed as follows: A common silicate mineral (with the cations of K^+, Na^+, Ca^+, Mg^+ represented by *M*) plus carbonic acid yields clay, a carbonate mineral (in solution), and minor hydrated silica. The important product is insoluble clay minerals.

4.16 Explain hydrolysis.

Some water (H_2O) dissociates into hydrogen ion (H^+) and hydroxyl ion (OH^-). The H^+ in essence replaces cations such as K^+, Na^+, Ca^{2+}, and Mg^{2+} (see Table 2.2) from silicate minerals, liberating them, and helps produce clay with a general qualitative formula of HAlSiO.

4.17 When a biotite granite weathers in a temperate climate, what products remain to form the soil?

Quartz sand and silt, iron oxides (hematite and/or limonite), and clay minerals. Organic matter is added by plants.

4.18 When a biotite granite weathers in a temperate climate, what cations may be carried away to the ocean?

Largely K^+, Na^+, Ca^{2+}, and Mg^{2+}.

4.19 Are all the ions listed in the previous problem carried away to the ocean and hence lost to the soil?

Fortunately, no. Many ions are taken up by clay minerals, from whence they can be released as necessary nutrients for plants.

4.20 What is the order of chemical stability of the common igneous minerals?

The order is the same as Bowen's reaction series of Fig. 3.3. The first minerals to crystallize from a magma are the first to weather. This is known as the Goldich Mineral Stability Series, after S. S. Goldich.

4.21 Which igneous mineral is the most stable, both mechanically and chemically?

Quartz. It has a three-dimensional crystal structure with strong covalent bonds and is therefore very hard and resists abrasion. The strong chemical bonds also make it resistant to chemical attack.

4.22 Which mineral, quartz or olivine, should be most abundant in beach sands?

Quartz. (See the previous problem.) Olivine breaks down by oxidation, as described in Problem 4.11. Olivine sands do occur in Hawaii, but the olivine is a product of geologically recent volcanism and time has not been sufficient to allow for its weathering.

4.23 Which rock should be most resistant to chemical weathering—gabbro or granite?

Granite, because it consists of the more stable minerals quartz, orthoclase, and a little biotite or hornblende (see Figs. 3.3 and 3.4). Gabbro consists of olivine, pyroxene, and intermediate plagioclase containing both calcium and sodium.

4.24 What is the most important factor in the production of good soils—the climate, the topography, or the rock type?

Climate is the most important, as it controls the amount of chemical weathering. Chemical weathering is necessary for the production of clay minerals. Topography is not as important, but it is nevertheless a big factor because if the relief is appreciable, rock material may be eroded before it can be weathered chemically. Rock type is also a factor; for example, a soil formed on a granite will contain the nutrient potassium (from orthoclase), whereas a soil formed on gabbro will not contain much potassium as gabbro contains no potassium-rich minerals.

4.25 Why are the best clay minerals, and hence the best soils, formed in temperate climates?

Because the climate is warm enough and moist enough to promote chemical reactions, but the rainfall is not so high that all the essential nutrients that the plants need (e.g., K^+, Na^+, Ca^{2+}, and Mg^{2+}) are leached away and carried to the oceans. Many of the ions are thus available to become part of the clay mineral structure.

4.26 Why are the soils in the tropics generally poor soils?

Because most of the valuable cations are leached away and the humus content of the soil is low because the plant matter in the soil decays 12 months a year and does not accumulate.

4.27 Why are clay minerals essential for a good soil?

Because many of the cations listed in Problem 4.25 are easily removed from the sheet structures of the clay minerals by plant roots. In minerals that have three-dimensional, strongly bonded structures (such as the feldspars), the cations are not easily available to the plants.

4.28 What are the products of chemical weathering in a subtropical to tropical climate?

With a high temperature and a high rainfall, chemical weathering is intense. Oxidation proceeds rapidly and iron oxides are formed. Clay minerals are formed as well, but they are cation-deficient types because of leaching. If the groundwaters have a high pH (low acidity), even the silica of the clay minerals will be leached away, leaving $Al_2O_3 \cdot H_2O$, which is the mineral bauxite. The soil is obviously lacking in the cations that plants require.

4.29 Why are so many soils yellow, brown, or red in color?

The iron oxides produced during chemical weathering color the soil. A small amount of hematite makes a soil red, whereas a small amount of limonite will make a soil some shade of yellow or brown.

4.30 Why are some soils black?

Because of a high content of incompletely decayed organic matter.

4.31 Where in the world are the high-protein plants grown and why?

In the temperate climatic zones where the clay minerals still contain the essential cations. For example, wheat with as much as 18 percent protein grows in temperate climates, whereas it will not grow well in subtropical or tropical hot climates. And, even if it could grow there, it would not be a nutritious wheat because the soil would not possess the proper cations. Less nutritious foods such as cassava (tapioca) with only 1.9 percent protein or bananas with only 1.2 percent protein both grow well in the tropics, but they are low in protein.

4.32 If subtropical and tropical climates produce poor soils, can any good come of the intense chemical weathering? (See Problem 4.28.)

Yes. Mineral deposits can result from extreme chemical weathering. Iron oxides, if sufficiently concentrated, can make up iron-bearing layers rich enough to mine as iron ore. The ore of aluminum is bauxite ($Al_2O_3 \cdot H_2O$). Gold can be freed by chemical weathering of the rock in which it resides. Nickel is commonly concentrated in residual soils.

4.33 What is *soil?*

Soil is a mixture of clay minerals, other mineral particles, and organic material.

Supplementary Problems

4.34 Describe a typical soil profile.

As shown in Fig. 4.5, a soil profile contains A, B, and C horizons.

4.35 Which rock should be most resistant to chemical weathering in, a temperate climate—a granite consisting of the igneous minerals quartz, orthoclase, and biotite or a red shale consisting of clay minerals, iron oxides, and minor quartz silt?

The red shale. Its components are already the products of chemical weathering at the earth's surface. The granite will weather much more rapidly. However, if the groundwater is alkaline (see Problem 4.28), even the clay can be broken down.

4.36 Explain the difference between weathering and erosion.

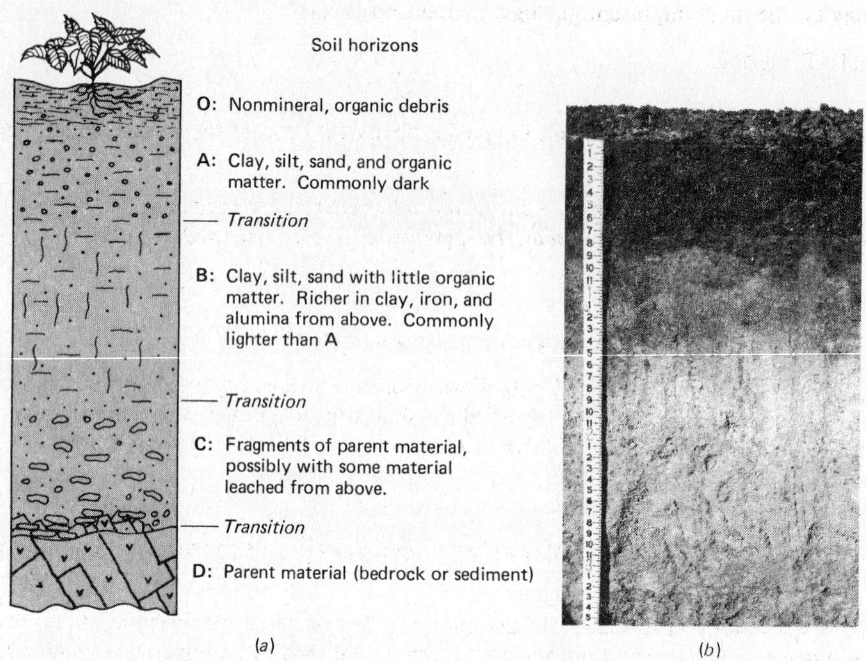

Fig. 4.5 Idealized soil profile in a temperate climate showing the A, B, and C horizons. (*From Ojakangas and Darby, 1976.*)

Weathering is the breakdown of preexisting rocks into smaller pieces and new products. Erosion is the removal of the materials from the sites at which they formed.

4.37 If you want your tombstone to survive a long time as a legible monument to your greatness, should you request that it be made of marble (a metamorphic rock consisting of calcite), or slate (a metamorphic rock consisting largely of fine-grained muscovite and quartz), or steel? (Request burial in a temperate climate.)

Request slate. The marble will slowly be dissolved by acid rainwater and organisms, even in an unpolluted area. The steel will rust. On neither will the inscription be legible for long. The slate is made of resistant minerals. (See Problem 4.20.)

4.38 What is *spheroidal weathering?*

The tendency of a rounded boulder to become smaller by the spalling (splitting) off of layers or shells of chemically weathered rock. The clay minerals produced by weathering in the outer layer of the rock into which air and water can penetrate have greater volumes than the original silicate minerals. This volume expansion in the outer layer causes it to spall off as the rock is weakened.

4.39 What is *exfoliation?*

The process whereby large curved slabs of rock break off of rock bodies of a homogeneous nature, such as a granite. This is probably partly the result of "unloading" of overlying rock cover and partly the result of chemical weathering as in the previous question.

4.40 Weathering affects the topography of the earth's surface in a broad and important way. How?

By breaking rocks down, mechanically and chemically, the broken rocks and the new products are available for removal by the main agents of erosion—running water, glacial ice, and wind. Thus the elevation of the land is lowered, and even mountains are worn away.

4.41 What may be the most important geologic process on earth?

Chemical weathering.

4.42 What mineral group is most abundant in igneous rocks, and why is its abundance so important to people?

The feldspars. It is important because this group weathers to form the major part of the clay minerals in soils. Note that the feldspars contain all the ingredients of clay (HAlSiO) plus the valuable cations (K^+, Na^+, and Ca^{2+}).

4.43 Which other elements, besides iron, are commonly oxidized?

Mn (manganese), U (uranium), and S (sulfur). Manganese ions have electric charges of +2 and +4, uranium of +4 and +6, and sulfur of +4 and +6. During oxidation, there is a loss of electrons (see Problem 2.24) and thus a gain in positive electric charge. The oxidized state of iron is +3; the reduced state is +2.

4.44 Why is the order of weathering of the common igneous minerals the same as Bowen's reaction series (see Problem 4.20)?

Because of the energy of formation of the minerals. Those that crystallized at the highest temperature (olivine, pyroxene, and calcium plagioclase) are most stable at high temperatures and pressures because of the greater stability of the chemical bonds and crystal structures under those conditions. Conversely, quartz crystallizes at a lower temperature and the chemical bonds and the crystal structure are stable at lower temperatures and pressures. Conditions at the earth's surface obviously include lower temperatures and lower pressures than at depth in the earth's crust.

4.45 How does the industrial revolution, with its output of CO_2 (carbon dioxide) from the burning of coal and petroleum affect chemical weathering?

The increased CO_2 output causes an increase in the amount of carbonic acid which will then result in an increase in the chemical weathering process of carbonation. (See Problems 4.13, 4.14, and 4.15.)

4.46 How does a gabbro weather in a temperate climate?

Mechanical weathering will certainly proceed, as will chemical weathering. A gabbro consists of olivine, pyroxene, and plagioclase (see Table 2.3). The iron will oxidize; the aluminum and silica will form clay; some of the potassium, sodium, calcium, and magnesium ions will go into solution and some will become part of clay minerals; and minor silica will go into solution. Therefore, the soil will consist of clays and iron oxides. (There will probably not be quartz silt and sand present, as gabbros do not generally contain quartz.)

4.47 What is the origin of acid rain, and how does it relate to weathering?

Sulfur dioxide (SO_2) from the burning of fossil fuels (especially coal), and nitrous oxides (NO_3) from automobile combustion combine with humidity in the air to form sulfuric acid (H_2SO_4) and nitric acid (H_2NO_3), which make the rainwater acidic. These acids attack rocks, just as the carbonic acid of Problems 4.13, 4.14, and 4.15.

4.48 What are the three main soil groups as classified on the basis of the minerals present in the A and B horizons?

Pedalfers: Those containing an abundance of aluminum and iron oxides and clay minerals. These are the common soils in eastern United States.

Pedocals: Those containing an abundance of calcium carbonate (calcite). These are the common soils in western United States.

Laterite: Those containing an abundance of iron oxides and aluminum oxides (bauxite). These are abundant in subtropical and tropical areas.

4.49 What are the three main clay groups that form by chemical weathering?

Kaolinite: Without appreciable cations other than H^{+1}, Al^{+3}, and Si^{+4}.

Smectite (montmorillonite): Expandable clays which expand as water and cations are incorporated between the sheets of ions in the crystal structure. Various ions are present. (This is the best clay for agriculture.)

Illite: With potassium ions (K^{+1}) between sheets of ions in the crystal structure.

Chapter 5

Sedimentary Rocks

INTRODUCTION

About 75 percent of the earth's land surface is covered by sedimentary rocks (Fig. 5.1), although by volume they make up only about 5 percent of the earth's crust. They are very important, for by understanding the origin of sediment and sedimentary rocks, much of earth's history can be deciphered. Sedimentary rocks also contain fossils (traces or remains of prehistoric life), and therefore the development and changes in life through time can be studied. Sedimentary rocks can also be of great commercial value, for nearly all earth's petroleum and groundwater are in sediments or sedimentary rocks, as are many valuable commodities such as uranium, gold, aluminum, and iron.

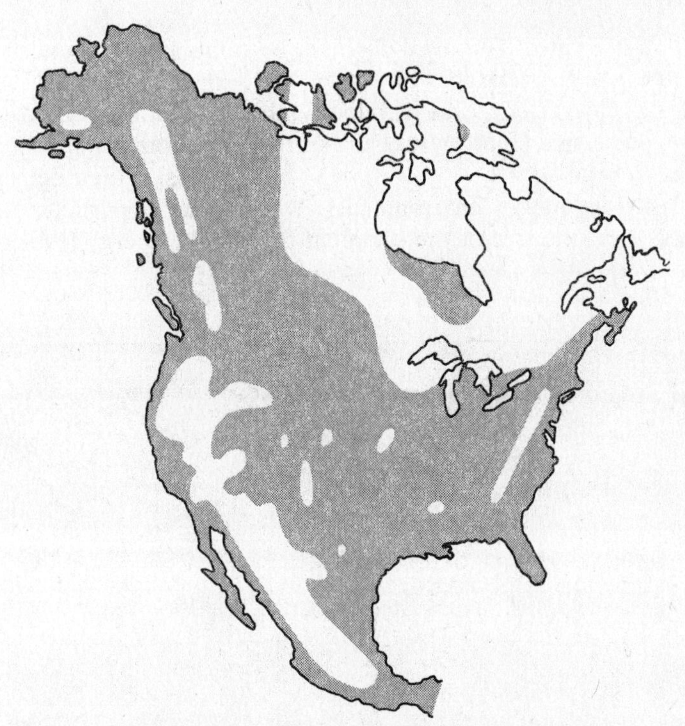

Fig. 5.1 Generalized map of North America showing the portion underlain by sedimentary rocks (shaded). The unshaded portions are underlain by igneous and metamorphic rocks. (Actually, the amount of sediment is even higher, for some of the metamorphic rocks were sedimentary rock types prior to metamorphism.)

SEDIMENT

Sedimentation is the process whereby particles (sediment) settle out of suspension. *Sediment* consists of fragments of minerals and rocks, as well as particles that are precipitated out of solution. The particles are classified according to size (Table 5.1). Practically all sediment is in one way or another a product of weathering, as discussed in Chapter 4.

Table 5.1 Wentworth Scale of Sediment Sizes (in millimeters)

Boulder	256
Cobble	64–256
Pebble	4–64
Granule	2–4
Sand	1/16–2
Silt	1/256–1/16
Clay*	< 1/256

*Note that the term *clay* has two meanings—grain size, as used here, and the name of a group of minerals with sheet structures.

Most sediment, once formed, is moved downhill by running water, under the influence of gravity. Some is moved by wind, glacial ice, or currents in lakes and oceans. Wind and water deposit their loads when they undergo a decrease in velocity; this is generally true for glacial ice as well, but the temperature is an additional factor here. During transportation to the depositional site, or during movement within the depositional site, *sorting* of sediment by size (and density) occurs. Small particles are deposited in low-energy environments, whereas large particles are deposited in high-energy environments. Thus the size of the sediment reveals the energy level at that site during deposition.

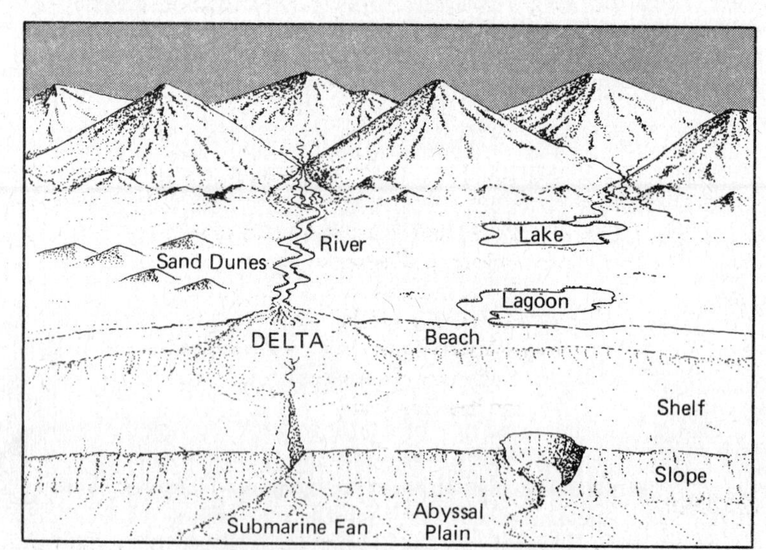

Fig. 5.2 Various environments in which sediment can be deposited. The ultimate site of deposition is the ocean.

Most sediment ultimately reaches the ocean (Fig. 5.2), there to eventually become sedimentary rock after burial, cementation, and compaction. However, the journey to the ocean may take a long time; for example, a grain of sand in a river may "bar hop" from sandbar to sandbar all the way to the ocean, taking hundreds or thousands of years, or more, to complete the trip.

Abrasion rounds particles of sediment. Boulders, cobbles, and pebbles are rounded easily due to chipping upon impact in high-energy environments, whereas sand rounds with difficulty because of the "cushion" provided by water. Wind is the best rounding agent, because air does not provide such a cushion. A sand consisting of sorted and rounded quartz, the most durable common mineral, is mineralogically and texturally *mature*.

Three main principles provide the framework to understanding the origin of sedimentary rocks. The *principle of uniformitarianism* can be broadly stated as "the present is the key to the past." That is, the observation of sedimentational processes occurring today helps in the understanding of the origin of older sedimentary rocks, and hence of earth history. The *law of superposition* states that in a sequence of sedimentary layers (strata), the oldest layer or bed is on the bottom and the youngest is at the top. That is, they were deposited bed by bed, one after the other (Fig. 5.3). The *principle of original horizontality* states that sediment, and hence sedimentary rocks, is generally deposited in horizontal or nearly horizontal beds. Therefore, if beds are tilted or vertical, geologists know that deformation has occurred.

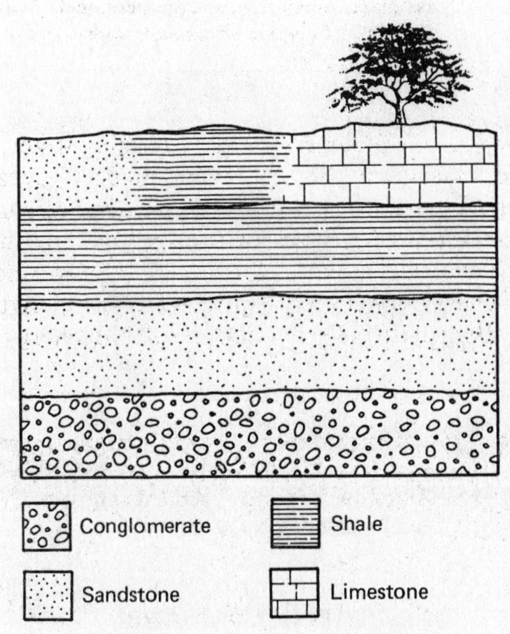

Fig. 5.3 Diagram of a cross-section of a sequence of sedimentary rock layers (strata) such as might be exposed in a cliff. Note that the different symbols indicate different sedimentary rock types. Also note the tree for scale.

SEDIMENTARY ROCKS

Sediment, when lithified (i.e., hardened), becomes sedimentary rock. Cementation and compaction are the two main lithification processes. Calcium carbonate ($CaCO_3$) and silica (SiO_2) are the two main cements, and they are precipitated out of water moving through the sediment.

There are three main types of sedimentary rock by volume: clay-rich or muddy rocks (50 percent or perhaps as much as 80 percent), sandstone (30 percent), and limestone (20 percent). All other sedimentary rock types, and there are many (mostly chemical precipitates), are volumetrically minor, comprising less than 1 percent of the total. Mud rocks include flaky-textured *shales* and massive-textured *mudstones*. Sandstones include *quartz sandstones, arkose sandstones* (with an abundance of feldspar in addition to quartz), *lithic sandstones* (with an abundance of rock fragments in addition to quartz), and *graywackes* (with an abundant clayey matrix in addition to rock fragments, feldspar, and quartz). *Limestone,* consisting of the mineral calcite, usually has a biochemical origin. Some limestone is altered during *diagenesis* (changes after deposition of the sediment) by magnesium-rich waters and becomes *dolomite* (dolostone) containing the mineral dolomite $CaMg(CO_3)_2$. Most dolomitized limestones become dolomitic limestones rather than pure dolomites.

SEDIMENTARY STRUCTURES

Bedding is the most obvious sedimentary structure. There are four main types—cross-beds, graded beds, laminated beds, and massive beds. There are many other sedimentary structures, including ripple marks and mudcracks. Such structures are of importance in deciphering the environment of deposition. Cross-beds and some other structures are *paleocurrent* (ancient current) *indicators* (Fig. 5.4). The assemblage of sedimentary structures is used to reconstruct the paleogeography and earth history.

Fig. 5.4 Tilted beds of cross-bedded sandstone. Note geologist for scale.

Solved Problems

5.1 How are sediments formed on the earth's surface?

This question could be answered with one word—weathering. Because minerals, and thus also rocks which are made of minerals, are stable only in the environment in which they formed, any change of environment will affect mineral stability. Igneous and metamorphic rocks formed at depth are unstable at surface conditions of 20°C, 1 atm of pressure, and abundant air and water. Even lava flows are unstable because they crystallized at a temperature of many hundreds of degrees. Few minerals are truly stable at surface conditions—quartz, which has a wide stability range, and clay, which forms at surface conditions, are the two most common stable minerals. However, even the clay can be changed to a different clay or to bauxite.

5.2 Name the three most abundant types of sedimentary rocks in order of decreasing abundance and tell why each is so abundant.

Mud rocks, sandstone, and limestone. The mud rocks are abundant because clay is a very abundant weathering product. Sandstone is abundant because quartz is the most resistant (chemically and mechanically) common mineral and because other minerals and rock fragments may be incompletely weathered. Limestone is abundant because calcium ions (Ca^{2+}) are freed in abundance during chemical weathering and combine with CO_2 to form calcite ($CaCO_3$).

5.3 What is the genetic difference between the two main classes of sediments—clastic and chemical?

Clastic ("broken") sediments are those composed of mineral and rock particles (clasts) which were eroded from other rocks or sediments. Chemical rocks are precipitated out of solution. Some limestones don't easily fit this dual classification, for although they are chemical or biochemical precipitates, they are clastic in that the clasts were derived from limey sediments which precipitated elsewhere in the basin.

5.4 How widespread are sedimentary rocks and sediments on the earth's surface?

Sedimentary rocks cover three-fourths of the earth's land surface (Fig. 5.1), with the thickness of the sedimentary rock column ranging from a few meters to a few tens of kilometers. In addition, the ocean basins are nearly everywhere covered with at least a thin cover of sediment over sedimentary rock or basalt.

5.5 Name the four main clastic sediment and sedimentary rock types on the basis of texture alone. State the size boundaries, in millimeters. (This answer will be obtained from Table 5.1.)

Sediment	Sedimentary Rock	Grain Size (mm)
Gravel	Conglomerate	2 mm and up
Sand	Sandstone	2–1/16
Silt	Siltstone	1/16–1/256
Clay	Shale, mudstone	Less than 1/256

5.6 How are loose sediments transformed (lithified) into hard sedimentary rock?

Upon burial beneath other sediments, compaction and cementation (generally by calcium carbonate or silica) take place.

5.7 Why is most sediment and sedimentary rock deposited in horizontal or near-horizontal layers or beds?

The agents of transportation—water, wind, and glacial ice—under the influence of gravity—rework and deposit sediment as horizontal or nearly horizontal beds. This is the principle of *original horizontality*. Locally, as near the base of a mountain, some sediment may be deposited with an initial dip of a few degrees.

5.8 Where is sediment deposited? That is, what are the environments of deposition? (See Fig. 5.2.)

There are many different environments in which sediment is deposited. On the continents, they are mainly at the bases of mountains as broad alluvial fans or plains, in river valleys, in lakes, and in swamps. In the oceans, they can be deposited on the shallow continental shelves, on the deeper slopes, and to a lesser extent, in the ocean deeps. At the boundary between continent and ocean, transitional environments include deltas, lagoons, estuaries, and beaches.

5.9 Why is the principle of uniformitarianism so important to sedimentary geologists?

By observing modern sediments and sedimentational processes, geologists can learn how ancient sedimentary rocks were formed.

5.10 (*a*) Define and explain the significance of the *law of superposition* which was first expounded upon by Nicholas Steno, a Dane, in 1669.

In a sequence of horizontal sedimentary rock layers, the younger layers are on the top and the older layers are on the bottom. To put it another way, each bed is younger than the bed beneath it. This realization was highly significant when first proposed, for it supported the concept of a gradual development of the rock column rather than a one-shot instantaneous creation. Geologists today use it as the basis for a bed-by-bed analysis of geologic history in a given area. (In the special case where a sequence of beds has been overturned by deformation of the earth's crust, superposition is not directly applicable.)

(*b*) What are the relative ages of the beds shown in Fig. 5.3?

The conglomerate is the oldest, then the sandstone, then the mudstone, and the youngest is the top layer that consists of three rock types.

5.11 (*a*) Why are cross-beds (Fig. 5.4) very important sedimentary structures?

Because they provide evidence about the environment of deposition and on the directions which the depositing currents (i.e., the paleocurrents) were moving, thus yielding data for paleogeographic interpretations. In areas where the beds have been folded and deformed, they can be useful as "top indicators" (i.e., which side of the bed was originally the top?), thereby identifying overturned beds.

(*b*) Which way were the currents that deposited the cross-bedded sand of Fig. 5.4 moving—left to right or right to left?

Left to right. Cross-beds dip downward in the direction toward which the current was moving.

5.12 What type of bed is illustrated in Fig. 5.5(*a*)?

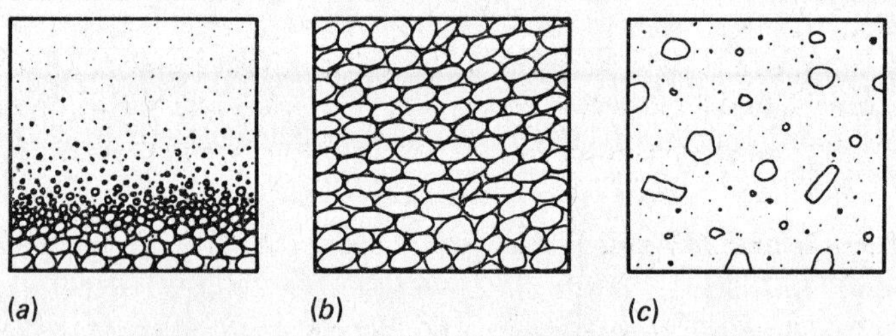

Fig. 5.5 Diagrams of three of the textures that are possible in sediments as a result of sorting. Different depositional agents are responsible for each texture.

A graded bed, in which there is a gradual decrease in grain size from bottom to top, for example, from coarse sand up to silt.

5.13 In Fig. 5.5, which sediment is most poorly sorted and which is best sorted, and why?

The sediment in Fig. 5.5(*b*) is the best sorted and that in 5.5(*c*) is poorly sorted. Poor sorting is generally the result of rapid deposition and little reworking, or deposition by a viscous medium such as a mudflow or a glacier. Running water sorts fairly well, but wind and wave action cause even better sorting.

5.14 How were thick sequences of abundant graded sandstone beds and mudstones (Fig. 5.6) probably formed?

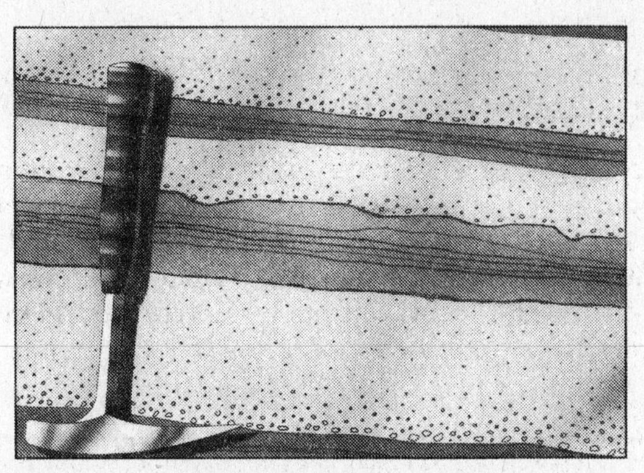

Fig. 5.6 A sequence of graded sandstone (graywacke) beds separated by mudstones.

By repeated turbidity currents which sporadically interrupted a continuous ''rain'' of mud (clay and silt) settling out of suspension onto the sea floor.

5.15 (*a*) Figure 5.7 includes a map and cross-section of a part of Grand Banks, Newfoundland, and the adjacent sea floor. On November 8, 1929, an earthquake occurred here, causing a huge submarine landslide and the subsequent breakage of several trans-Atlantic telegraph cables on the sea floor. Most geologists believe the landslide incorporated water as it moved downslope and became a turbidity current. Calculate the approximate speed of the current between points H and I, I and J, J and K, and K and L, in nautical miles per hour (i.e., knots). (1 nmi = 1.15 mi = 1.85 km.) Please read the figure legend before attempting to answer the question. You will need a ruler to accurately measure the distance along the scale at the bottom of the diagram.

Between H and I, 112 nmi is the measured distance. The elapsed time is 2 h and 4 min, or 2.07 h. Then, 112/2.07 = 54.1 knots, or 54.1 nmi/h. Since 1 nmi = 1.15 mi, this is 62.2 mi/h.

Between I and J—121 nmi, 5 h and 58 min, or 5.97 h. Then, 121/5.97 = 20.27 knots. Times 1.15, this is 23.3 mi/h.

Between J and K—21 nmi, 1 h and 17 min, or 1.28 h. Then, 21/1.28 = 16.41 knots. Times 1.15, this is 18.9 mi/h.

Between K and L—33 nmi, 2 h and 59 min, or 2.98 h. Then, 33/2.98 = 11.07 knots. Times 1.15, this is 12.73 mi/h.

(*b*) Decades after this event, oceanographers obtained samples of the sediment on the sea floor where the sediment finally came to rest to determine whether or not a turbidity current had indeed been responsible for the cable breaks between points H and L. This was proven to most geologists' satisfaction. What must they have found in the sediment to prove this?

A graded bed, such as the type deposited by turbidity currents.

5.16 The four main types of sandstone are illustrated in Fig. 5.8. Which is which, and why?

Figure 5.8(*a*) is a quartz sandstone—more than 90 percent of the sand grains are composed of quartz. Figure 5.8(*b*) is an arkose sandstone—25 percent or more of the sand grains are feldspar.

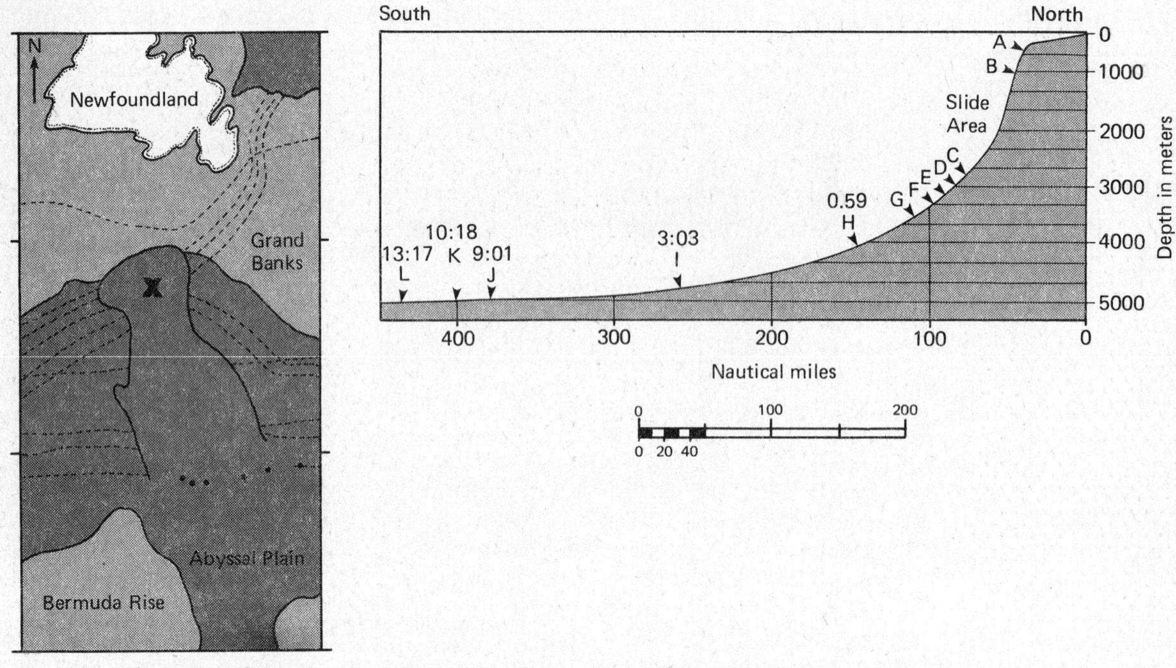

Fig. 5.7 Diagrammatic map and cross-section of a portion of the Grand Banks of Newfoundland showing the location of the submarine landslide and turbidity current of November 18, 1929. On the map, the white depicts Newfoundland, the gray areas depict the shallow shelf or "banks," and the light gray shade depicts deeper ocean. The slide originated at point X as a result of an earthquake, and it became a turbidity current as it incorporated water while moving downslope. The dotted lines are trans-Atlantic telegraph cables, labeled A through K. Dots are locations of core samples. The numbers above points H through L show the times at which the cables stopped transmitting signals.

Figure 5.8(*c*) is a lithic sandstone—more than 25 percent of the sand grains are rock fragments of various types.

Figure 5.8(*d*) is a graywacke—more than 15 percent of the rock consists of clayey matrix, it contains an abundance of feldspar and/or rock fragments as well as quartz, and it is dark-colored.

Note that these are "end members," and many sands and sandstones will have compositions somewhere between them. For example, a sandstone might contain 15 percent feldspar, 10 percent rock fragments, and 75 percent quartz, and it would not be a quartz sandstone, an arkose, a lithic sandstone, or a graywacke. It might be called a "feldspathic-lithic quartzose sandstone." But let's not get too involved in terminology.

5.17 Contrast the histories of a well-rounded, well-sorted quartz sandstone [Fig. 5.8(*a*)] and a poorly sorted graywacke with angular grains of various compositions [Fig. 5.8(*d*)].

The quartz sandstone is composed of only the most resistant common mineral and has undergone a long history of weathering in the source area and abrasion during transportation and reworking. Perhaps several cycles of erosion and deposition were involved, for it is difficult to make such a quartz sandstone. The graywacke sandstone had a history of incomplete weathering in the source area, rapid erosion and transportation, and burial soon after deposition so that reworking by waves and currents could not occur. Or to put it another way, a quartz sandstone is "well washed" whereas a graywacke sandstone is "dumped."

5.18 Where are modern turbidity current deposits (turbidites) generally found? See Figs. 5.2 and 5.7.

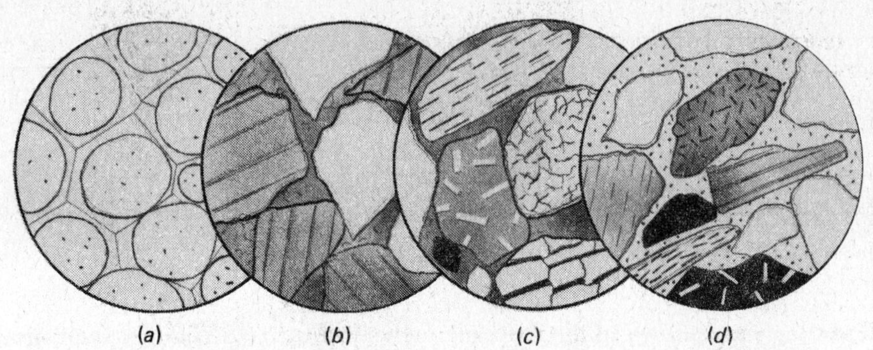

Fig. 5.8 Sketches of microscopic views of the four main types of sandstones: (*a*) quartz sandstone, (*b*) arkose sandstone, (*c*) lithic sandstone, and (*d*) graywacke sandstone. Each field of view is about 2 mm in diameter. In (*a*), all grains are quartz, cemented by more quartz. In (*b*), the darker grains with cleavage (straight lines) are altered feldspar grains and the white grains are quartz; all are cemented by calcite. In (*c*), the dark grains with various internal textures are rock fragments and the white grains are quartz. The material between grains is calcite cement. In (*d*), quartz, feldspar, and rock fragments are all present, and there is an abundance of clay (stippled) between the sand grains.

On and at the bases of continental slopes, with the turbidite deposits commonly forming submarine fans. The coarser and thicker turbidites are in channels on the fans, whereas thinner and finer turbidites and the mud that accumulates during the time between turbidity currents are common on interchannel areas.

5.19 Some sandstones are said to be "mature." What does this mean, and what does it reveal about the history of the sandstone?

Maturity is an expression of the texture and composition of a sandstone and is related to the amount of energy expended on the sand prior to its final deposition. A quartz sandstone made of well-rounded, well-sorted quartz grains [as in Fig. 5.8(*a*)] is mature, whereas the arkose of Fig. 5.8(*b*) with its angular feldspar, the lithic sandstone of Fig. 5.8(*c*) with its rock fragments, and the graywacke of Fig. 5.8(*d*) with angular, poorly sorted grains of variable composition are all immature.

5.20 What kind of igneous rock is a likely source rock for an arkose sandstone?

A granite, because an arkose contains more than 25 percent feldspar with the remainder being mostly quartz. Recall that a granite consists of feldspar, quartz, and minor biotite or hornblende.

5.21 Why wouldn't gabbro be a likely source rock for an arkose sandstone?

Although a gabbro could supply an abundance of feldspar, it could not supply the abundant quartz of an arkose. Besides, the calcium-sodium plagioclase of a gabbro is more susceptible to weathering than is orthoclase (see Problem 4.46).

5.22 The precipitation of calcite can be illustrated by this equation:

$$Ca(HCO_3)_2 = CaCO_3 + H_2O + CO_2$$

What would cause this reaction to move to the right (i.e., to precipitate calcite) if the seawater were saturated with respect to calcium bicarbonate?

The loss of carbon dioxide would mean that more calcium bicarbonate would dissociate to produce carbon dioxide. This could be caused by plants using it for photosynthesis, by the warming up of water (warm water, like warm soda, holds less dissolved gas than does cold water or cold soda), or by agitation of the water.

5.23 In the above equation, what is the result of the loss of CO_2?

The precipitation of $CaCO_3$ as the mineral calcite (or aragonite).

5.24 Two limestones are shown in microscopic views in Fig. 5.9. Contrast their histories.

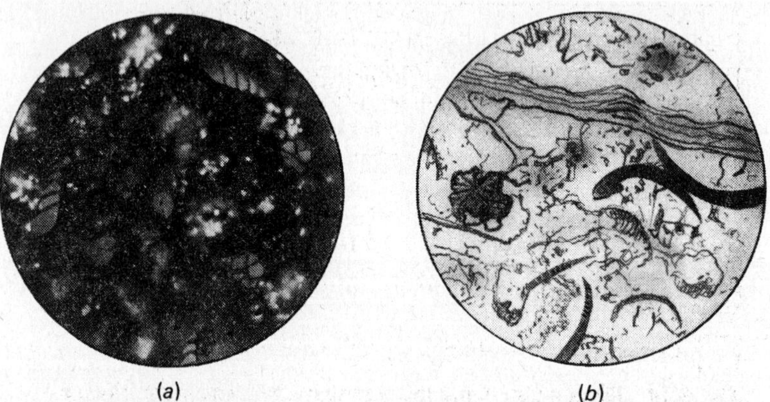

Fig. 5.9 Photomicrographs of (*a*) chalk and (*b*) fossiliferous limestone. Note the fossils in both. Each field of view is about 2 mm in diameter. Calcite cement is present in both.

The limestone of Fig. 5.9(*a*) consists of clay-sized particles of calcium carbonate and represents deposition in a low-energy, quiet environment. (Otherwise, the small particles could not have been deposited.) The limestone of Fig. 5.9(*b*) consists of shell fragments that have been abraded and sorted by higher-energy currents and later cemented.

5.25 What does the presence of a limestone indicate about the nature of the area in which it was formed?

It shows that not much land-derived sand, silt, and clay were being deposited at this locality, or else the carbonate would have been overwhelmed and could not form a bed of essentially pure calcium carbonate. Thus limestone beds have generally formed far from land, or if land is present, it must have been a very low-lying landmass.

5.26 How are sediments commonly sorted according to size so that pebbles are in one place, sand in another, silt in another, and clay elsewhere?

By variations in the current velocities of the transporting mechanisms—water, air, and ice. For example, as a fast river loses its velocity, the pebbles will come to rest first, the sand will be deposited further downstream where the current is moving less rapidly, and the silt and clay may end up in a lake or ocean off the mouth of the river.

5.27 Explain sedimentary facies. (See Fig. 5.10.)

The lateral variations in sediment type in a sedimentary formation are called *sedimentary facies*. They are the result of variations in current velocity, as in Problem 5.26, and other variables in the basin of deposition.

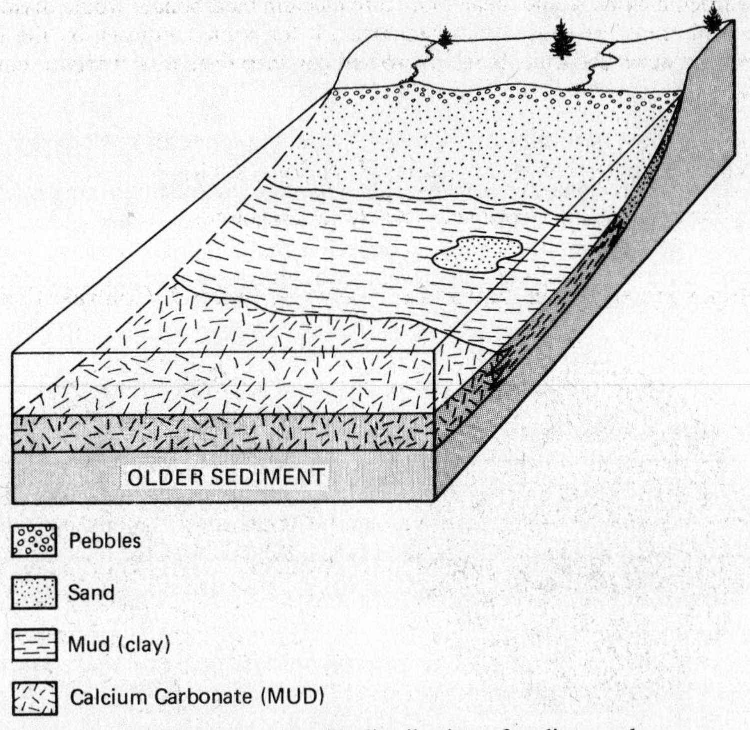

Fig. 5.10 Diagram showing the distribution of sediment along a seashore. Note that the legend gives the sediment types. These lateral variations in sediment types are sedimentary facies.

5.28 Why is the study of sedimentary rocks of more than academic interest?

Because the earth's fossil fuels (petroleum and coal), and much of its uranium and gold, and nearly all its iron ores, aluminum ores, salts, and many other economic minerals as well, are found in sedimentary rocks. In addition, groundwater is found in loose sediment and in sedimentary rocks.

5.29 Why do geologists study sandstones under the microscope? (See Figs. 5.8 and 5.9.)

Because identification of the sand-sized minerals and rock fragments may lead to identification of the source areas, and study of the texture (sorting and roundness) provides clues to the nature and length of the geologic history.

5.30 What are the main cations (positively charged ions) and anions (negatively charged ions) in seawater?

Cations	Anions
Na^+	Cl^-
K^+	CO_3^{2-}
Mg^{2+}	SO_4^{2-}
Ca^{2+}	

Note that two are "complex anions," ions consisting of a cation and oxygen bound together electromagnetically and with net negative charges.

5.31 How are beds of salt formed?

By evaporation of seawater which contains about 3½ percent salts by weight.

5.32 If a body of seawater is totally evaporated, what sequence of rock types results?

The salts in solution precipitate in a given order, based on solubility. The least soluble precipitate is first, and the order is calcite, gypsum, anhydrite, halite, and potassium and magnesium salts. Halite makes up the greatest part of the sequence. Repeated addition of salty water, as into a shallow lagoon or salt pan through a narrow and/or shallow entryway, is necessary for the production of the thick salt formations of the geologic column.

5.33 What mineralogical connotation do the terms *sand* and *sandstone* have?

They should not have mineralogical connotations, for sand and sandstone are names based on the size of the particles, between 1/16 and 2 mm in diameter (see Table 5.1). Any mineral and any relatively fine-grained rock type can occur as sand grains. Even gypsum, which has a hardness of 2 on the hardness scale and which is quite soluble, exists as sand at Whitesands National Monument in New Mexico. Unfortunately, many people associate sand and sandstone only with the mineral quartz, which is indeed the most common resistant mineral.

5.34 What is an explanation for the fact, experimentally proven, that quartz sand grains are rounded much faster by wind than by a river?

Grains become rounded as their angular corners are broken off by impact. Water cushions the impact of the sand grains, and they do not hit each other hard enough to easily chip off corners. Wave and surf action on a beach, which drags grains across other grains, may also round sand grains faster than the running water of rivers.

5.35 Why do boulders, cobbles, and pebbles round with only a few kilometers of transport whereas sand is difficult to round and silt hardly ever becomes rounded?

Because of the greater mass of boulders, cobbles, and pebbles, water has little cushioning effect as described in the previous question. Conversely, silt grains are so small that they rarely hit with enough impact to cause chipping of corners, and hence don't become rounded.

Supplementary Problems

5.36 What is *dolomite* or dolostone?

A rock composed primarily of the mineral dolomite, $CaMg(CO_3)_2$. While some may be precipitated directly from very salt-rich waters (brines), most is formed from limestone by chemical alteration.

5.37 How is most dolomite formed?

Magnesium-rich brines permeating limestone can replace half of the calcium ions in the crystal structure of calcite with magnesium ions.

5.38 What do mudcracks in a sedimentary rock indicate?

That wet mud was dried and cracked, and the cracks were filled with silt or sand. This indicates an environment with alternating wet and dry periods, exposure, and shallow water when wet.

5.39 What does the size of cross-bedding (Fig. 5.4) indicate about the origin of the sedimentary rock in which it is found?

The larger the scale of the cross-beds, the higher the velocity of the current that formed them.

5.40 What does a conglomerate indicate about the energy conditions in the environment of deposition?

High energy (strong current) was necessary to carry pebbles, cobbles, or boulders.

5.41 Do ripple marks preserved in a sedimentary rock indicate deposition in shallow water?

Wave (symmetrical) ripple marks are indicators of relatively shallow water, shallow enough for the bottom sediment to be affected by waves. Storm waves can disturb the sediment at depths as great as 100 m. However, current (asymmetrical) ripple marks have been found at depths of several thousand meters.

5.42 How are sedimentary rocks classified into types?

By texture and composition, as are igneous and metamorphic rocks.

5.43 Much calcium carbonate is precipitated biochemically. How?

Many organisms, especially marine organisms, secrete calcareous shells or other hard parts. The most obvious large examples of such skeletal structures are reefs, which are the products of a complex of organisms, many of which secrete hard skeletal material. This is clearly a biochemical process. Also, marine plants such as algae use carbon dioxide during photosynthesis, and calcium carbonate may be precipitated. (See Problem 5.22.)

5.44 Muddy rocks, sandstones, and carbonates make up about 99 percent of all sedimentary rocks. What are some other sedimentary rock types?

Conglomerate, iron-formation, phosphorite, chert, and coal.

5.45 Who is given credit for the *principle of uniformitarianism* (the present is the key to the past)?

Sir James Hutton, the Father of Geology, a Scot who published the first geology textbook in 1795. By the observation of natural surficial processes over several decades, he came to this conclusion.

5.46 Some sequences of sedimentary rocks are tens of thousands of meters thick. Does this thickness indicate that some great catastrophe occurred?

No. Such great thicknesses are generally the result of sedimentation over a long period of time. The site of deposition must have been subsiding during this deposition, essentially keeping pace with the sedimentation.

5.47 In Chapter 4, "clay minerals" are discussed. Yet, in Problem 5.5, clay has a size connotation. Does the word *clay* have a dual meaning?

Yes. There are many different clay minerals, and any particles smaller than 1/256 mm are clay-sized particles. However, much of the clay-sized material does consist of clay minerals.

5.48 Explain, in general terms, *sedimentary differentiation*.

Sedimentary differentiation is the fractionation of an original source rock into many chemically unique products via the processes of chemical and mechanical weathering (the breakdown of preexisting rock into smaller pieces and the formation of new minerals, especially clays and those minerals that precipitate out of solution), transportation resulting in size-sorting and abrasion, and diagenesis (changes after deposition). Thus, for example, aluminum-rich clays are deposited at one place, iron oxides at another, and quartz sand at yet another.

5.49 Carbonate mud becomes hard limestone in large part by the process of recrystallization. Explain.

The fine-grained calcium carbonate of carbonate mud is largely the mineral aragonite when originally precipitated. Aragonite is unstable and recrystallizes into interlocking calcite crystals (see Fig. 5.9). (*Note:* Both aragonite and calcite have a chemical composition of $CaCO_3$ but they have different crystal structures.)

5.50 How do clastic sedimentary rocks reveal clues to the climate in the source area from which the sediment was derived?

The clay minerals in muddy rocks offer one clue. Kaolinite indicates a climate of severe and thorough weathering, whereas illite and montmorillonite indicate a more temperate climate. If the feldspar sand grains are fresh (without any alteration to clay), it suggests that chemical weathering may not have been an important process; desert or polar climates may have prevailed.

5.51 What is chert?

Chert is a rock consisting of fine-grained sedimentary silica. It is either precipitated from seawater as bedded chert or replaces limestone during diagenesis (postdepositional changes) to form chert nodules.

5.52 What is chalk?

Chalk is a fine-grained limestone, originally composed of carbonate mud and usually an abundance of microfossils. If it contains clay minerals as well, it is called *marl* (see Fig. 5.9).

5.53 Where are evaporite minerals such as salt forming today?

In some lakes in semiarid regions (e.g., Great Salt Lake), in the Caspian Sea, the Dead Sea, and on tidal flats of the Persian Gulf.

5.54 What is coal?

Coal is a sedimentary rock consisting largely of carbon, formed by the decay of ancient plant material.

5.55 What is the most abundant chemically (or biochemically) precipitated rock?

Limestone.

5.56 Are limestones being formed today?

Yes, calcium carbonate is being precipitated within about 20° of the equator on shallow marine platforms. The best-studied area is the Bahaman platform, an area of about 60,000 mi^2 (155,000 km^2), where the shallow, warm water is supersaturated with respect to calcium carbonate. Another site is the northeastern coast of Australia.

5.57 What are deep sea oozes?

Deep sea oozes are sediments composed of SiO_2 and $CaCO_3$ that are accumulating at very slow rates. The carbonate oozes are the most important and consist of the shells of $CaCO_3$-secreting microorganisms, dropped into the deep water from the life-rich sunlit top layer of the oceans.

5.58 Where are fossils found?

In most types of sedimentary rocks. However, some sedimentary rocks are too old to contain fossils, as noted in Chapters 18, 19, and 20.

5.59 It has been stated that sediment is moved downhill from the site of weathering by running water (ice), or other agents of transportation, under the influence of gravity. Is there an exception to this "downhill" rule?

Yes, wind can move sediment from lower to higher places. Even sand can be moved upslope by wind (as up and over a sand dune), but silt- and clay-sized particles are most easily moved to all elevations.

5.60 What is probably the major control on sedimentation?

Tectonism (earth movements) controls sedimentation. Movements in the source areas govern the relief of the source areas. The gradients of the streams are determined in part by tectonism; high relief will result in higher velocities and greater carrying capacity for sediment. Movements in the sites of deposition govern the amount of subsidence that occurs and hence control the total thickness that may accumulate.

5.61 Which rock unit in Fig. 5.3 diagrammatically illustrates sedimentary facies?

The top layer, as the rock types change laterally within that layer.

Chapter 6

Metamorphic Rocks

INTRODUCTION

Metamorphism means a change in form. *Metamorphic rocks* are rocks that have undergone changes in texture or mineralogy or both, under the influence of changed temperature and/or pressure, usually in the presence of chemically active fluids. These changes occur because minerals, and hence rocks, are stable in the environment in which they formed (see Fig. 2.6). Preexisting rocks are thus changed texturally and/or mineralogically. Metamorphism does not usually involve melting, as it generally occurs at temperatures below the melting point of the rock.

MINERALOGICAL CHANGES

Perhaps the easiest example to visualize when studying metamorphism is the case where the clay minerals of shale or mudstone, which formed at the earth's surface at about 25°C and 1 atm of pressure, are subjected to a temperature of several hundred degrees and a pressure of 1000 atm (1 kbar) or more. Such conditions exist at depths of several kilometers. So, if shale is buried to that depth, the clay minerals would, for example, change to muscovite.

Conversely, a basalt that crystallized at a temperature of about 1000°C at the earth's surface will not be stable at the above cited temperature and pressure conditions and will change to a rock called *greenstone,* with the plagioclase, pyroxene, and olivine changing to the green minerals epidote, actinolite (an amphibole), and chlorite.

Even though new minerals are formed during metamorphism, the process is usually isochemical, with no major additions or subtractions of material, other than the loss of H_2O and CO_2. (This can be compared to a group of students in a classroom rearranging their seating order, as into different crystal structures, but with no one coming into or leaving the room.) On the other hand, if solutions are moving in and out of a rock, adding and/or removing ions, the chemical composition can be changed; this is called *metasomatism*. The common metamorphic minerals are the same as the common igneous minerals of Fig. 3.3.

TEXTURAL CHANGES

A rock undergoing metamorphism may simply recrystallize into a coarser-grained rock, with no overall changes in mineralogical or chemical composition. Examples are limestone recrystallizing to marble and quartz sandstone recrystallizing to quartzite. Such a recrystallization may be the result of burial pressures due to the weight of overlying rock, and the new rock is quite *massive* in texture [Fig. 6.1(*d*)]. However, if directional pressures are involved, the result of deformation of the crust, then an arrangement of minerals into planes will occur, forming *foliation*. See Figs. 6.1(*a*) (an unmetamorphosed shale) and 6.1(*b*) and 6.1(*c*), which are products of the metamorphism of shale. *Slaty cleavage* in slate and *schistosity* in schists are two examples of foliation. The flat minerals, such as the micas, are oriented perpendicular to maximum pressure as they grow (Fig. 6.2).

MINERAL STABILITY

Metamorphic rocks can be studied where they have been exposed by uplift and erosion. Field observations, especially the tracing of an original rock type into a metamorphic rock, are supplemented

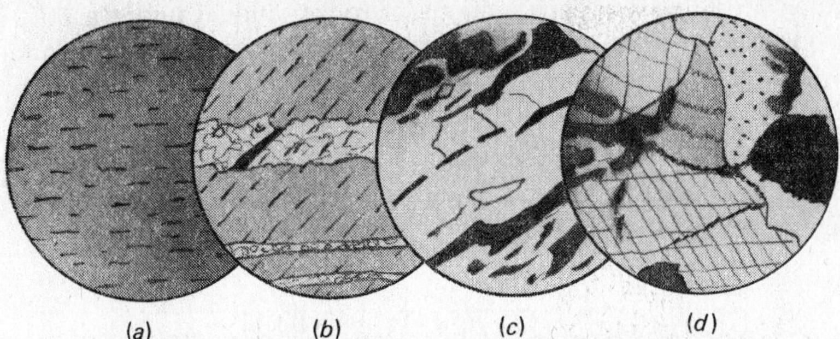

Fig. 6.1 Sketches of microscopic views of: (*a*) unmetamorphosed shale; (*b*) slate with horizontal bedding and small aligned mica crystals resulting in a cleavage at about 30° to the bedding; (*c*) schist with larger micas than in slate, so large that they are easily visible to the naked eye; and (*d*) marble made of large crystals of calcite. The first three can be used to illustrate the progressive effects of metamorphism under directional pressures. Marble is a metamorphosed limestone, with the major characteristic being a coarsening of grain size. Fields of view are about 2 mm in diameter.

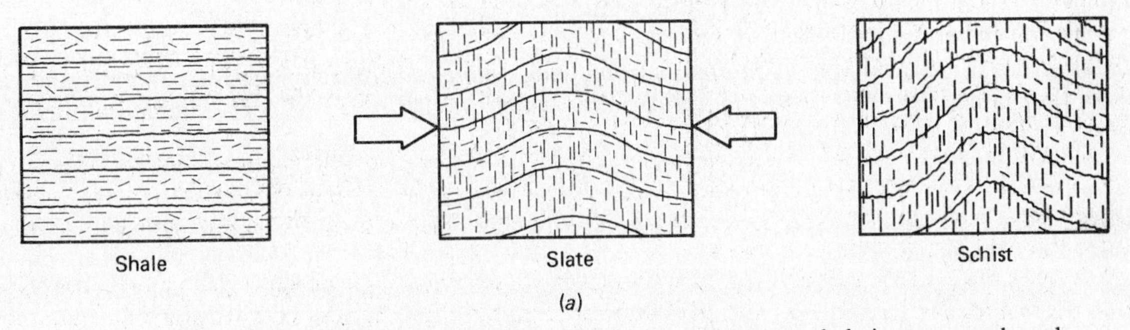

Fig. 6.2 (*a*) Diagrammatic representation of the changes as mudstone or shale is metamorphosed. With the increased pressure during folding, the clay minerals become unstable and recrystallize into microscopic micas oriented perpendicular to the principal stress (pressure). With continued or increased changes in pressure and temperature, the micas become large enough to be seen without a microscope (i.e., macroscopic). The arrows indicate the principal stress directions. (*b*) Sketch of a rock outcrop showing original bedding folded into a broad fold now cut by vertical slaty cleavage. The arrows represent the principal stress direction caused by compression. In reality, the cleavage planes will not be perfectly vertical and parallel; on the flanks or sides of the fold, they will flare outward, toward the upper left on the left side of the fold and toward the upper right on the right side. (See Figure 16.3.)

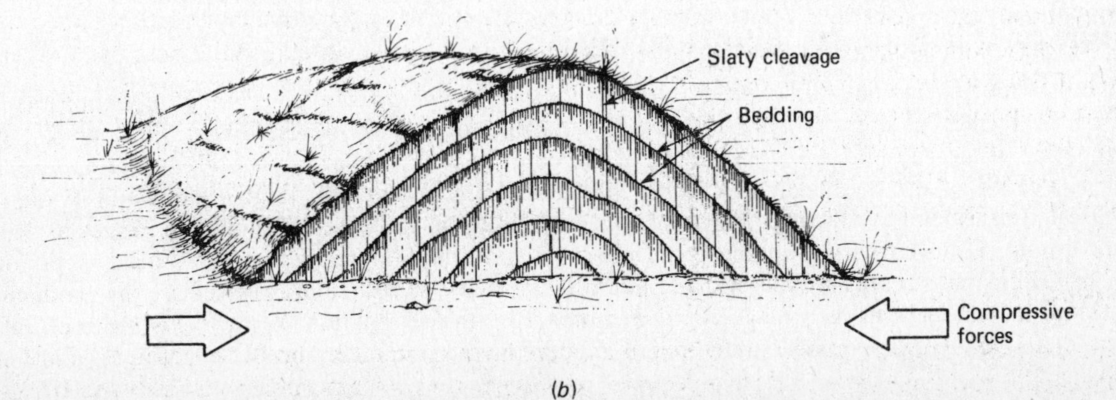

by laboratory studies of the rocks. Thin sections of rocks studied under a polarizing-light microscope commonly show that one mineral was "caught in the act of changing" to another mineral (Fig. 6.3); such preservation is probably due to a change of pressure and temperature conditions before the original mineral was completely changed to the new metamorphic mineral.

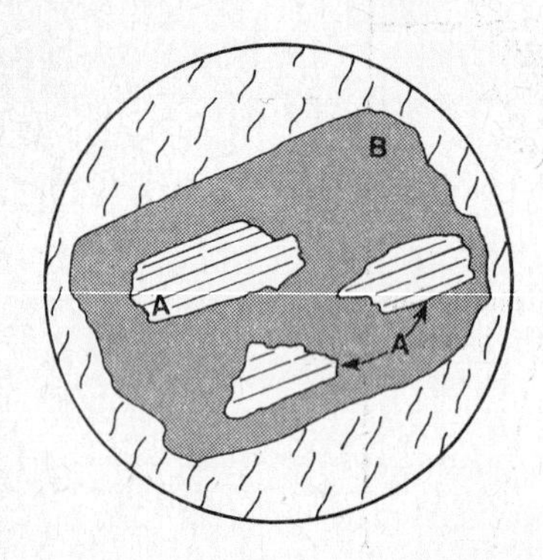

Fig. 6.3 Sketch of a microscopic view of a metamorphic rock showing that mineral A was changing to mineral B. It was "caught in the act of changing" when the temperature and pressure conditions changed before the reaction was complete. Note that cleavage (straight lines) in mineral A are all parallel, indicating that the pieces of A were once a single crystal.

Laboratory experiments in high-temperature and high-pressure "bombs" (containers) reveal the stability fields in terms of pressure and temperature of each metamorphic mineral.

Several minerals are indicators of *metamorphic grade* (i.e., particular temperatures and pressures) as in Fig. 6.4 and are called *metamorphic index minerals*. Assemblages of minerals, and the rocks they constitute, define *metamorphic facies* which are definitive of certain temperatures and pressures (Fig. 6.5).

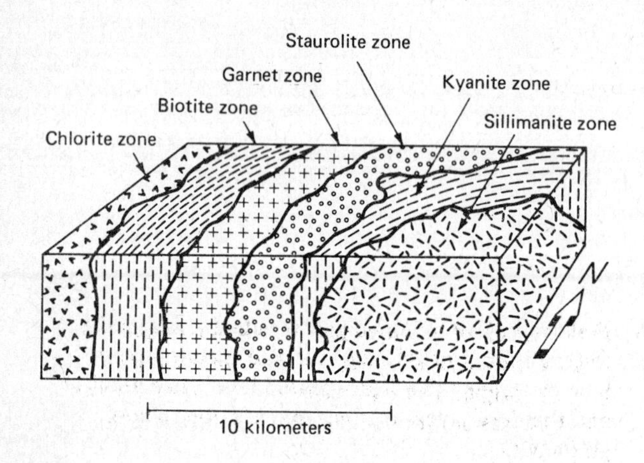

Fig. 6.4 Diagrammatic representation of metamorphic zones proceeding from the low-grade metamorphism zone (chlorite) to the high-grade metamorphism zone (sillimanite). The mineral assemblages in each zone differ, but each zone has the same chemical composition. These are metamorphic index minerals that indicate the temperature and pressure in each zone. (*From Ojakangas and Darby, 1976.*)

CONTACT AND REGIONAL METAMORPHISM

Most metamorphism can be classified as either *contact metamorphism* or *regional metamorphism*, based on the scale of the changes. Changes adjacent to igneous flows, dikes, sills, stocks, or batholiths, at least in part a simple baking of the country rock, are examples of *contact metamorphism*, caused by the hot magma. *Regional metamorphism* occurs over large regions and is due to regional stresses and elevated temperatures, as are present in the cores of mountain ranges as they are forming. Deep burial will increase the pressure and temperature and cause "burial metamorphism." Regional stresses are directional, and foliated slates, schists, and gneisses are the products. Extreme metamorphism can result in *granitization* (the transformation of a preexisting rock into granite) and *migmatites* (mixed banded rocks of both igneous and metamorphic parentage, as in Fig. 6.6).

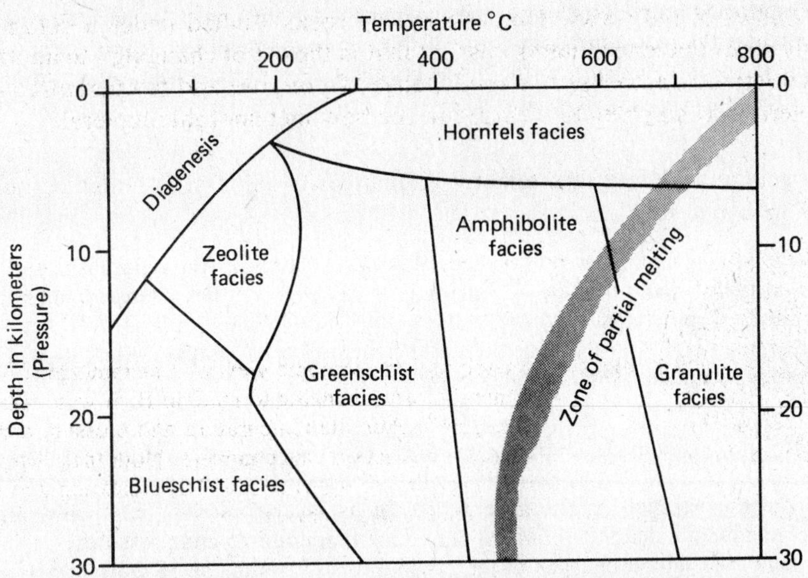

Fig. 6.5 Generalized metamorphic facies diagram, showing relationships of metamorphic facies to pressure and temperature. The boundaries between facies are gradational. Partial melting of rocks occurs to the right of the shaded zone. (*After Judson, S., Kauffman, M. E., and Leet, L. D., Physical Geology, Prentice-Hall, Englewood Cliffs, N.J., 1987.*)

Fig. 6.6 Field sketch of a veined gneiss called a migmatite. Such a rock probably has a mixed metamorphic-igneous origin. The light-colored layers are granitic and the dark layers contain much biotite and/or amphibole.

Solved Problems

6.1 What are the metamorphic equivalents of the following sedimentary rocks: (*a*) mudstone or shale, (*b*) limestone, and (*c*) quartz sandstone?

(*a*) Slate or schist. (*b*) Marble. (*c*) Quartzite.

6.2 Explain what happens as a mudstone becomes a schist during metamorphism [Figs. 6.1(*a*), 6.1(*b*), 6.1(*c*), and 6.2].

As temperature and pressure increase, the clay minerals in the mudstone become unstable. Micas which are stable under the new conditions form. If the micas are small (microscopic) and aligned parallel to each other, the rock is a slate. As the micas grow larger with increased temperature and pressure, the resultant rock is a schist.

6.3 How do geologists know that mineral A changes to mineral B under certain pressures and temperatures (Fig. 6.3)?

Under the polarizing microscope, thin slices of rock reveal the spatial relationships of minerals. Mineral A may be partially altered to mineral B along its outer edges. Other grains of mineral B may retain the crystal outlines of mineral A.

Also, remnants of A in B are usually in the same crystallographic orientation. Finally, laboratory experiments may simulate the natural conditions under which A changes to B.

6.4 How do metamorphic facies (Fig. 6.5) differ from sedimentary facies (see Problem 5.27)?

Metamorphic facies, such as the greenschist facies or the amphibolite facies, are assemblages of metamorphic minerals that are definitive of certain temperatures and pressures. Sedimentary facies, on the other hand, are lateral changes in the original type of sediment or sedimentary rock within a rock unit, with the changes due to depositional differences.

6.5 Does metamorphism occur only when rocks are subjected to higher temperatures and pressures?

No, it also occurs when rocks are subjected to lower temperatures and pressures than those at which they formed. Remember, minerals and rocks are stable at conditions under which they formed (see Fig. 2.6).

6.6 Name some common metamorphic minerals.

The common igneous minerals of Fig. 3.3—olivine, pyroxene, hornblende, biotite, plagioclase, orthoclase, muscovite, and quartz—are the common metamorphic minerals as well.

6.7 Name some minerals which generally form only under metamorphic conditions.

Staurolite, kyanite, sillimanite, asbestos, and talc are examples. Garnet is usually a metamorphic mineral.

6.8 Which rock—an impure limestone or a quartz sandstone—would be most likely to yield new minerals during metamorphism?

A quartz sandstone would simply recrystallize into larger crystals. In an impure limestone, there may be clay minerals, and hence Al, Si, and other cations, to react with the $CaCO_3$. So, new minerals would be more likely in the impure limestone.

6.9 Which kind of metamorphism—contact or regional—is illustrated in Fig. 6.7?

Contact metamorphism, as it is found only adjacent to the igneous rock body.

6.10 Refer back to Fig. 3.1. Near which kind of igneous rock body would contact metamorphism be the most pronounced?

Near the batholith, which would probably stay hot longer and release more solutions into the country rock than would the smaller igneous bodies.

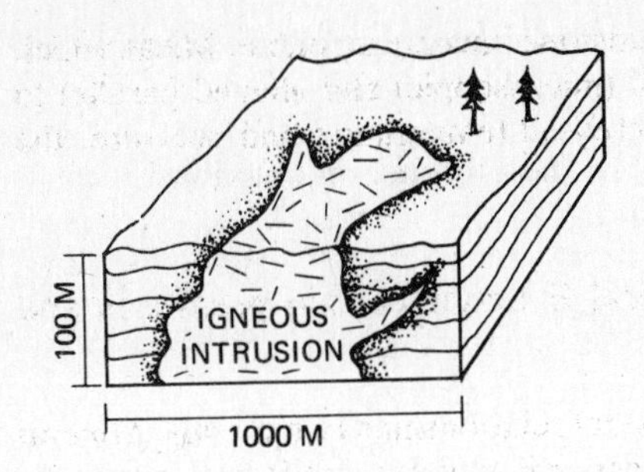

Fig. 6.7 Contact metamorphism (dotted zone) occurs next to an igneous intrusion, largely the result of the increased temperature and solutions from the intrusion. Obviously, the intrusion and the metamorphic rock have been exposed by uplift and erosion. (*From Ojakangas and Darby, 1976.*)

6.11 Why does slate possess a good rock cleavage and break into flat pieces? (See Figs. 6.1 and 6.2.)

Because of the aligned microscopic micas which impart planes of weakness to the rock.

6.12 What is the difference between rock cleavage (slaty cleavage) and cleavage in minerals?

Rock cleavage is caused by the orientation of minerals in the rock. Cleavage in minerals is related to the crystal structures, with the minerals tending to break along planes of ions. (See Problems 2.31 and 2.32.)

6.13 On a geology exam, one question read as follows: "During metamorphism, rocks may be melted at temperatures as high as 1000°C. Comment." OK, comment.

Yes, rocks can melt at high temperatures. (That is how all magmas are formed!) During metamorphism, most changes take place without melting. The ions move around and reorganize into different crystal structures at temperatures below the melting temperatures of the rock. Once melting occurs, a magma is being formed, and the processes are both igneous and metamorphic.

6.14 Where does regional metamorphism generally occur?

In the bowels of mountain ranges, when temperatures and compressive stresses are increased.

6.15 Where does contact metamorphism occur?

Adjacent to intrusive bodies or lava flows. The heat of the magma "bakes" the country rock. In some cases, depending upon the amount of volatiles in the magma, solutions also move through the country rock, adding or removing ions.

6.16 During metamorphism, what happens to the size of crystals?

They get larger.

6.17 What is foliation?

Foliation is a texture found in metamorphic rocks. It is a parallelism of mineral grains and mineral bands, formed during metamorphism in the presence of directional pressures. The rocks break along the foliation planes.

6.18 Name the metamorphic rock that contains bands of different minerals, for example, plagioclase and quartz in the light-colored bands and biotite and hornblende in the dark-colored bands.

A gneiss.

6.19 What does foliation reveal about the conditions under which it formed?

The presence of foliation indicates that the rock was metamorphosed under a directional stress, with the pressures stronger in the direction perpendicular to the foliation.

6.20 Temperature and pressure are commonly cited as the causes of metamorphism. What is another factor?

Chemically active fluids. Metamorphism proceeds more rapidly in the presence of solutions because ions can move around more readily in solutions than in a dry state.

6.21 What is a mica-rich metamorphic rock called?

Schist.

6.22 Name three types of mica that are common in metamorphic rocks.

Muscovite, biotite, and chlorite.

Supplementary Problems

6.23 What is a greenschist or greenstone? (See Fig. 6.5.)

It is a green rock that is commonly formed by the metamorphism of basaltic rocks at temperatures of a few hundred degrees and pressures of a few kilobars (a few thousand times the pressure at the surface of the earth). It is green because of the presence of green metamorphic minerals, including chlorite, epidote, and actinolite (a variety of amphibole).

6.24 What is granitization?

Granitization is the metamorphic process whereby a preexisting rock is converted to granite. The former rock is *not melted* during granitization. Usually, K^+ and Al^{3+} are added by solutions.

6.25 Figure 6.8 shows two granite bodies. What evidence indicates that one crystallized from a magma and the other is the product of granitization?

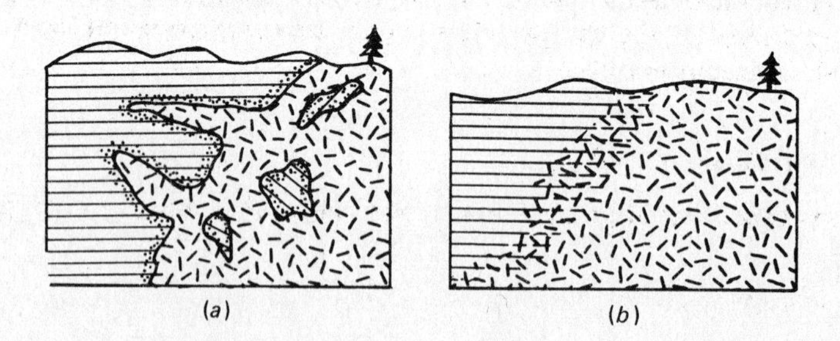

Fig. 6.8 (*a*) A magmatic granite. Note the sharp contacts, the contact metamorphism (shown by the dotted pattern), the dikes of granite in the sedimentary country rock, and the rotated blocks of metamorphosed sedimentary rock within the granite. (*b*) A granite which is the product of granitization. Note the gradational contact, the lack of dikes, and the lack of rotated blocks. Although local melting may occur during granitization, the process does not generally involve any melting. (*From Ojakangas and Darby, 1976.*)

The granite in (*a*) has sharp contacts with the country rock, has caused contact metamorphism, has sent dikes into the country rock, and has engulfed blocks of the country rock, some of which have rotated in the magma. This is a true igneous granite. In contrast, the granite in (*b*) has gradational contacts with the country rock, lacks dikes, and does not contain rotated inclusions. It is the result of granitization.

6.26 What is retrograde metamorphism?

"Metamorphism" generally brings to mind conditions of increased temperature and pressure, and this is usually the case. However, minerals that formed at these increased temperatures and pressures may be unstable under new conditions of waning temperatures and pressures, and new minerals that are stable under these succeeding conditions may form, especially if the new conditions exist for a long enough period of time for equilibrium to be attained. Such changes are "retrograde" (reverse) changes as opposed to the more usual "prograde" changes. Retrograde changes are not common, or all prograde metamorphic products would be obliterated.

6.27 What is a hornfels?

A *hornfels* is a fine-grained, hard, metamorphic rock that forms during contact metamorphism. It is generally equigranular and lacks foliation.

6.28 Metamorphism occurs at depth. How, then, can geologists study metamorphic rocks?

Uplift and erosion have exposed the metamorphic rocks.

6.29 Why are some metamorphic rocks massive, rather than foliated?

Foliated rocks formed under conditions of directional pressures (stresses). If a metamorphic rock is massive, only the pressure caused by the weight of the overlying rock affected it, and this "overburden pressure" is equal in all directions.

6.30 What are the sources of water that may be present during metamorphism?

The water may be present in some of the preexisting minerals (such as the micas or clays), or it may be supplied by nearby magmas. (It will be present as water vapor rather than liquid water when the temperature is above the boiling point.)

6.31 What are the sources of heat for metamorphism?

The heat may be geothermal (the temperature of the earth's crust increases with depth), it may be produced by deformational pressures (temperature increases with pressure), or it may be from a nearby magma or still-hot igneous body.

6.32 What is hydrothermal alteration?

This is alteration caused by hot waters. It can be thought of as one aspect of metamorphism.

6.33 What is a blueschist?

A special metamorphic rock formed at low temperatures and high pressures, as indicated on Fig. 6.5. It is blue because of the presence of blue, sodium-rich amphiboles. It forms in subduction zones, where a part of the earth's crust is moved downward beneath other crust. (Subduction zones will be discussed in Chapter 17.)

6.34 What is a migmatite (Fig. 6.6)?

A "veined gneiss" of mixed igneous and metamorphic origin. It is the product of extreme metamorphism, during which some melting may have occurred. However, it is also possible that the granitic layers between the foliated mineral layers may have had a source from a nearby magma.

6.35 Who first established the principles of metamorphic facies?

Professor Pentti Eskola of Finland, in the 1930s.

6.36 Name two main characteristic minerals of each of the three main metamorphic facies of Fig. 6.5.

Greenschist facies: Chlorite, epidote, quartz

Amphibolite facies: Hornblende, garnet, quartz

Granulite facies: Pyroxene, sillimanite

6.37 What is burial metamorphism?

Burial metamorphism is metamorphism caused by the weight of overlying rock. This is also a reflection of the geothermal gradient, the increase of temperature with increasing depth.

6.38 At what temperature does melting of preexisting rocks occur? (See Fig. 6.5.)

Depending upon the pressure and the presence of volatiles, from temperatures of about 650 to 800°C.

6.39 Metamorphism occurs at depth. Since geologists cannot see this going on, can they apply the axiom of "the present is the key to the past"?

Yes, by utilizing laboratory experiments in which the temperatures and pressures (and solutions) at depth are simulated.

6.40 During metamorphism, most chemical reactions are isochemical. Explain.

This means that the bulk chemical composition does not change. For example, if all the students in a classroom exchange places with each other, the arrangement would be different (i.e., different crystal structures would be present), but the total makeup of the room (i.e., the bulk chemical composition) has not changed.

6.41 When the chemistry does change, what two ingredients are most commonly involved?

Water and carbon dioxide. At the temperatures of metamorphism, these volatiles can be driven off.

6.42 What happens to a conglomerate during metamorphism under directional pressures?

The pebbles or boulders are commonly flattened and elongated in planes perpendicular to the principal stress. The result is a "stretched pebble conglomerate."

6.43 A special type of a metamorphic rock is a *cataclastic rock*. How is it formed?

By movement along faults or other zones of shear. Granulation of the rock is the result. This is the one type of metamorphism in which the grain size becomes smaller rather than larger. *Mylonite* is commonly used as a synonym for a cataclastic rock, but to some specialists, a mylonite involves a recrystallization at very high pressures and temperatures in shear zones.

6.44 Figure 6.4 shows metamorphic zones related to the temperature of mineral formation. Note the north arrow on the block diagram, and then suggest where the heat source that caused these mineralogical changes was located relative to the map of the surface.

Probably in the southeast corner or to the southeast of the area. The zones curve, and if the diagram were larger, a "bulls-eye" would be formed with the sillimanite zone at the center.

Chapter 7

Fossil Fuels and Economic Minerals

INTRODUCTION

Mineral resources can be classified as metallics, nonmetallics, and fuels. The fossil fuels (oil, gas, and coal) technically are not minerals because they lack crystallinity and are not inorganic, having formed by the decomposition of organic material.

Many minerals—in fact, more than 100—are of economic importance, being needed for various uses by modern civilization.

ECONOMIC MINERALS

These valuable minerals, like the more common less valuable ones, are formed by igneous, metamorphic, or sedimentary processes. Some minerals, such as chromite (the main ore mineral of the metallic element chromium) and nickel sulfide minerals, crystallize out of mafic magmas. Most high-grade massive sulfide deposits have a volcanic origin and contain the main ore minerals of copper (chalcopyrite and chalcocite), zinc (sphalerite), and lead (galena) as well as other sulfide minerals.

Large feldspar crystals in pegmatites (see Problems 3.15 and 3.16) are mined for use in ceramics. Several rare elements such as tantalum, tin, and uranium are found in pegmatites which crystallize from the last solutions of a cooling magma and which contain many elements with the wrong ionic size and charge to be incorporated in the common rock-forming minerals.

Asbestos, a much used industrial mineral before its carcinogenic attributes became known, commonly forms by the metamorphism of ultramafic rocks containing the mineral olivine. Talc and the massive variety known as soapstone have a similar origin.

The main iron ore minerals, hematite and limonite, are sedimentary minerals, precipitated out of solution. The iron was dissolved during weathering or placed into solution by volcanic and/or hydrothermal ("hot water") fluids.

Hydrothermal quartz veins, common in rocks affected by igneous or metamorphic processes, may contain native gold. This gold may be released during weathering and then concentrated by placering (a sorting out of heavy minerals) during sedimentation because of its great density (19 +) compared to common quartz and feldspars with densities of about 2.7.

The main ore mineral of aluminum (bauxite) is the end product of chemical weathering of feldspar-rich rocks under subtropical to tropical conditions.

Gypsum is used for making plaster of paris and wallboard. Phosphate rock (containing phosphorus as PO_4 combined with other elements) is used as a fertilizer. Graphite is used as a lubricant. Garnet, corundum, and nongem-quality diamonds are used as abrasives because of their great hardness. Several minerals, notably diamonds, are used as gemstones.

Some common rocks and sediments, such as granite, limestone, gravel, and sand, can be very valuable if located close to markets. Granite, limestone, dolomite, and marble are quarried for building stones. Limestone is used for making agricultural lime and cement, which when mixed with gravel, makes concrete. Basalt, limestone, and other rocks are commonly used as crushed rock for various construction projects, including highways. Gravel and sand are used in various types of building projects.

FOSSIL FUELS

Coal, petroleum, and natural gas, together known as the fossil fuels, are composed of the remains of organisms.

Coal is formed in fresh water swamps from incompletely decayed plants. Swamps have poor circulation, and the oxygen is used up by bacteria before they can digest all of the plant material. If peat (an accumulation of plant material) is buried and compacted beneath overlying sediment, then water, oxygen, nitrogen, and other components are expelled, and the residue is carbon-rich material called *coal*. Coal is abundant; world reserves are estimated at 8.5×10^9 tons (8.5 billion tons!); nearly all is in North America, Europe, and Asia (Figs. 7.1 and 7.2).

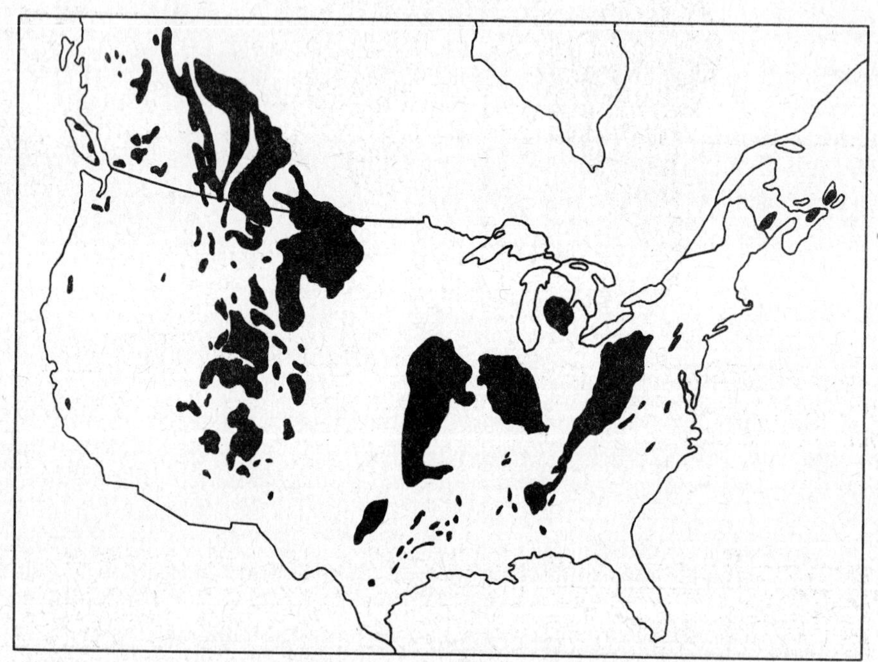

Fig. 7.1 Map showing the distribution of coal in North America. The eastern coals are higher grade (they produce more heat) but generally contain higher amounts of sulfur than western coals and are thus sources of pollutants when burned.

Crude oil (a liquid) and natural gas (a gas that is mostly methane—CH_4) are the components of petroleum. It is generally thought to have formed from organisms (animals and plants), mostly microscopic species. Petroleum is a hydrocarbon, consisting of carbon and hydrogen, plus sulfur, nitrogen, oxygen, and other minor elements. The "source beds" are black organic-rich muds, mostly marine. "Heavy oils" (large organic molecules with high molecular weights) form first; with burial, the temperature and pressure increase, and the large molecules are "cracked" into lighter, more mobile molecules. No two oils are exactly alike.

Petroleum migrates into reservoir beds that are porous (i.e., have pore spaces or cavities) and permeable (i.e., the pores are connected, allowing for the movement of fluids). Reservoir beds are usually sandstones (59 percent), porous carbonate units such as reefs (40 percent), or fractured rocks (1 percent). The reservoir bed must be overlain by an impermeable cap rock, and a trap must exist in which the petroleum can be concentrated (Fig. 7.3).

Burial to depths of 1500 m (5000 ft) seems to be required before petroleum can form.

Only a small amount of petroleum occurs at depths of less than 1200 m (4000 ft); most (85 percent) occurs between 1200 and 3000 m (10,000 ft). The first petroleum exploration was directed at oil seeps on the surface; then anticlines were drilled (see Fig. 7.3), and now geophysical techniques

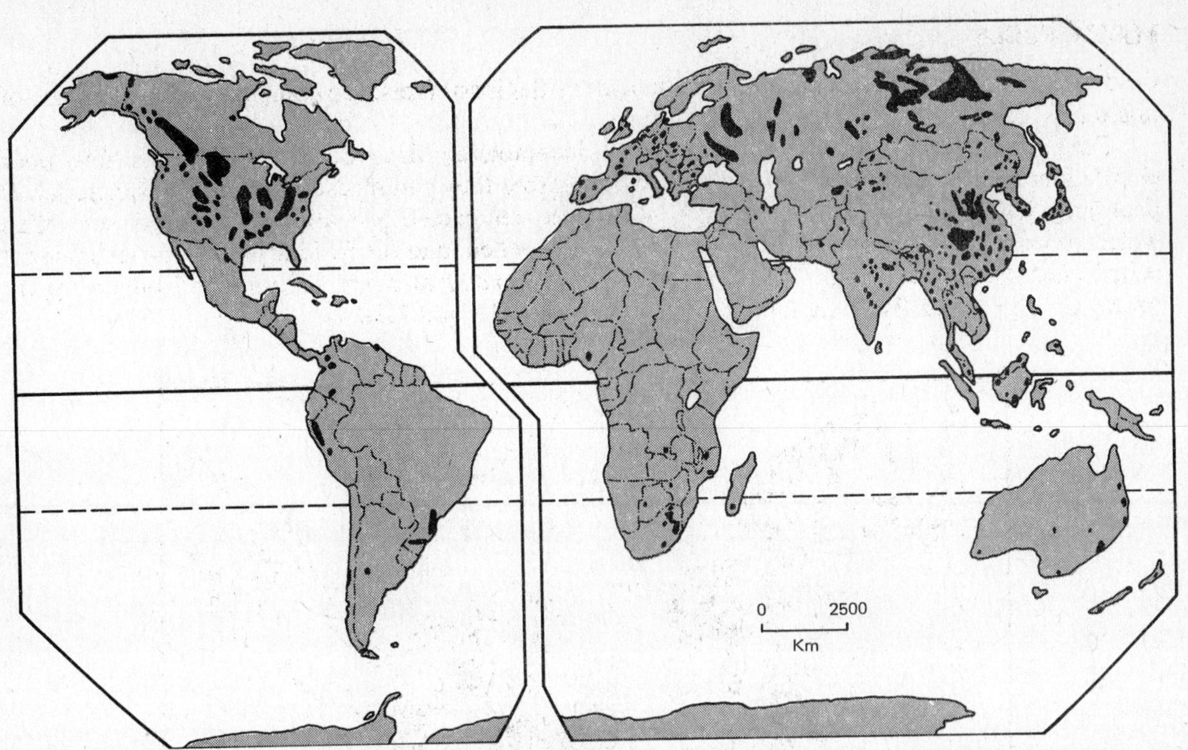

Fig. 7.2 Map showing the distribution of coal in the world. (*After Skinner, B. J., Earth Resources, Prentice-Hall, Englewood Cliffs, N.J., 1969.*)

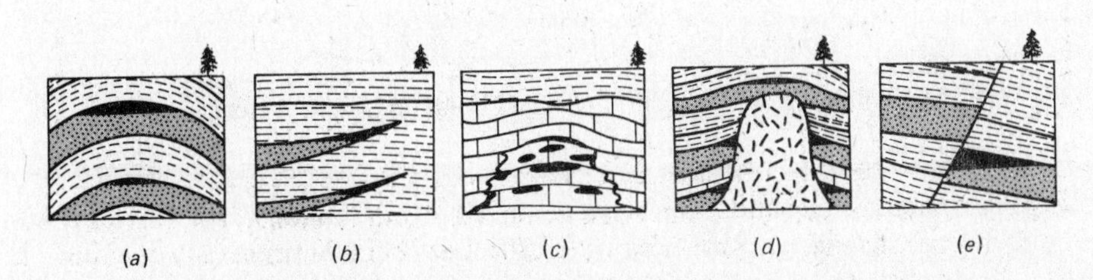

Fig. 7.3 Types of oil traps. (*a*) Anticline. (*b*) Stratigraphic. (*c*) Porous limestone reef. (*d*) Salt dome. (*e*) Fault. (*From Ojakangas and Darby, 1976.*)

(seismics, in which sound waves are reflected off of rock layers) are necessary to find most of the buried traps shown in Fig. 7.3.

Petroleum is widely distributed (Fig. 7.4), but the nations of the Middle East have the greatest reserves. Petroleum is indeed a "fossil" fuel; it has been estimated that 15 percent is Paleozoic (57 to 225 million years old), 27 percent is Mesozoic (65 to 225 million years old) and 58 percent is Cenozoic (less than 65 million years old). Thus it is essentially a nonrenewable resource and must be used wisely.

Oil shales are shales that contain solid hydrocarbons (kerogen). At temperatures of a few hundred degrees Celsius, the kerogen becomes fluid and can be removed from the rock. In the United States, large reserves of oil shale exist in the tristate area of Colorado, Wyoming, and Utah in lake beds of Eocene age (about 50 million years old) (Fig. 7.5). The petroleum reserves in oil shales are great. The total amount may equal all the crude oil that the United States has ever had. Average oil shales yield about 25 to 30 gal of petroleum per ton of shale.

Tar sands are sandstone units that contain a heavy hydrocarbon residue (tar) left after the escape of lighter elements. The rock must be mined and then treated with steam to make the petroleum

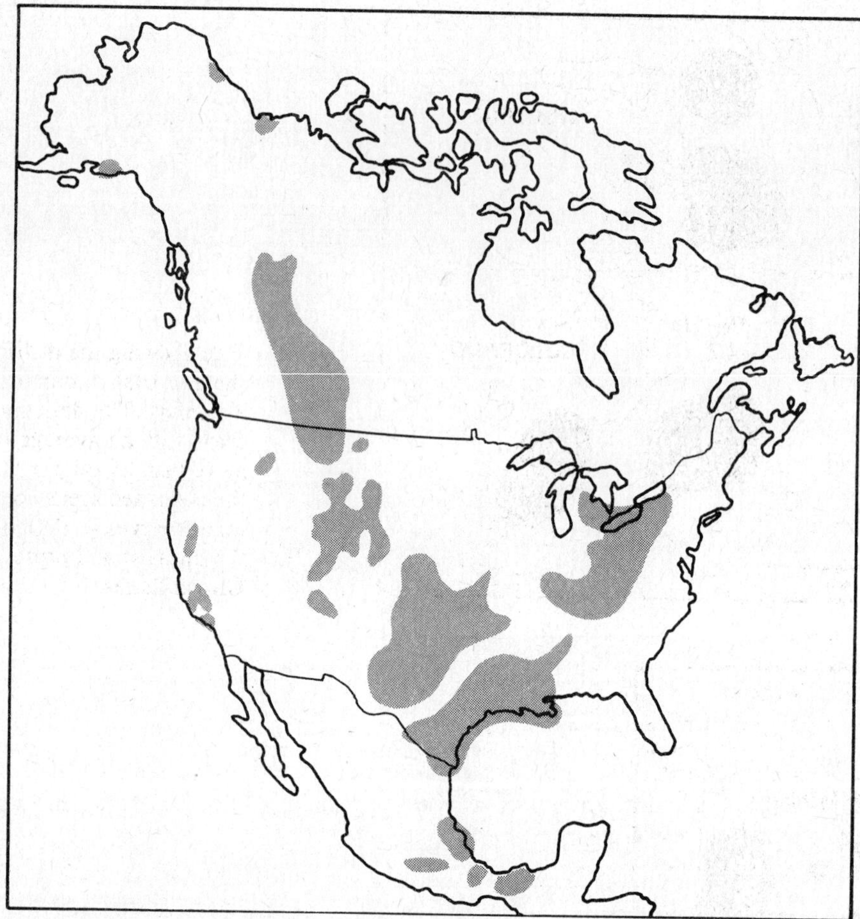

Fig. 7.4 Map showing areas with oil and gas fields in North America.

flow out of the pore spaces. Tar sands are present in Alberta, Canada (Fig. 7.6) in a 60-m (200-ft) thick sandstone formation found over an area of 77,700 km^2 (30,000 mi^2).

NUCLEAR FUELS

Uranium is the only element used as a nuclear fuel. Uranium ores include more than a hundred minerals, but uraninite (pitchblende) with the formula UO_2 is by far the most important. There are numerous types of deposits formed by igneous (e.g., in pegmatites), metamorphic, and sedimentary processes. The three most important types are in ancient quartz-pebble conglomerates (as in South Africa and Ontario), in sandstones (as in Colorado, Wyoming, and a few other western states), and along unconformities (buried surfaces of erosion) between ancient sandstones and even older basement rocks (as in northern Australia and in northern Saskatchewan).

SUMMARY STATEMENT

The search for metallic and nonmetallic minerals, fossil fuels, and nuclear fuels is the primary objective of most geologists and geophysicists. These resources are indeed valuable, but perhaps the most valuable resources of all are soils and water. How can a value even be estimated for these basic necessities of life? Without them, the search for other minerals would be meaningless. In fact, without them, there would not be people on this earth to search for minerals.

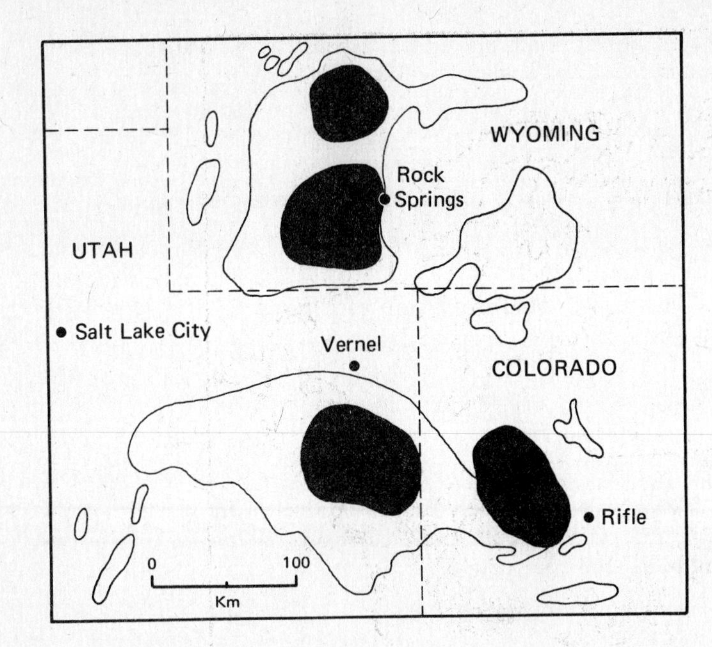

Fig. 7.5 Map showing the distribution of oil shale in Utah, Colorado, and Wyoming. The dark shading shows areas with an average oil content of more than 25 gal per ton of shale; the unshaded areas contain less. The total reserves of oil may equal all the oil, past and present, in the United States.

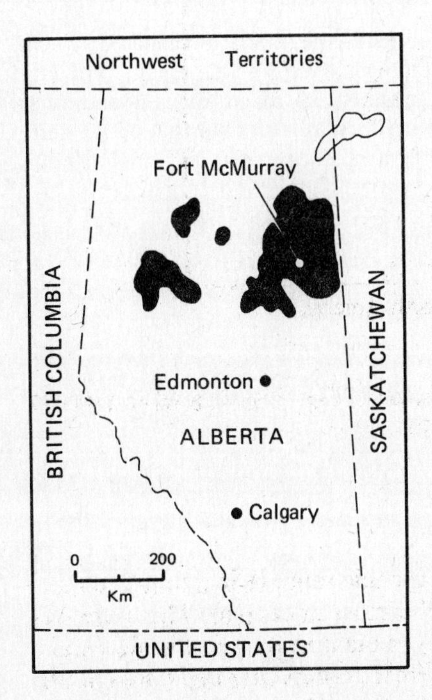

Fig. 7.6 Map showing the location of tar sands in Alberta, Canada.

In Chapter 25, People and the Earth, these economic resources will be discussed further in the context of environmental problems and supply problems.

Solved Problems

7.1 Name and define the three major classes of mineral resources.

Metallic mineral resources are mined for their metallic elements. Dozens of different metals are used by modern civilizations; the most important are iron, copper, aluminum, lead, zinc, silver, and gold. Nonmetallic mineral deposits include stone, clay, gravel, sand, phosphate, gypsum, salts, and gems. Fuels include both fossil fuels (coal, oil, and natural gas) and nuclear fuel (uranium).

7.2 Why are coal, oil, and natural gas called "fossil fuels"?

Because they are composed of the remains of organisms that lived as long ago as 500 million years. Some fossil fuels are even older.

7.3 Why don't coal, oil, and natural gas fit the accepted definition of a mineral? (See Problem 2.21.)

They are neither crystalline nor inorganic.

7.4 What are peat, lignite, and coal?

Peat is slightly decayed plant material. *Lignite* is a low-grade coal, the product of further decay and compaction. *Coal* is made up largely of carbon, with the bulk of the other original constituents of the plant matter driven off as volatiles during burial and compaction.

7.5 How was coal formed?

As thick plant growth in swamps died, the organic matter accumulated in the poorly oxygenated swamp waters where bacterial destruction of the organic matter was incomplete. *Peat* was the result. Burial by sand and mud provided pressure and heat that helped drive off the water, hydrogen, oxygen, and other volatiles, forming coal, both *lignite* and the higher-rank *bituminous* types. Additional heat and pressure of metamorphic processes transformed bituminous coal into higher-rank *anthracite coal* which is composed mainly of carbon and only very few volatiles.

7.6 Is coal in short supply?

No. Coal is very abundant. World reserves are estimated at 8.5 billion tons.

7.7 What are the four basic requirements for an oil or natural gas field?

Source rock, reservoir rock, impermeable cap rock, and a trap.

7.8 Name the five types of oil and gas traps shown in Fig. 7.3.

Anticlines, sedimentary pinchouts (stratigraphic traps), porous limestone reefs, salt domes, and fault traps.

7.9 *Why* do oil and natural gas accumulate in the traps shown in Fig. 7.3? (See Fig. 7.7.)

Petroleum and gas move with water through rocks. Gas is less dense than oil (petroleum) and oil is less dense than water, and they accumulate in the upper parts of the traps.

7.10 What is the likely source rock of petroleum and natural gas?

Organic-rich black muds. The organic material decays to hydrocarbons that are molecules of carbon and hydrogen combined in numerous ways. The simplest of the hydrocarbons is methane (CH_4).

7.11 Where in North America are petroleum and natural gas found (Fig. 7.4)?

Thirty-seven of the states in the United States and four of the Canadian provinces produce oil and gas. Offshore production from the continental shelves comes mainly from the Gulf Coast, with some from

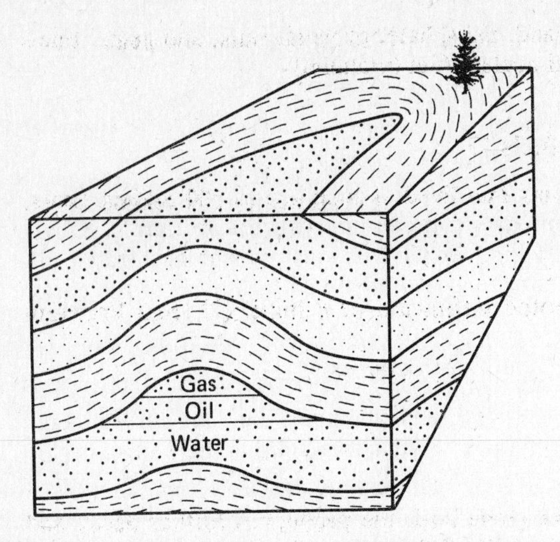

Fig. 7.7 Generalized diagram showing the locations of water, oil, and gas in an anticlinal trap.

the West Coast. (Very little has been found off the eastern coast, in spite of a great amount of exploration.)

7.12 What thickness of cover rock is generally thought necessary to generate petroleum and natural gas from organic material?

About 1500 m (5000 ft). Of present U.S. production, only about 1 percent comes from depths less than 1200 m (4000 ft); about 85 percent is from depths between 1200 m (4000 ft) and 3000 m (10,000 ft).

7.13 What are the geologic ages of the petroleum and natural gas deposits in the United States?

About 15 percent is Paleozoic (225 to 570 million years old), 27 percent is Mesozoic (65 to 225 million years old), and 58 percent is Cenozoic (less than 65 million years old). Although migration certainly occurs, most of the oil and gas is probably found in rocks of similar age to those in which it was formed.

7.14 In which rock types are oil and gas found?

Nearly 60 percent is in sandstones, and most of the remainder is in carbonate rocks (limestone and dolomite).

7.15 What is oil shale?

Oil shale is a dark, organic-rich shale containing solid hydrocarbons called *kerogen*. The kerogen can be liquefied at high temperatures, and the oil can be driven out of crushed shale. A normal yield is 25 to 30 gal of oil per ton of shale.

7.16 Where are the big reserves of oil shale in the United States? (See Fig. 7.5.)

The big deposits are in the tristate area of Utah, Wyoming, and Colorado, in lake beds of Eocene age (about 50 million years old).

7.17 What are tar sands, and where are the earth's largest known deposits? (See Fig. 7.6.)

Tar sands are sandstones with a very heavy crude oil that could even be called *asphalt* (tar). It is too viscous to flow naturally, so it is heated with steam. Tar sands probably represent the heavy residue left after volatile fractions escaped from the rock. The largest known deposits are in northern Alberta.

7.18 Which processes—igneous, metamorphic, or sedimentary—form valuable mineral resources?

All of them do.

7.19 What is an ore?

Ore is generally defined as rock that contains useful minerals that can be extracted at a profit. Obviously, most rock is not ore, and some natural concentrating mechanisms must have concentrated the materials. In some cases, a certain mineral may be essential and will be mined regardless of the cost.

7.20 Give an example of a mineral formed by primary igneous processes.

The ore minerals of chromium (chromite), nickel (pentlandite), and platinum (e.g., the rare mineral sperrylite, with a formula of $PtAs_2$) are interpreted to be minerals that formed during the first stages of crystallization and settled in the magma chamber.

7.21 Is iron ore a resource that is in short supply?

No, it is one of the most abundant of the metals, constituting 5 percent of the earth's crust.

7.22 Where are iron deposits found?

The large deposits of hematite and magnetite occur on all the inhabited continents (Fig. 7.8).

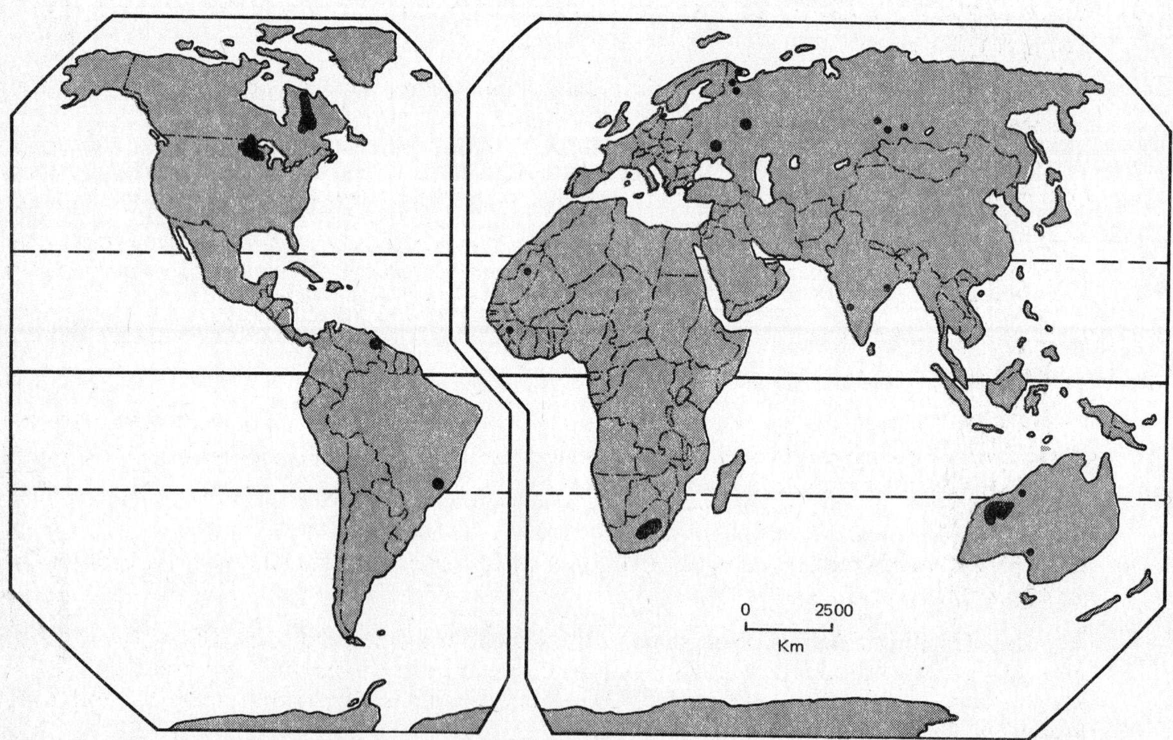

Fig. 7.8 Map of the world's major iron ore deposits. They are between 1800 million years and 2400 million years old. (*After Skinner, B. J., Earth Resources, Prentice-Hall, Englewood Cliffs, N.J., 1969.*)

7.23 What is the main ore mineral of iron, and does it have an igneous, sedimentary, or metamorphic origin?

Hematite (Fe_2O_3). It is a sedimentary mineral, formed by the oxidation of iron as it combines with oxygen under surficial conditions.

7.24 What is the main ore mineral of aluminum, and what is its origin?

Bauxite ($Al_2O_3 \cdot H_2O$), which is really a mixture of three hydrated aluminum oxides, has a sedimentary origin, being the end product of the severe weathering of rocks containing aluminum-bearing minerals, especially the feldspars. Under subtropical to tropical conditions, where chemical weathering is intense, most of the cations are removed and leached away, forming kaolinitic clays (HAlSiO). If the leaching waters are alkaline, even the silica goes into solution, leaving Al_2O_3. (See Problems 4.28 and 4.32.)

7.25 Name a main ore mineral of copper, and is it igneous, sedimentary, or metamorphic?

Chalcopyrite ($CuFeS_2$). It may be igneous, as in early crystallized mineral of a mafic magma, or it may be volcanogenic or hydrothermal in origin.

7.26 Name the main ore mineral of zinc and give its origin.

Sphalerite (ZnS). It is volcanogenic or hydrothermal in origin.

7.27 Name the main ore mineral of lead and give its origin.

Galena (PbS). It is volcanogenic or hydrothermal in origin.

7.28 Where are the rare earth elements such as zirconium, lithium, beryllium, and columbian-tantalum found?

In pegmatites, which crystallize from the last solutions of a magma and form veins and pods near the main granitic intrusion. (See Problems 3.15 and 3.16.)

7.29 Why do quartz and calcite veins sometimes contain ore minerals such as native gold?

These veins are commonly hydrothermal in origin. The hot waters, as they move at depth under pressure, dissolve (leach) elements from the rock through which they are passing. As the waters reach cooler upper crustal levels, they cool and precipitate as veins with ore minerals.

7.30 How are placer gold deposits formed?

As rock containing native gold is weathered, the unweathered native gold is carried off with other sediment. It has a high density (19 +) compared to the density of quartz and feldspar (2.7) and is trapped in irregularities on the bottoms of stream channels.

7.31 Why can't common rock be mined for elements such as copper? That is, why must high-grade deposits be found?

Most metallic elements are present in very small amounts in the earth's crust, and the cost of processing vast quantities of rock is prohibitive. For example, copper makes up only 0.0058 percent of the crust by weight. Fortunately, igneous and hydrothermal processes have formed copper ores with as much as 5 percent copper.

7.32 Metals such as manganese, nickel, cobalt, tungsten, and vanadium are also necessary in our modern civilization. What is their main use?

As additives to iron and carbon in the steel-making process, thus forming steels with different properties.

7.33 What is the nuclear fuel of the nuclear industry?

Uranium. Pitchblende and uraninite (both with the formula UO_2) are the main uranium minerals, but many others are also utilized, including carnotite (a complex hydrated oxide) that is mined in several western states.

7.34 Where does most silver come from?

It is commonly a by-product of the mining of copper, zinc, and other sulfides. It occurs as minor silver sulfides or silver arsenides, with pyrite, chalcopyrite, and sphalerite. Some native gold also contains some native silver.

7.35 What is gypsum, how does it form, and what is its main use?

Gypsum is $CaSO_4 \cdot 2H_2O$. It is formed by the evaporation of seawater, and it is used primarily in the manufacture of plaster of paris and plasterboard.

7.36 Why are phosphates an important nonmetallic mineral resource?

The element phosphorus is an essential constituent of a fertile soil and is a critical element in the diet of animals. Phosphorus is present in apatite—$Ca_5(PO_4)_3OH$—a mineral that has igneous, sedimentary, and metamorphic origins. The large deposits, however, are sedimentary, deposited where cold phosphate-rich marine waters well up onto continental shelves.

Supplementary Problems

7.37 What minerals, other than native gold, are found as placer deposits?

Any heavy mineral can be placered (i.e., the concentrating of heavy minerals) by the sorting action of water in streams or on beaches. Recall that feldspar and quartz have densities of about 2.7, relative to water which has a density of 1.0. Following are some examples with their densities also given:

Magnetite	3.0–3.2
Ilmenite (a titanium-iron mineral)	4.7
Zircon	4.7
Diamond	3.5
Ruby and sapphire (corundum)	4.02
Native gold	15.0–19.3

7.38 How do diamonds form?

Diamonds are found in the ultramafic rock kimberlite, which occurs as small pipelike intrusions. The weathering of kimberlite can free diamonds which can than be placered. It is thought that they originally form under high temperatures and pressures in the upper mantle, perhaps at depths of 100 km, and are carried to the upper crust by ultramafic magma.

7.39 What is a possible future source of manganese?

Manganese nodules that form on the sea floor. These nodules, usually only a few cm in diameter, also contain rarer elements such as copper, cobalt, and nickel. An average composition for the nodules is 20 percent manganese, 6 percent iron, 1 percent copper, and 1 percent nickel.

7.40 What is the origin of the large iron ore deposits of the world? (See Fig. 7.8.)

Iron-formation is formed by the precipitation of various iron minerals out of water. Iron-formation consists of about 30 percent iron (mostly as the minerals hematite, limonite, and magnetite) and 70 percent silica (chert), and is called *taconite*. Taconite is mined today, but in the past only higher-grade natural concentrates were economical. In these, silica was dissolved by alkaline waters, leaving a concentrate of iron that made up 50 to 55 percent of the rock as the minerals hematite, limonite, and magnetite.

7.41 How many times richer is copper ore of (*a*) 5 percent copper, compared to the average crustal abundance of 0.0058 percent and of (*b*) 0.25 percent copper, the lowest grade mined?

(*a*) 862 times. (*b*) 43 times. To get these answers, simply divide the average abundance into the ore values (i.e., 5/0.0058 = 862, and 0.25/0.0058 = 43). Obviously, nature was an efficient and inexpensive concentrator.

7.42 What are the three main metals that are mined in large quantities?

Iron, copper, and aluminum.

7.43 What is a porphyry copper deposit?

A low-grade copper ore in porphyritic felsic to intermediate intrusive rock bodies (stocks). Although low grade (as low as 0.25 percent copper, mostly as chalcopyrite—$CuFeS_2$—disseminated through the rock), such deposits have large tonnages and can be mined profitably.

7.44 What are probably the two most important mineral resources which must be conserved because of their necessity to life?

Soil and water. They are so valuable that their monetary worth cannot even be estimated. How could we put a price on them?

7.45 Describe supergene enrichment (secondary enrichment) of a pyrite- and chalcopyrite-bearing rock by weathering processes (Fig. 7.9).

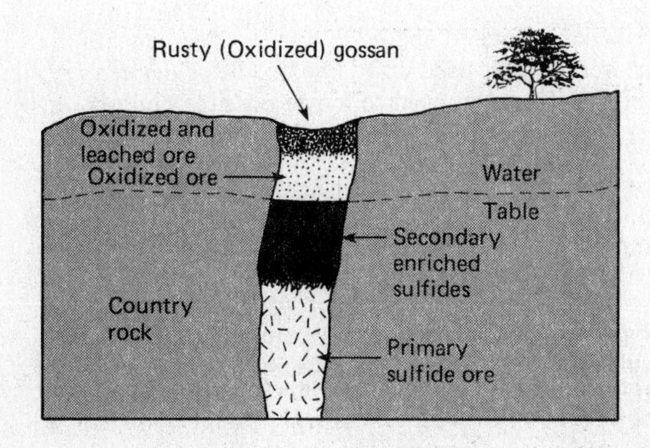

Fig. 7.9 Diagram of secondary enrichment by weathering of a low-grade, sulfide-bearing deposit.

Water percolating down through the rock dissolves the pyrite, forming sulfuric acid. This in turn leaches the copper from the chalcopyrite and precipitates it below the water table as an enriched ore zone. Such "supergene caps" are valuable.

7.46 Calculate the amount of iron (weight percent) to the nearest whole percent in each of the following iron minerals: Magnetite (Fe_3O_4), hematite (Fe_2O_3), goethite [FeO(OH)], and siderite ($FeCO_3$).

Magnetite = 72 percent

Hematite = 70 percent

Goethite = 63 percent

Siderite = 48 percent

To make the calculations, divide the atomic weight of the contained iron of a given mineral by the atomic weight of the total mineral, given these atomic weights: Fe = 55.8, C = 12, O = 16, H = 1.

For example, magnetite's weight is 3 × 55.8 (because there are three Fe's) or 167.4, plus 4 × 16 (because the atomic weight of oxygen is 16 and there are four of them), or 64, for a total of 231.4. Divide the weight of the three Fe's (167.4) by the total weight of 231.4 to get the percentage of iron in magnetite.

For hematite, the answer is obtained by dividing the weight of two Fe's (111.6) by the total weight of hematite (111.6 plus three oxygens or 3 × 16 = 48), or 159.6.

For goethite (limonite), the answer is obtained by dividing the weight of one Fe (55.8) by the total weight of goethite (55.8 plus two oxygens, or 32, plus one hydrogen, or 1, for a total of 88.8).

For siderite, the answer is obtained by dividing the weight of one Fe (55.8) by the total weight of siderite (55.8 plus one C, or 12, plus three oxygens, or 48, for a total of 115.8).

7.47 Where are the biggest gold deposits in the world, and what type of deposits are they?

In South Africa in the Witwatersrand Basin. These 2300- to 3000-million-year-old quartz-pebble conglomerates contain what is interpreted by most geologists to be ancient placer deposits, concentrated on alluvial fans. These deposits produce about 60 percent of the non-Communist world's gold. The same rocks also yield about 20 percent of the non-Communist world's uranium.

7.48 What is the role of an economic geologist?

To find mineral deposits that can be developed into economic ventures. (The actual extraction methods are the realm of the mining engineer, and the metallurgical processing techniques are the responsibility of the metallurgist. Obviously, accountants and economists analyze the overall viability of the project.)

7.49 Why is it more difficult to find mineral resources today than it was in the first half of the twentieth century?

The deposits exposed at the surface were relatively easier to find compared to those buried beneath the surface. The easily found deposits were limited and were largely used up by the middle of the century. Today, geophysical techniques (see the following problems) are utilized to locate possible deposits of, e.g., petroleum or sulfide minerals. Finally, however, drilling is necessary in order to determine whether a deposit is actually present or whether other features "acted like" deposits.

7.50 How is geophysics used in the search for petroleum?

Seismic surveys, in which sound waves are sent into the earth by explosions or vibrations, are used to detect buried anticlines, faults, and stratigraphic traps (see Fig. 7.3). The waves are reflected off of the different strata, and the time they take to travel down and back provides a measure of depth. Thus, many readings can delineate a buried structure.

7.51 How is geophysics used in the search for mineral deposits?

Magnetic surveys can locate buried bodies of magnetite. Electromagnetic surveys measure the conductivity of buried rock units; massive sulfide bodies are better conductors than are the silicate minerals.

Gravity surveys measure the densities of buried rock bodies, thereby helping to identify the types of rocks present; many ore bodies are composed of denser minerals.

7.52 How is geochemistry used in mineral prospecting?

Chemical tests of soils, plants, and waters can reveal anomalous amounts of different elements (e.g., copper). Then geophysical methods are utilized to locate the source rocks that released the abnormal quantities of ions.

7.53 What is the chemical composition of petroleum?

Petroleum includes both crude oil (a liquid) and natural gas (gaseous). It consists mainly of C and H, plus S, N_2, O_2, and other very minor ions. Oils range from "heavy oil" with a high molecular weight due to the large organic molecules, to "light oils" which formed when larger, heavier molecules were broken or "cracked" during burial under high temperatures and pressures. Heavy oils grade into asphalt or tar.

7.54 What are the differences in the combustible properties of lignite, bituminous coal, and anthracite coal?

Lignite contains more moisture (43 percent) and less carbon (38 percent) than bituminous coal (5 percent moisture and 54 percent C) and anthracite (3.2 percent moisture and 96 percent C). Anthracite is the highest-grade coal and provides about twice the heat of lignite.

7.55 Where is most of the world's petroleum located?

Most is in the Middle East. Slightly different estimates have been made, and range from 50 to 75 percent of the world's resources.

7.56 What country produces most of the world's diamonds?

The Congo. Also, very important are South Africa, Ghana and the U.S.S.R.

7.57 What country produces most of the world's other gemstones (ruby, sapphire, and others)?

Brazil.

7.58 What countries produce most of the world's bauxite?

Australia, the African nations of Guinea and Cameroon, southern Europe, Jamaica, Surinam, and Guyana

7.59 What countries produce most of the world's copper?

U.S.A., U.S.S.R., Chile, Zambia, and Canada.

7.60 Which type of petroleum trap was most easily discovered in the early days of petroleum exploration? (See Fig. 7.3.)

Anticlines, as they may have a surface expression.

Chapter 8

Erosion and Mass Movement

THE HYDROLOGIC CYCLE

The *hydrologic cycle* (Fig. 8.1) summarizes the travels of water on or near the earth's surface. It can be stated in words as follows: Precipitation = runoff + infiltration + evaporation + transpiration. The water cycle is one of the most important of the earth's many cycles, for it sustains life on earth by recycling and hence replenishing the water supplies on or near earth's surface. Water that is precipitated on the 21 percent of earth's surface that is land rather than ocean moves downhill immediately after falling as rain if it doesn't infiltrate (soak into the ground), or later if it was precipitated as snow. This water that moves downhill, or runs off, is important, as water is the great leveler of the earth's surface.

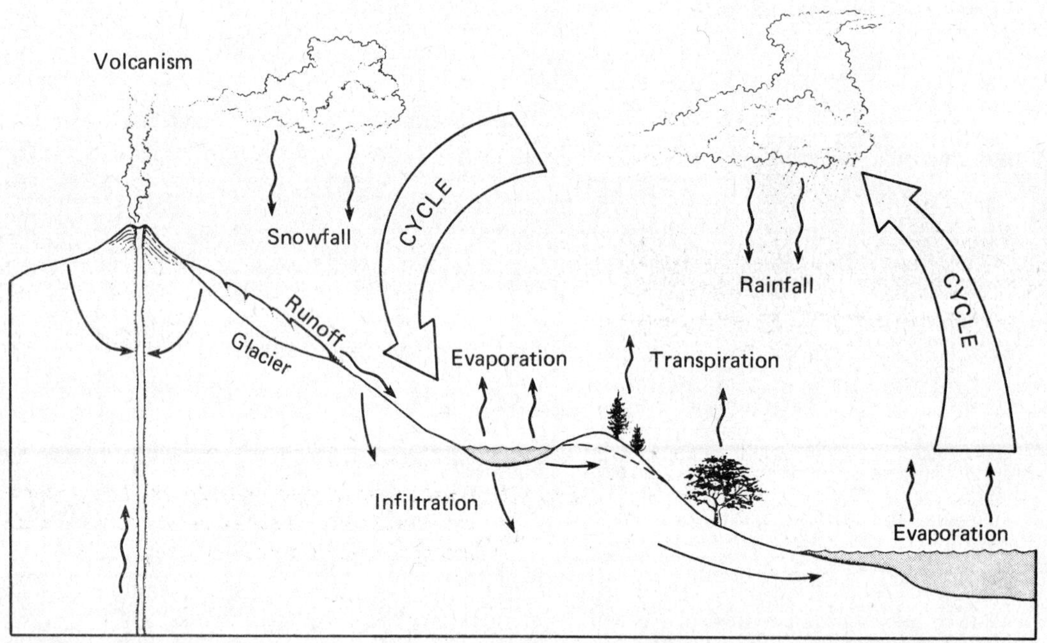

Fig. 8.1 The hydrologic cycle, showing the route that water takes as it is recycled over and over again.

INTRODUCTION TO EROSION

The products of weathering—clays, iron oxides, and quartz sand and silt—rarely stay where they were formed. This is because most of the earth's surface is not horizontal, but has some slope. Therefore, the agents of erosion, especially running water but also glacial ice (Fig. 8.1), tend to move the particles downslope, toward *base level* (the lowest level of erosion) which is ultimately sea level. Sea level is thus the "bottom of the hill" for all the earth's surface that is above sea level.

To paraphrase the author of Isaiah 40.4, the hills and the mountains will be leveled and the valleys will be filled. Landscapes are in continual change, a never-ending contest between tectonism (earth movements) and erosion. Running (moving) water and glacial ice act under the influence of gravity, and even gravity alone can move material to lower elevations, as when a boulder falls down a cliff.

Wind also moves material, but not necessarily only downslope, for fine clay- and silt-sized particles can be carried up to altitudes of thousands of meters and transported around the world. (This is especially true of fine material put into the air by volcanic eruptions.) Currents in lakes and oceans can move sediment downslope into deeper waters (e.g., turbidity currents as in Problems 5.14 and 5.15), but they also commonly move sediment parallel to the continental slopes and parallel to the shorelines. In the latter situation, the movement of sediment is accomplished in spite of the pull of gravity rather than because of it. Human activity is responsible for much movement of material as well, a process we might call "people erosion."

All the above-named agents of erosion deposit their loads when there is a decrease in velocity. When the air stops moving, its load falls. When a fast mountain stream reaches the base of the mountain, it drops much of its load. When a turbidity current slows, it starts dropping its load, and when it finally stops, it deposits the remainder of its load. When a continental glacier reaches a warmer latitude, or when a valley glacier reaches a lower and warmer elevation, it deposits its load.

Nevertheless, the products of weathering, and in some cases entire soil profiles, escape erosion and are preserved *in situ*. These *regoliths,* or zones of weathered rock (Fig. 8.2), can even be preserved in the rock record beneath or between layers of sedimentary rock. There are several examples of regoliths as old as 2200 million years. Old regoliths are especially interesting to geologists, as they can yield clues to the nature of the climate and the atmosphere during the time they formed (see Problem 4.24).

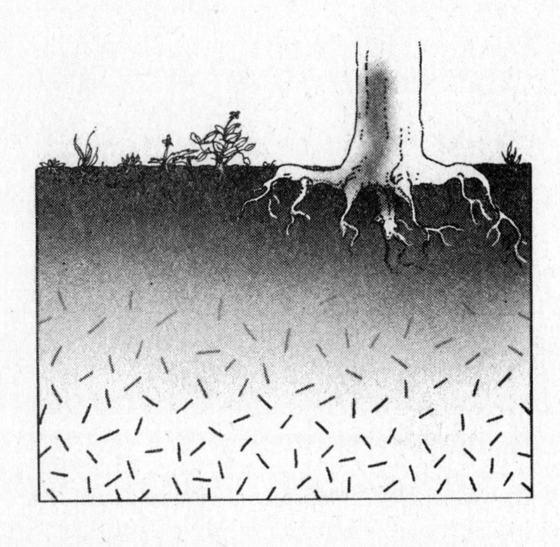

Fig. 8.2 Regolith developed on bedrock. The upper part of the regolith may be a true soil profile if weathering has proceeded far enough, as in Fig. 4.5.

MASS MOVEMENTS

The next five chapters deal with both erosion and deposition by running water, glacial ice, wind, shoreline processes, and underground water. In this chapter, the emphasis will be on *mass movement processes,* those controlled largely by gravity. During mass movement, material moves downslope "toward the center of the earth." The earth exerts a force of attraction on any mass of material on its surface, as defined by the universal law of gravitation (see Problem 1.30). The material can fall down vertical slopes or slide down less steep slopes.

Mass movements can be classified into two types, rapid and slow, but there is obviously a gradation between them. *Rapid mass movements* include mudflows, landslides, avalanches, and fast subsidence. Some occur in subaerial environments whereas others can occur in subaqueous environments. For example, a landslide can occur on land or beneath the ocean. Soil creep and slow subsidence are *slow mass movements*. Both rapid and slow mass movements are most abundant in areas with considerable relief and steep slopes. Dry material can move under the influence of gravity, but water commonly acts as a lubricant, and freezing and thawing action can also aid in the downward movement. Whereas running water can move particles virtually across a continent, mass wasting processes move material over only relatively short distances.

Solved Problems

8.1 What is the hydrologic cycle, and how can it be expressed in words?

The route that the earth's water takes, from precipitation onto the earth's surface, to its movements on or near the surface, and finally, to evaporation-transpiration back into the air. It can be expressed as precipitation = runoff + infiltration + evaporation + transpiration.

8.2 What is transpiration?

Transpiration is the release of water to the atmosphere through the leaf surfaces of plants.

8.3 What is the difference between weathering and erosion?

Erosion is the removal, transportation, and deposition of earth materials, generally from topographically higher to topographically lower locations. Weathering is the breakdown of rock material into smaller pieces or into ions in solution, as described in Chapter 4.

8.4 What are the agents of erosion?

Water, wind, glacial ice, gravity, and people.

8.5 What is the most important agent of erosion?

Running water.

8.6 When and where do the agents of erosion deposit their loads?

When there is a decrease in velocity, as when a mountain stream slows upon reaching the base of the mountain.

8.7 What is a regolith (Fig. 8.2)?

A *regolith* is a layer of weathered rock, sometimes including soil, that is essentially in place (*in situ*). It may be on the surface or be buried beneath layers of sediment or sedimentary rock.

8.8 Beneath a 500-million-year-old lava flow is a 3-m-thick regolith of decayed rock fragments, red iron oxides, and cation-deficient clay minerals. Why is the geology student who discovered this excited about his discovery?

Because it shows that this was the earth's surface sometime before 500 million years ago and that the climate was probably wet and warm in order to produce the cation-deficient clay. (See Problem 4.28.)

8.9 What is base level?

Base level is the level below which a stream cannot erode. Ultimate base level is sea level, but local base levels are the result of lakes or hard rock units which affect the depth of erosion.

8.10 How do talus piles (talus cones) form? (See Fig. 4.4.)

By blocks falling off of cliffs to form coarse piles at their bases.

8.11 What are the two classes of mass (gravity) movements?

Slow movements and rapid movements.

8.12 What is a rock glacier, and why does it move?

A *rock glacier* is a mass of large rock fragments that moves downslope under the influence of gravity, aided by ice between fragments that acts as a lubricant. The rock mass thus moves somewhat like a glacier does, and hence the name. Rather steep slopes are necessary.

8.13 Why do mudflows move?

Mudflows are masses of largely muddy material (clay- and silt-sized), lubricated with water, that move downslope under the influence of gravity.

8.14 What is the angle of repose (the steepest angle at which a pile of sand will stand) of loose dry sand?

About 35°.

8.15 What is solifluction?

Solifluction is soil creep on slopes in cold climates. The soil may move only a few centimeters or meters per year and only during the summer when ice in the surficial material melts and the underlying ground remains frozen. The water-saturated top layer moves.

8.16 Does soil creep operate in warmer climates as well?

Yes. On steep slopes, soil creeps or moves downhill at a slow rate, usually detected only by bent trees or tilted posts (Fig. 8.3). The steeper the slope, the more rapid is the creep.

Fig. 8.3 Soil creep on a steep slope.

8.17 What generally displaces the blocks on a cliff, forming talus piles?

Frost wedging. (See Problem 4.9.)

8.18 What is colluvium?

Colluvium is a general term for material on or at the bases of slopes, moved there by mass movements under the effect of gravity.

8.19 Are avalanches (snow slides) mass movements?

Yes. The mineral ice is moved rapidly downhill under the influence of gravity. Avalanches can do considerable damage and are especially dangerous in ski areas. They commonly carry loose rocks from higher to lower elevations.

8.20 What are landslides?

Landslides are masses of rock material that move rapidly down steep slopes.

8.21 Briefly describe the Gros Ventre landslide that occurred in 1925 near Jackson Hole, Wyoming.

Melting snow and heavy rains caused this landslide of 37 million cubic meters of rock and soil to slide down a 20° dipslope (rock bedding surface). Sandstone, unsupported because of erosion of the gently dipping sandstone by the river on the valley floor, slid on the underlying water-saturated shale layer. The slide acted as a dam in the river valley, and a lake quickly formed behind it. The lake overflowed 2 years later, eroding the dam and causing a flood further down the valley.

8.22 How does freeze-thaw action aid downslope movement of material?

As water under mineral grains or rock fragments freezes, it raises the particles perpendicular to the slope. As the ice melts, the particle moves downward vertically under the influence of gravity (Fig. 8.4).

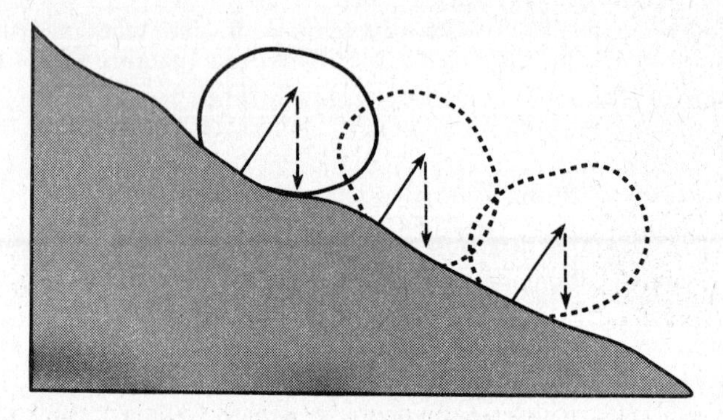

Fig. 8.4 Freeze-thaw action can move particles downslope. The arrows pointing upward show the movement due to freezing and the arrows pointing downward (perpendicular to the horizontal) show the movement when the ice melts.

8.23 What causes material that was previously stable to move downslope under the influence of gravity, either slowly or rapidly?

The usual cause is water saturation of soil and rock material on slopes. Also, slopes that become oversteepened due to human activity or natural erosion at the base of the slope may move under the influence of gravity. Other causes are deforestation by cutting or fires, or a decrease in vegetation by overcultivation.

8.24 Whereas oversteepening makes slopes unstable, what actually triggers rapid mass movements?

There are several causes. Earthquakes or heavy rainfall can cause a lack of cohesion of particles, thus initiating landslides or debris flows. For example, the Madison River, Montana, slide of August 17, 1959, was triggered by an earthquake; it moved 160 km/h (100 mi/h), killing several campers. It formed a dam that was breached with an artificial channel to reduce flood danger.

8.25 How fast can rapid mass movements move?

Fast! In Peru in 1970, a mass of rock, snow, and ice slid down Mt. Huascare at speeds as fast as 400 km/h (240 mi/h); an estimated 20,000 to 30,000 people were buried within 4 min.

Supplementary Problems

8.26 How can one distinguish mass movement deposits from water-moved material?

The products of mass movement are poorly sorted, or visually unsorted, whereas water sorts the material, resulting in layers of different grain sizes.

8.27 In Fig. 8.5, the highway department is clearing a mudflow off a highway. What will probably happen next?

Fig. 8.5 A highway crew will soon be clearing a mudflow off a highway.

Recall that most mass movements occur because of oversteepening of slopes. Removal of the toe of the slide will result in oversteepening, and a heavy rainfall or an earthquake, or even the weight of the sediment itself, will result in another mudflow moving onto the highway. (A steady job!)

8.28 What kind of mass movements might be expected in areas underlain by limestone or dolomite, as in Florida?

Collapse of rock into underlying cave systems. Sinkholes form in this manner. Similar collapse of rock into old underground mine systems also can occur.

8.29 See Fig. 8.6. Your good friend has the opportunity to buy this riverside land in a beautiful canyon at a good price. She wants to build a log cabin on the site. Knowing some geology, what should you tell her?

There are two problems with this land. One is that blocks may fall down the cliff and demolish the cabin and perhaps even squish your friend. An even larger problem is that there is danger of a slide that could be triggered by heavy rainfall or an earthquake. Note that the cabin site is located upon material that was probably brought there by a previous slide.

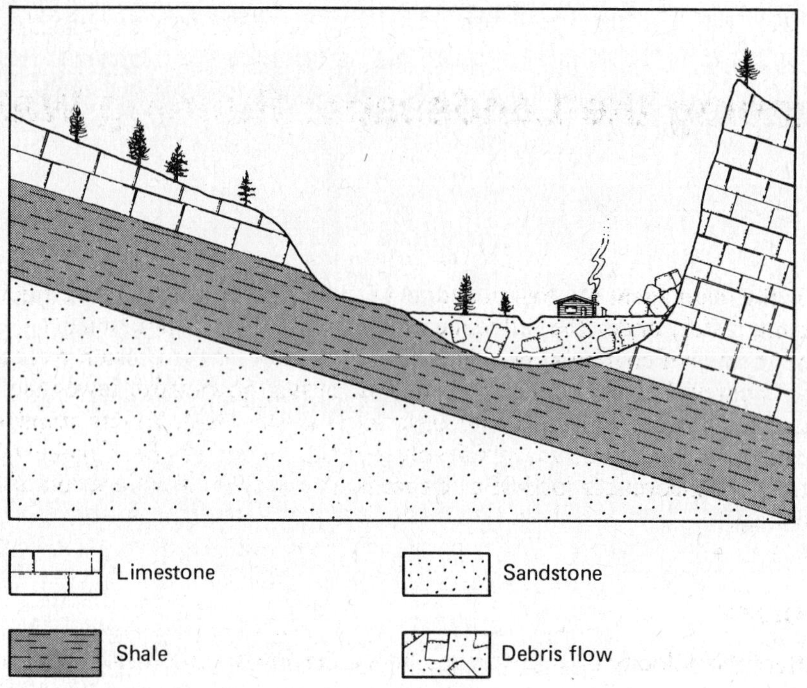

Fig. 8.6 Isn't this a beautiful site for a cabin?

8.30 Many massive debris flows are related to volcanism, as in 1985 when a volcanic eruption in Columbia caused debris flows that buried a village and about 20,000 people. Why are volcanoes often the sites of massive debris flows?

Because hot ash and gases can melt snow and ice on high volcanoes. In addition, the volcanism commonly causes rainfall over the volcanoes which adds moisture to volcanic ash and can cause debris flows.

8.31 Can mass movements be predicted?

Sometimes. Sensitive instruments and even animals can detect slight ground movements. Thorough studies by geologists or geological engineers may detect dangerous situations.

8.32 In a "Peanuts" cartoon strip, Lucy is decrying the erosion process, saying, "Billions of people scrape their shoes as they walk along the ground, day after day, wearing this ol' earth down to the size of a basketball." What is she obviously ignoring?

Material removed from one area must be deposited in another area. The total mass remains the same.

8.33 Could poor engineering of structures, such as dams, result in disasters?

Yes. For example, a dam in a steep valley in Italy was built despite the presence of weak layers of limestone and shale on steep slopes above the reservoir site, and previous landslides in the area. A huge landslide plunged into the reservoir in 1963 and created a huge wave that washed over the 875-ft-high dam and killed 2600 people downstream.

8.34 Give an example of a submarine landslide (submarine slump).

See Fig. 5.7 for details of the submarine landslide that developed into a turbidity current off the Grand Banks of Newfoundland.

Chapter 9

Shaping the Landscape: Running Water

EROSION

Running water is the main agent of erosion and includes both sheetwash and channel flow (streams). Erosion by running water, under the influence of gravity, can level mountains.

Rivers erode their own channels, but sheetflow does most of the erosion over large areas. The more visible rivers carry the sediment away, and, by cutting downward, lower the local base level and thus enhance further erosion. Ultimate base level is sea level. Rivers transport sediment as *bedload,* in *suspension,* and in *solution* (dissolved load); together these make up the *load* of a stream. The maximum theoretical load of a stream is its *capacity*. Most erosion and transportation is done at flood stage.

STREAM VELOCITY

Many factors affect the velocity of a stream and hence the amount of erosion. A generalized formula for stream flow is:

$$V \propto \left(\frac{SA}{RP}\right)^{1/2}$$

where V is the velocity, S is the slope, A is the cross-sectional area of the channel, R is the roughness of the channel floor, and P is the wetted perimeter (the total length of that part of the channel sides and bottom in cross-section that is under water). So, the velocity of a stream is proportional to the square root of SA divided by RP. The formula shows that stream flow is complex, and a change in one factor can cause changes in the other factors or in the stream velocity. And only some major factors are given in the formula. Others, such as total sediment load and discharge, have been omitted for simplicity.

The amount of water discharged by a stream is given by the formula $Q = WDV$, where Q is the discharge, W is the channel width, D is the channel depth, and V is the velocity.

A stream is in equilibrium, or is a *graded stream,* if there is *no* net erosion or deposition. That is, any erosion is exactly balanced by deposition. A graded condition is only temporary, for changes in any of the factors mentioned in the previous two formulas disrupt equilibrium.

DEPOSITION

Streams eventually deposit the sediment that they are transporting; this occurs when there is a significant decrease in velocity. Some is deposited on the stream's *floodplain* (the part of the stream valley that is under water during floodstage), usually to be reeroded and retransported later. Where a stream leaves a rugged or mountainous terrane and suddenly loses velocity, it deposits its load as an *alluvial fan.* Where a stream enters a lake or an ocean, a *delta* may be formed if the river's load is too large to be rapidly removed by currents in the lake or ocean.

EVOLUTION OF VALLEYS AND LANDSCAPES

Stream valleys change or evolve through time, progressing from youth to maturity to old age. *Youthful valleys* are V-shaped, have steep gradients, and contain rapids and waterfalls. A *mature valley* is wide, constituting the floodplain of the stream, with the actual stream channel occupying only a part of the valley. The stream generally winds or meanders across its floodplain, changing position by lateral erosion; such broad loops or *meanders* that are isolated or cut off by stream migration form *oxbow lakes*. An *old-age stream valley* is very wide with low valley walls and extensive development of meander loops and oxbow lakes.

A topography or landscape theoretically follows a similar evolution from youth to maturity to old age. After initial uplift of a broad area, streams will cut downward to reach a new base level; an area with broad and flat interstream areas between youthful deep stream valleys is in *youth*. As erosion by sheetwash and streams, accompanied by mass-wasting (see Chapter 8), proceeds, the flat interstream areas are transformed into a hilly terrane with sloping surfaces, and the area is said to be in the *mature* stage. As the hills are eroded down to local base level, the area is transformed into a *peneplain* (''almost a plain'') with a few remnant hills or *monadnocks*. This progression of a landscape from youth to maturity to old age is commonly called the *erosion cycle*.

Each stream has a *drainage basin,* the total area that contributes water to the main stream via its tributaries and sheetwash. If there is no major structural control by underlying rocks on stream flow, the stream network is likely to be *dendritic,* with a stream pattern similar to the veining on a leaf; the V's formed by the junction of any two streams point downstream. Bedrock control can result in other geometries, such as a *rectangular* drainage pattern on an area where the bedrock has joint or fault sets at right angles.

Solved Problems

9.1 ''Running water, or surface runoff as in Fig. 8.1 of the previous chapter, means river water.'' Is this a correct statement?

It is partly correct. The term *running water* includes all runoff water, both channel flow and unchanneled flow that is more commonly known as ''sheetwash.''

9.2 When does precipitation become surface runoff (Fig. 8.1)?

When the amount of precipitation exceeds the infiltration capacity of the soil and the loss through evapotranspiration.

9.3 What is sheetwash or sheet flow?

When rain falls on any surface, it accumulates and forms thin sheets of water that move downslope under the influence of gravity; such flow is *sheetwash* or *sheet flow*. Irregularities on the surface result in small trickles of water, which merge to form larger rivulets, then streams, and eventually join to form large rivers.

9.4 Compared to rivers, do these thin sheets of water contribute significantly to erosion?

Yes! Every square inch of the earth's land surface can be subjected to erosion by sheetwash. It is sheetwash that starts the erosion process and initially moves particles.

9.5 What is the major role of rivers in the erosion process?

Rivers are important to the erosion process in two major ways. They act as the ''freight trains'' that transport loose sediment out of a given area. Also, rivers with their large volumes of water, can, under certain circumstances, cut their channels deeper and deeper, even into hard rock. This lowers the local base level (see Problem 8.9), resulting in more erosion of the nearby land as sheetwash and small streams strive to cut down to the lowered base level. We could say the sheetwash does the work and rivers haul away the debris.

9.6 What could you reply to a friend who says, "Surely the Mississippi River does more erosion than does sheetwash!"

You could say that that is not a valid statement. Virtually all the water in the Mississippi was sheetwash before it reached the river. Only that insignificant fraction of precipitation that fell directly from the clouds into the river channel was not sheetwash.

9.7 Is the amount of sediment removed from the land surface by erosion significant?

Yes. It has been estimated that running water removes an average of 1 m of sediment (including soil) every 30,000 years from the entire land surface of North America. (Obviously the actual amount varies from place to place, depending upon the amount of rainfall and many other factors.)

9.8 Calculate the annual rate of erosion, based on the data in the previous question.

One meter, or 100 cm, divided by 30,000 years gives 0.0033 cm (or 0.033 mm) per year, or 1 cm per 300 years. Is this a trivial amount? It may seem so, but erosion continues over great lengths of time, and the net result is truly significant. Mountains can be worn completely away, as has happened time and time again in the earth's history.

9.9 Several factors affect the amount of erosion by running water that can occur in a given area, but two are generally the most important. Name them.

The amount and intensity of precipitation and the relief or ruggedness of the area. (Much rain falling on a low-lying featureless area will not cause significant erosion.)

9.10 What types of sediment constitute the total load of a stream?

Coarse sediment is moved along the river bottom as *bedload*; fine sand, silt, and clay are carried in the water as *suspended load*; and ions in solution are carried as *dissolved load*.

9.11 What is the difference between the load and the capacity of a stream?

The *load* of a stream is the total amount of material that the stream is carrying, whereas the *capacity* is its theoretical maximum load.

9.12 What is the competence of a stream?

The largest particle a stream can move is a measure of its *competence*.

9.13 The *gradient* of a stream can be defined as the drop in stream elevation along its course and is expressed in meters per kilometer or in feet per mile. A related concept is the *slope* of a stream, which is defined as the vertical drop of a stream divided by the horizontal distance over which the drop takes place. What is the slope of stream A that drops 264 ft/mi? What is the slope of stream B that drops 200 m/km (1 mi = 5280 ft; 1 km = 1000 m)?

$$\text{A: } \frac{264}{5280} = 0.05$$

$$\text{B: } \frac{200}{1000} = 0.20$$

9.14 Stream velocity is complex. (*a*) What would happen to the velocity of a stream at a given point on the stream if the slope decreased? (*b*) If the angle of the slope increased? (*c*) If the roughness of the bottom increased? (See the formula on page 93.)

The velocity would: (*a*) Decrease. (*b*) Increase. (*c*) Decrease.

9.15 Refer back to Problem 9.13. Which stream, A or B, should have the highest velocity and erode the most if the other factors are equal?

Stream B. Note that in the formula for velocity in the introduction to this chapter, as the slope S increases, the velocity V also increases.

9.16 The previous problem suggests that a stream's velocity might change frequently. What if the velocity remained rather constant, as it commonly does, during a given period of time?

If any factor changes and the velocity remains the same, then some other factor or factors must change to compensate. The formula shows that stream flow is extremely complex.

9.17 How does vegetation affect the amount of runoff?

A good cover of vegetation decreases the amount of water that runs off a surface. It holds the water and therefore aids infiltration, which commonly can be slow. Arid areas with less vegetation thus have more runoff and more erosion. For example, in the drainage basin of the Colorado River in southwestern United States, more than three times as much material is eroded away per unit area as in the Mississippi River drainage area where there is more vegetation. A good vegetative cover will, of course, result in more evapotranspiration of water to the atmosphere.

9.18 What is the concept of a graded stream?

A stream is said to be *graded,* or in equilibrium, if it is neither eroding nor depositing. (It is an idealized concept, and the previous few problems indicate that a stream's velocity, and hence its ability to carry sediment, is always in a state of change along its entire course because there are so many variables that are constantly changing.)

9.19 When do rivers do most of their work of transporting sediment?

During flood stage when the water discharge and velocity are the greatest.

9.20 What would happen to the sediment load of a stream if the velocity of the stream decreased considerably?

The stream would deposit its bedload and part or all of its suspended load, the coarsest sediment first and the finest last, perhaps further downstream where the velocity might be less and where the gradient is likely to be less.

9.21 What would happen to the dissolved load of a stream if the velocity decreased?

Nothing. The amount of material in solution is not dependent upon the velocity. It will be deposited (i.e., precipitated) only when the chemical conditions are right.

9.22 What is the discharge of a stream?

The total volume of water passing a given point in a given time is the *discharge*. For example, 100,000 l/s (about 23,640 gal).

9.23 Imagine you are trying to cross a fast stream that has a constant width. The water is moving at the same velocity along the straight stretch which you must cross. The stream seems to be waist deep and the water is COLD! After scouting, you find a shallow place where it appears that you can wade across with the water only about ankle-deep. So you do it. What happens? (Remember the formula for stream discharge that is given in the text.)

The discharge would probably have remained constant all along the straight stretch, and since the width and velocity are constant, only the depth can and must vary. It may *appear* to be shallow, but somewhere along the way you are crossing, the depth must increase drastically. You may be in over your head! (Sorry to say, this actually happened to the author of this book!)

9.24 How does the discharge of a stream affect the amount of sediment that the stream can carry?

As the discharge increases, the increased water volume means that the stream can carry more sediment, all other things being equal.

9.25 At what point along a stream's course should the discharge be the greatest, and why?

At its mouth, because tributaries all along its course are adding water to the main stream.

9.26 Figure 9.1 is a generalized longitudinal profile of a river, with elevations above sea level plotted against distance from the mouth of the river. What does it show?

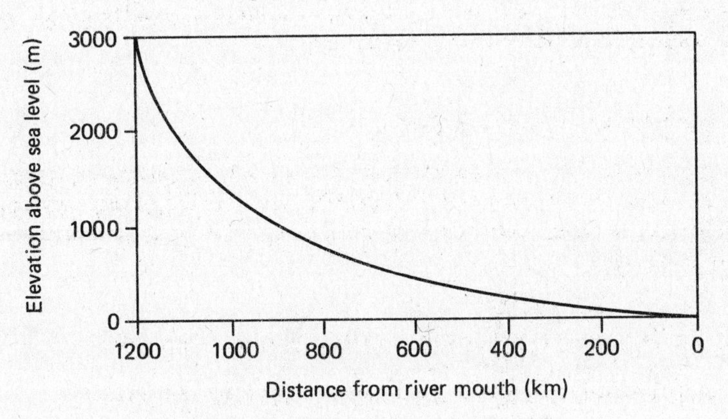

Fig. 9.1 Idealized longitudinal profile of a stream.

It shows that the slope S of a river decreases downstream.

9.27 In Problem 9.25, it was stated that the discharge of a stream increases downstream. Since $Q = WDV$, would the velocity increase downstream as well?

Generally not, because the slope is less, and this results in a lower velocity. (See Problem 9.26.)

9.28 If the velocity does not increase downstream, as stated in the previous question, what will have to change to accommodate the greater volume of water (i.e., discharge)?

The width and the depth of the channel will be greater. Indeed, this is the case—streams are wider and deeper at their mouths than they are elsewhere along their courses.

9.29 How are stream valleys formed?

Streams erode their own valleys. (However, preexisting lines of crushed or weathered rock, etc., may dictate the actual course that a stream takes.)

9.30 How do stream channels become longer?

Streams can make their channels longer in three ways: by headward erosion at the source, by building a delta to flow across at the stream mouth where it enters a lake or ocean, or by establishing meanders (bends) in the river channel.

9.31 What is a drainage basin?

The total area that contributes water to a main stream via its tributaries and sheetwash is the *drainage basin* of that major stream.

9.32 (*a*) What are drainage divides and continental divides?

A *divide* is a height of land separating drainage basins; it can be shown on a map as a line. Precipitation that falls on opposite sides of the divide flows off in different directions. A *continental divide* is a divide of large scale, separating major drainage of a continent. For example, there is a divide in the Rocky Mountains that lies between eastward-and westward-flowing waters.

(*b*) On Fig. 9.2, how are continental divides and more local drainage divides shown?

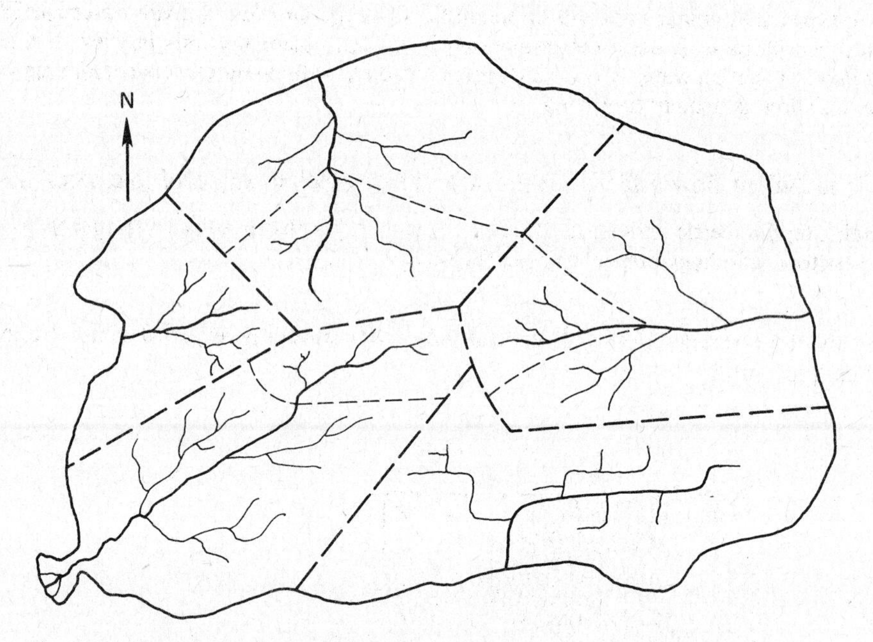

Fig. 9.2 Hypothetical continent with drainage basins, streams, and divides (dashed lines) between drainage basins.

The heavy dashed lines are continental divides, and the lighter dashed lines illustrate local divides.

9.33 How large is the drainage basin of the Mississippi River?

It extends from the Appalachians to the Rocky Mountains, and includes an area of about 3.2 million km^2, about 65 percent of the area of the 48 states.

9.34 Name and describe two types of flow in water (Fig. 9.3).

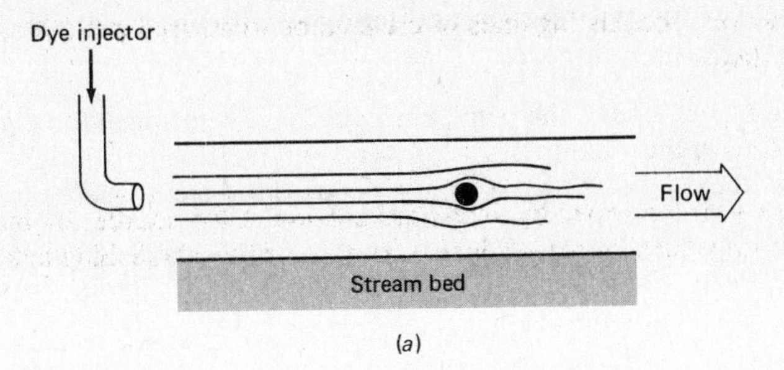

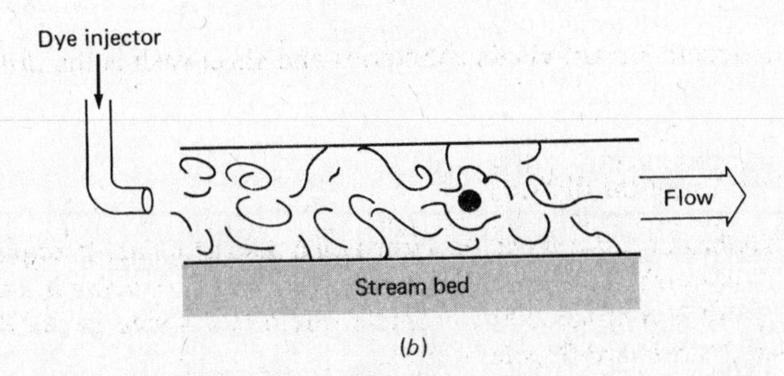

Fig. 9.3 Laminar flow (top) and turbulent flow (bottom). (*From Ojakangas and Darby, 1976.*)

The two types are laminar flow and turbulent flow. In *laminar flow,* a given water particle will follow a straight, smooth path. It is rare in nature and is present essentially only in very slow streams. In *turbulent flow,* the path a water particle follows is complex, with numerous eddies moving in all directions. In streams, flow is usually turbulent.

9.35 How is turbulent flow a factor in the suspension of clay, silt, and fine sand grains?

The grains tend to settle under the influence of gravity, but the turbulent water tends to keep the grains off the bottom, and hence traveling with the water.

9.36 Where on the cross-section of the stream channel shown in Fig. 9.4 is the place of maximum velocity and why?

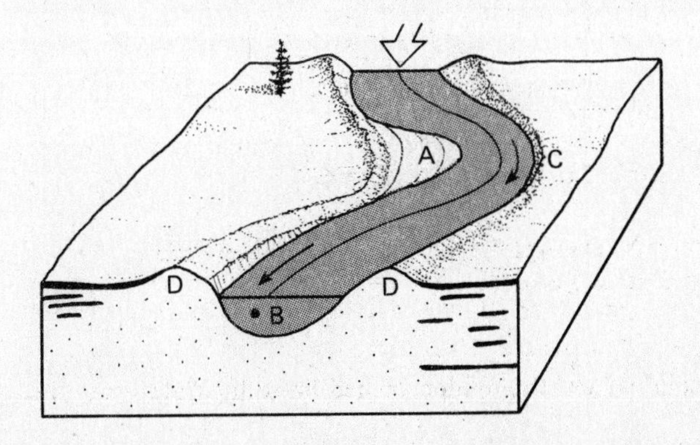

Fig. 9.4 A stream channel with a meander. The current flows in direction of the arrows.

In the upper part of the stream, at less than half the depth of the channel, because friction on the channel bottom slows the water considerably, whereas minor friction with the air at the top of the channel slows the water only slightly. The point of maximum velocity is marked as point B. It is offset from the center of the channel because of the bend in the river, which causes the current to swing outward against the bank.

9.37 In a bend in a stream, as in Fig. 9.4, where will the stream deposit sediment, and where will it erode sediment?

The water velocity will be fastest on the outer side of the bend, partly because centrifugal force swings the current outward, so erosion will occur in the vicinity of C. Conversely, the velocity is lowest in the vicinity of A, and deposition will occur there. (Recall that a stream deposits sediment where there is a decrease in velocity.)

9.38 How do river bends, or meanders, move or migrate?

Continued erosion on the outer side and continued deposition of the inner side of a channel, as in Fig. 9.4, can cause meanders to move over a broad valley floor and become a very exaggerated loop or meander.

9.39 (*a*) Do meanders migrate small distances or large distances?

They may migrate only a few meters a year, especially during flood stage. The distance can also be large; some meanders in the Mississippi River have migrated 6.5 km (4 mi) in 100 years.

(*b*) At what average rate did the above meanders migrate?

6.5 km or 6500 m divided by 100 years gives a rate of 65 m (about 200 ft) per year!

9.40 Do rivers erode laterally as well as downward?

Yes, the development of meanders illustrates lateral migration, whereas a steep V-shaped valley is the result of down-cutting.

9.41 Which feature on Fig. 9.5 is a braided stream? Describe its origin.

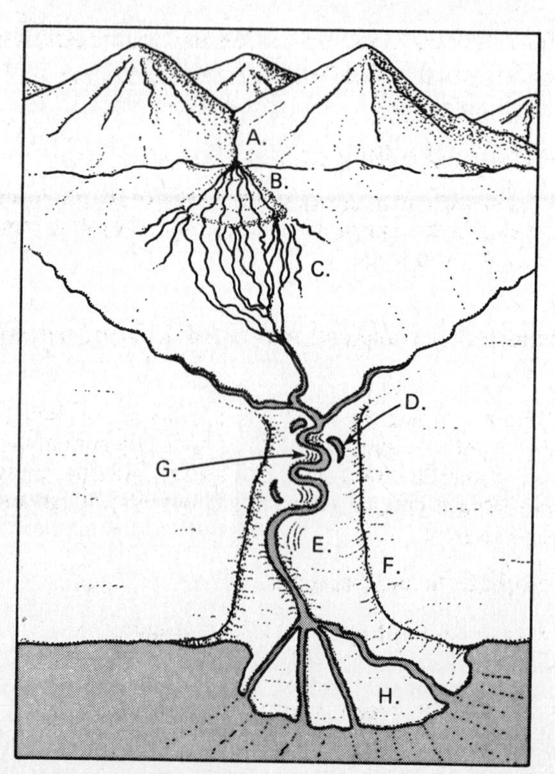

Fig. 9.5 Diagrammatic representation of mountains with drainage to the ocean (foreground).

Feature C. A stream with a large sediment load will have its channel filled with sediment, and the result is a series of small channels within the main channel, separating and rejoining each other, resulting in a *braided stream*.

9.42 Which feature on Fig. 9.5 is a meandering river? Describe its origin.

Feature G. A river which loops or meanders back and forth across its valley is a *meandering river*. Meanders are the result of lateral erosion as the stream moves across its floodplain.

9.43 Which feature on Fig. 9.5 is a floodplain, and how is it formed?

Feature E. A *floodplain* is that portion of a river valley that is covered with water during floodstage. Floodplains are floored by sediment deposited in meandering river channels (as in Problem 9.38) and by finer sediment deposited outside of channels when rivers overflow their banks.

9.44 Which feature on Fig. 9.4 is a natural levee, and how are natural levees formed?

Feature D. *Natural levees* are low ridges formed along the banks of rivers, the result of a velocity decrease and deposition of suspended sediment as sediment-laden streams overflow their banks during floodstage. In some cases, they allow the river to flow at a higher elevation than the surrounding floodplain.

9.45 Which feature on Fig. 9.5 is an alluvial fan, and how did it form?

Feature B. It is a result of deposition at the base of a steep slope where the stream velocity decreases rapidly. Alluvial fans are prominent where streams leave the mountains.

9.46 On an alluvial fan, where will the coarsest sediment, such as pebbles and boulders, be located?

Near the top end or apex of the fan where the velocity of a fast stream first decreased markedly. Finer material is carried on down the fan, with the finest deposited at the distal ends of the fan, or beyond.

9.47 Which feature on Fig. 9.5 is a delta, and how did it form?

Feature H. A *delta* is a depositional feature formed where a stream enters a lake or ocean, loses velocity, and deposits its load. However, if the currents in the lake or ocean are strong, they may remove the sediment as quickly as it is deposited, and no delta will form.

9.48 (*a*) What are the differences between dendritic, rectangular, and radial drainage patterns, and how does each form?

A *dendritic* drainage pattern is branching, like a tree and its branches; it forms on a rather uniform, gently dipping surface. In a *rectangular* drainage pattern, streams and tributaries join at right angles and may take right-angle bends; such patterns are controlled by tilted rock units of differing resistance to erosion or by joints or by faults. In a *radial* drainage pattern, streams flow outward or radiate from a central higher area such as a volcano or a domal uplift.

(*b*) On Fig. 9.2, which types of drainage patterns are present?

Four of the patterns are dendritic, and the one in the southeast corner is rectangular.

9.49 Figure 9.6 illustrates the theoretical evolution of a stream valley from its source to its mouth. The figure is highly compressed and diagrammatic. Which portion of the diagram illustrates the youthful stage? The mature stage? The old-age stage?

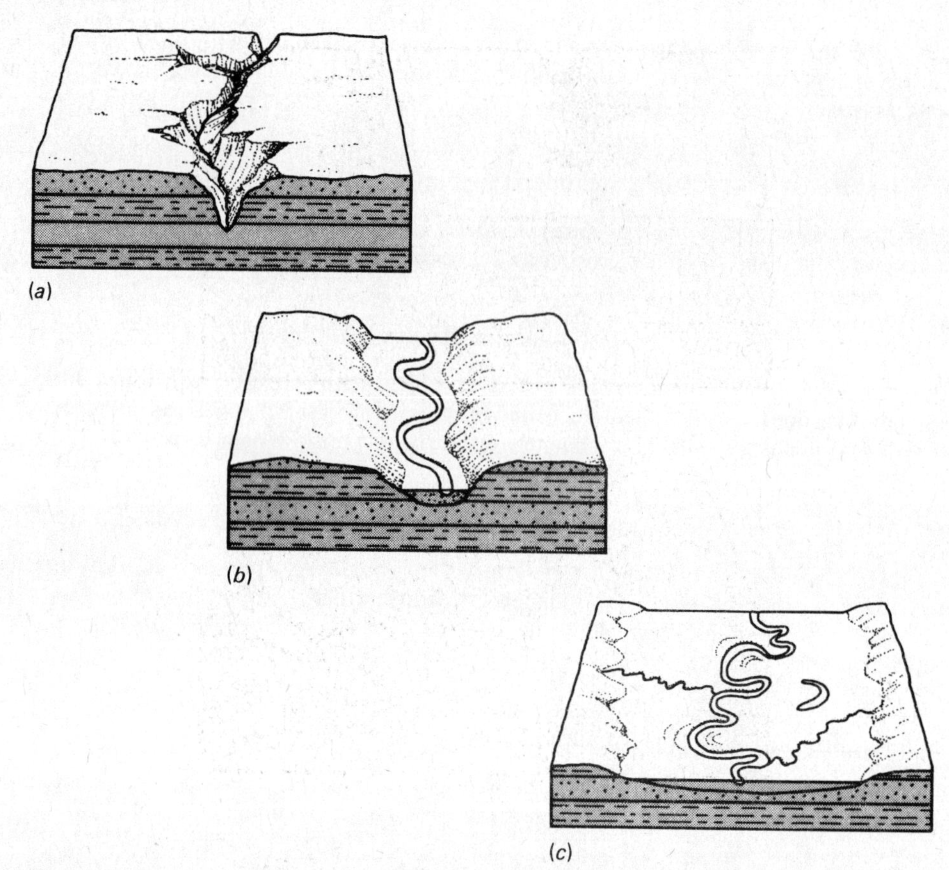

Fig. 9.6 Stages in the evolution of a stream valley.

The youthful stage is illustrated by A, the mature by B, and the old age by C.

9.50 In which stage (youth, maturity, or old age) in the evolution of a stream valley do each of the following occur? Refer to Fig. 9.6.

(*a*) Rapids and waterfalls
(*b*) Floodplains
(*c*) Meanders and oxbow lakes
(*d*) V-shaped valleys without floodplains.
(*e*) Natural levees
(*f*) Very wide valleys
(*g*) Very low banks
(*h*) Narrow floodplains

(*a*) Youth. (*b*) Maturity and old age. (*c*) Maturity and old age. (*d*) Youth. (*e*) Maturity and old age. (*f*) Old age. (*g*) Old age. (*h*) Maturity.

9.51 What are the progressive stages in the evolution of a topography or a landscape, as shown in Fig. 9.7?

Youth, maturity, and old age, the same terms as were used in the evolution of a stream valley. This is also known as the *erosion cycle*.

9.52 In which stage of evolution of a landscape are each of the following features most abundant? (See Fig. 9.7.)

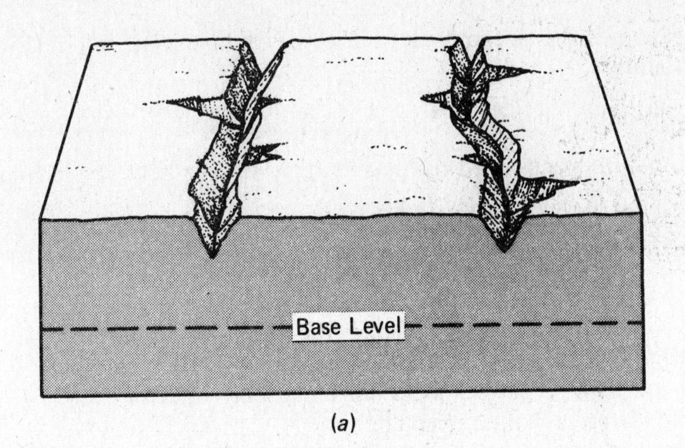

(*a*)

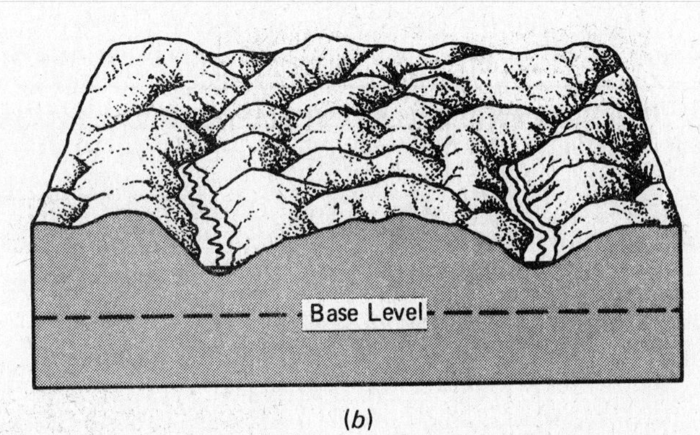

(*b*)

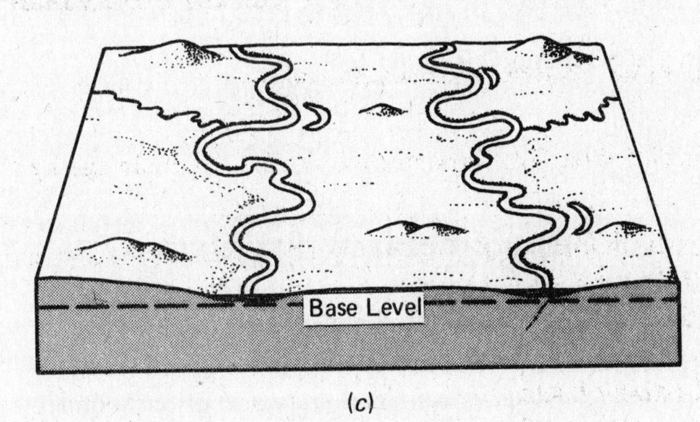

(*c*)

Fig. 9.7 Stages in the evolution of a landscape. (*From Ojakangas and Darby, 1976.*)

(*a*) Rivers with V-shaped valleys
(*b*) Rivers with falls and rapids
(*c*) Broad flat uplands or interstream areas
(*d*) Rivers with moderately wide floodplains
(*e*) Hillsides (slopes)
(*f*) Rivers with very wide floodplains
(*g*) Peneplain
(*h*) Rivers with many meanders

(*a*) Youth. (*b*) Youth. (*c*) Youth (*d*) Maturity. (*e*) Maturity. (*f*) Old age. (*g*) Old age. (*h*) Maturity or old age.

9.53 (*a*) Does the evolutionary "age" of stream valleys in a landscape always agree with the evolutionary "age" of the landscape itself?

Not always. For example, the stream valleys in a mature landscape, all in slope, may still be V-shaped and therefore youthful.

(*b*) Find such a situation in Fig. 9.7.

In Fig. 9.7(*b*), the mature stage, most of the tributary streams between hills are youthful.

9.54 How can a meandering river without a floodplain be explained?

Rivers meander on their floodplains. Therefore, if a river is meandering but has steep valley walls and no floodplain, it is a rejuvenated valley. That is, the river was undoubtedly flowing across a floodplain at one time, but the area was uplifted and the river was able to cut downward toward a new base level, keeping pace with the uplift. The meanders are thus "entrenched meanders." The Goosenecks of the San Juan River in Utah are a classic example.

9.55 In what stage of valley development is the Colorado River valley in the Grand Canyon area?

Youth. The Colorado River occupies essentially its entire valley floor and has many rapids. It is essentially a V-shaped valley. Actually, the river valley has been rejuvenated by uplift when the Colorado Plateau was uplifted.

9.56 In what stage of landscape development is the region around the Grand Canyon?

Youth. It has broad interstream areas between youthful streams, including the Colorado River. This indicates that the region has been geologically recently uplifted (i.e., rejuvenated).

9.57 In Fig. 9.7, which landscape (A, B, or C) is in youth? In maturity? In old age?

A is in youth, B is in maturity, and C is in old age.

Supplementary Problems

9.58 Refer to Fig. 9.8 which relates water velocity and grain size of the sediment.

(*a*) At what water velocity will a cobble 50 mm in diameter be eroded and at what velocity will it be deposited?

Eroded at about 300 cm/s and deposited at 150 cm/s. (Note that this is a log-log plot, and the distances between numbers cannot be divided into equal segments.)

(*b*) At what water velocity will a grain of coarse sand (2 mm in diameter) be eroded and at what velocity will it be deposited?

Eroded at 30 cm/s and deposited at about 12 cm/s.

(*c*) At what water velocity will a silt grain 0.02 mm in diameter be eroded, and at what velocity will it be deposited?

Eroded at about 30 cm/s and deposited at 0.15 cm/s.

9.59 Based on the previous problem, formulate a statement relating grain size, erosion, and deposition.

Larger particles are deposited and eroded at relatively comparable velocities, whereas fine particles (silt and clay) are easily deposited at very low velocities but are eroded only at very high velocities.

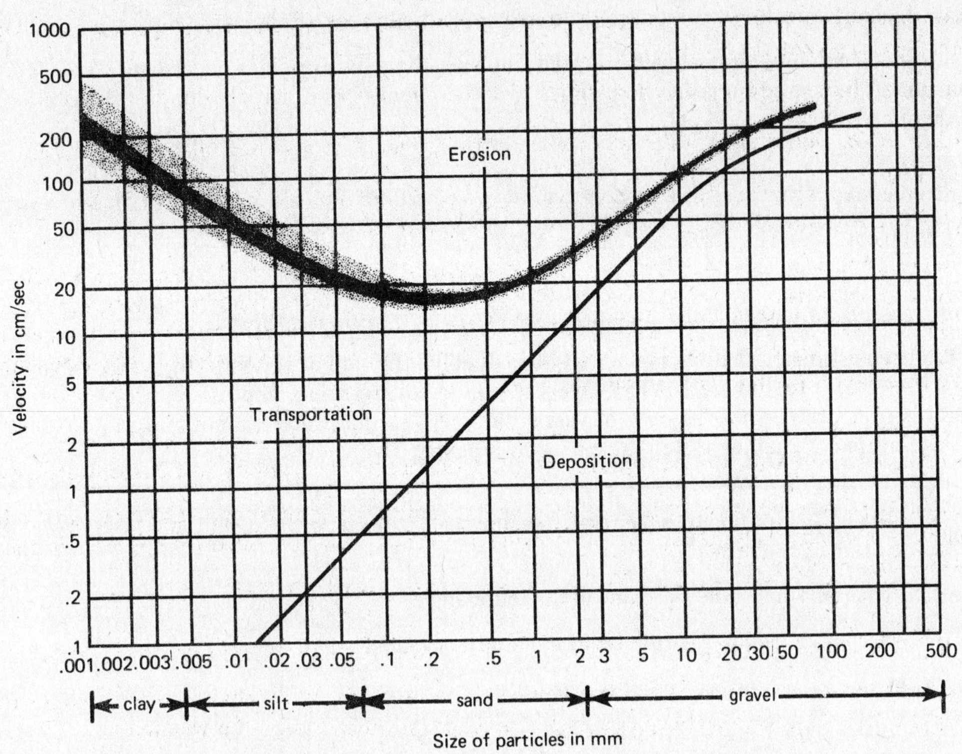

Fig. 9.8 Diagram illustrating relationships of water velocity and particle size. Areas of erosion, transportation, and deposition are shown. (*After Dunbar, C. O., and Rodgers, John, Principles of Stratigraphy, Wiley, New York, 1957.*)

9.60 "Large particles settle rapidly in water because their mass is so large relative to the density of the water. However, small clay-sized particles settle so slowly that a particle can take 100 years to reach the deep ocean floor from the surface." How can this statement be related to the distribution of clay-sized material on the earth's surface?

Since they settle so slowly, clay-sized particles are distributed worldwide and deposited in lakes, oceans, rivers, as well as on dry land.

9.61 What can be said about the energy of the environments of deposition of clay-sized particles?

From Fig. 9.8, it is clear that clay is deposited in places where the water velocities are very low, much less than 0.1 cm/s.

9.62 How do meanders start?

The reasons that meanders form initially are very complex but are related to irregularities or obstructions in the stream channels which divert the current toward a bank. Once started, they become more and more exaggerated. Even a straight "stream channel" on an experimental sand table will develop meanders. Meanders can be thought of as related to a stream's attempts to adjust to changes in various factors. For example, a stream's length is increased by the development of meanders, and thus the amount of slope (i.e., the drop in meters per kilometer or feet per mile) is lessened.

9.63 What is the difference in the ruggedness of a landscape in a humid versus an arid climate, and why?

In humid climates, the topography is more rounded because chemical weathering of rock leads to rounded features. The principle is illustrated on a smaller scale in Fig. 4.3.

9.64 How does urbanization with its concrete and asphalt surfaces affect runoff?

The runoff increases dramatically. A study in East Meadow Brook, Long Island, New York, showed that runoff increased by nearly four times over a 25-year period, largely due to urbanization. Flooding is therefore much more likely.

9.65 Calculate the discharge Q of a stream that has a velocity of 3 m/s, a width of 10 m, and a depth of 2 m.

See Problem 9.23. The discharge will be 3 m/s times 10 m times 2 m, or 60 m^3 of water per second.

9.66 How much does human activity contribute to erosion by running water?

It has been estimated that human activity has doubled the rate of erosion and perhaps even increased it by a factor of 3 to 10 times. Why? Overcultivation, overgrazing, and overcutting of forests are three reasons. Also, the cementing over of vast urban areas so that water cannot infiltrate but instead rushes off as torrents of running water causes much erosion.

9.67 (*a*) See Fig. 9.2. Formulate a general rule on the angle at which tributaries join major streams in dendritic drainage patterns.

Streams generally join other streams at an acute angle.

(*b*) Of what value would this rule be in the study of maps?

The acute angles will always point downstream.

Chapter 10

Shaping the Landscape: Glacial Ice

INTRODUCTION

Glacial ice today covers most of Antarctica and Greenland (10 percent of earth's land surface), but as recently as 10,000 to 20,000 years ago during the last Great Ice Age, it covered about 30 percent of the earth's land surface. Glaciers exist where snow accumulates faster than it melts, and the net result is that as the snow thickens, the lowermost snow recrystallizes into ice that moves or "flows" under the weight of the snow-ice pile. Once it is moving, it is a glacier. Large, continental-scale glaciers are called, logically, *continental glaciers*. Smaller glaciers in mountain valleys are termed *valley glaciers* or alpine glaciers; these may merge at the bases of mountain ranges to form *piedmont glaciers*.

GLACIAL MOVEMENT

Like running water, glaciers move downhill in mountainous terranes. Continental glaciers move "down their own slope," shearing or slumping off of the glacier front. In detail, glacier ice moves by sliding along ruptured planes of atoms within the ice (recall that glacial ice is a mineral with a crystalline structure) or by slipping over a wet subsurface beneath the glacier. Such slippage occurs at the bases of "warm glaciers"; "cold glaciers" are frozen to the substrate. The latter type exists usually in polar latitudes or seasonally at warmer latitudes.

GLACIAL EROSION

Each glacier can be subdivided into a zone of accumulation and a zone of ablation or wastage. A glacier is usually actively eroding beneath its zone of accumulation and depositing beneath its zone of wastage. Glacial ice contains "tools" of silt, sand, and larger particles; these actively abrade the bedrock over which the ice is moving, forming whalebacks, grooves, striations, crescentic gouges, chatter marks, and polished bedrock. Larger erosional features are commonly formed by valley glaciers and include cirques, aretes, horns, and U-shaped valleys. Active glaciers easily remove any soil, weathered rock, or sediment but abrade only limited amounts of solid rock except in mountainous terranes where the high gradients enhance erosion. It is a common misconception that glaciers of the past have removed entire mountain ranges, but the most effective agent of erosion has been running water.

GLACIAL DEPOSITION

As valley glaciers reach lower, warmer elevations and lose velocity, or as continental glaciers reach warmer latitudes and lose velocity, they deposit their loads and no longer erode. Debris deposited directly from glacial ice is unsorted and is called *glacial till*. Till may constitute glacial drumlins, and terminal, recessional, and lateral moraines. Broad, blanket-like deposits of till are called ground moraine; some contain *oriented clasts*, oriented by ice movement during deposition. Ground moraine deposited when a glacier stagnated and melted while standing still does not contain oriented clasts.

Sediments deposited by meltwater from a melting glacier are sorted and stratified and are collectively called *stratified outwash deposits* or *stratified drift*. Some of the material is deposited on glacial outwash plains in front of melting glaciers. Meltwater deposits may also be formed beneath the glacier as kames and eskers which, when the glacier finally melts away, stand up as rather equidimensional hills and elongate ridges, respectively. This sediment constitutes sand and gravel deposits.

Glacial lakes commonly form near melting glaciers and are sites for the deposition of fine-grained sediment (silt and clay) that settles very slowly. *Glacial varves* are couplets of silt and mud, interpreted to be seasonal deposits of summer silt in unfrozen lakes and winter clay in frozen lakes.

CAUSES OF GLACIATION

What are the ultimate causes of ice ages or glaciations? There have been several in earth's history, but the causes remain obscure. We know that today's major glaciers are in polar latitudes, so any landmass situated in a polar latitude may be the site of glaciation. (Small valley glaciers can exist even on the equator if the elevation of the mountains is such that the upper reaches of the mountains are at cold altitudes.) But there have been times when even the polar latitudes have not hosted glaciers. Why not? Certainly the reasons are climatic and related to periods of cooling.

What causes the earth to cool? There are several hypothetical explanations, including the elevating of landmasses, reduced solar heating, and changes in atmospheric circulation. And now humans are altering the atmosphere with the addition of carbon dioxide from the burning of fossil fuels, thereby creating the "greenhouse effect" which allows the sun's heat to enter but not readily escape. Thus the earth could warm up, and earth's present glaciers could melt. If sea level rose 30 or 60 m as a result, where would many of the world's cities be?

Solved Problems

10.1 Where is most of the world's fresh water?

It is tied up in glacial ice. Only about 2.8 percent of earth's water is nonmarine, and about 80 percent of this is in glacial ice.

10.2 Where is most of the glacial ice located today?

About 98 percent is in Antarctica (90 percent) and Greenland (8 percent). The remaining 2 percent is in hundreds of small valley glaciers in mountains on all continents except Australia (where the mountains are not high enough to reach cool altitudes).

10.3 What are the two main types of glaciers?

Continental glaciers (of continental scale) and valley glaciers. (Valley glaciers may merge at the base of a mountain range to form a piedmont glacier.) The big glacier on Greenland is commonly called an ice cap rather than a continental glacier, as Greenland is not a continent as is Antarctica.

10.4 What portion of the earth's land surface is covered by ice today, and how much was covered during the last Great Ice Age in the Pleistocene?

Today 10 percent is covered, as compared to about 30 percent during the Great Ice Age.

10.5 What is the major requirement for a glacier to form?

That more snow falls each year than melts, resulting in an increasing accumulation.

10.6 A friend of yours has heard that glaciers exist only in areas of extreme cold, colder than 30°F below zero. Is your friend correct?

No. If it is cold enough for precipitation to fall as snow (i.e., below freezing at 32°F or 0°C), and if it remains cool enough so that it does not all melt, then a glacier could form. (If the temperature is extremely low, it doesn't snow at all.) Sufficiently low temperatures, maintained low enough to prevent complete melting, today exist only in polar latitudes and on high mountains.

10.7 Are valley glaciers found only in polar latitudes?

No, some even occur near the equator on high mountains. One example of a mountain with glaciers is Mt. Kilimanjaro in Tanzania, Africa, which is located only 4° from the equator. They are found on all continents except Australia.

10.8 How does snow turn to ice?

The weight of a thick pile of snow compacts the lowermost snow, melting it at points of contacts of snow crystals. The water moves and refreezes. There may also be melting at the top or at the ground if the ground is warm. Some sublimes, or changes from a solid to a gas, and then condenses immediately and freezes. This new ice recrystallizes into little spheres or "BBs" of ice, called *firn*. Old snowbanks and downhill ski areas in the spring contain a similar material commonly called corn snow. Eventually at a depth, say, deeper than 30 m (100 ft), it will become a denser, solid, bluish mass of recrystallized, interlocking ice crystals.

10.9 See the previous problem. Is the result of that process a glacier?

No. It must move before it is called a glacier. It usually doesn't move until it is 45 to 60 m (150 to 220 ft) thick.

10.10 How does ice move, thereby becoming a glacier?

It moves by gravity. It slides "down its own hill" (i.e., its own slope of tens or hundreds of meters) by moving along countless failed layers of atoms within ice crystals deformed by the weight of the ice. This latter movement may be termed "plastic flow." Some glaciers slip across the surface beneath the ice, whereas others are frozen to the underlying surface and can only deform plastically. Water at the bases of some glaciers apparently acts as a lubricant.

10.11 How fast do glaciers move?

It varies, from feet to miles (meters to kilometers) per year. Sometimes glaciers "surge" or move rapidly. A valley glacier in Kashmir moved 8 mi (13 km) in 2½ months in the summer of 1953, and Hubbard glacier in Alaska surged 12 m a day. A glacier in India moved 110 m a day.

10.12 Describe the difference in the behavior of an advancing glacier, a stagnant glacier, and a retreating glacier.

An advancing glacier moves downslope by the processes described in Problem 10.10. A stagnant glacier has stopped moving and simply sits still and melts. A retreating glacier "retreats" because its front end is melting back even though the glacial ice within the glacier may be standing still or even moving forward.

10.13 Two mountain climbers in the Swiss Alps were caught in a large avalanche of snow which came to rest on an active glacier. What do you think happened to them?

They were in a deep freeze. A geologist, based on his studies of the movement of the glacier, predicted that they would reappear at the lower, melting end of the glacier in about 40 years. He was quite right, for 41 years later they appeared, dead.

10.14 Figure 10.1 is a generalized longitudinal section of a valley glacier. Label and explain the zone of accumulation and the zone of wastage or ablation.

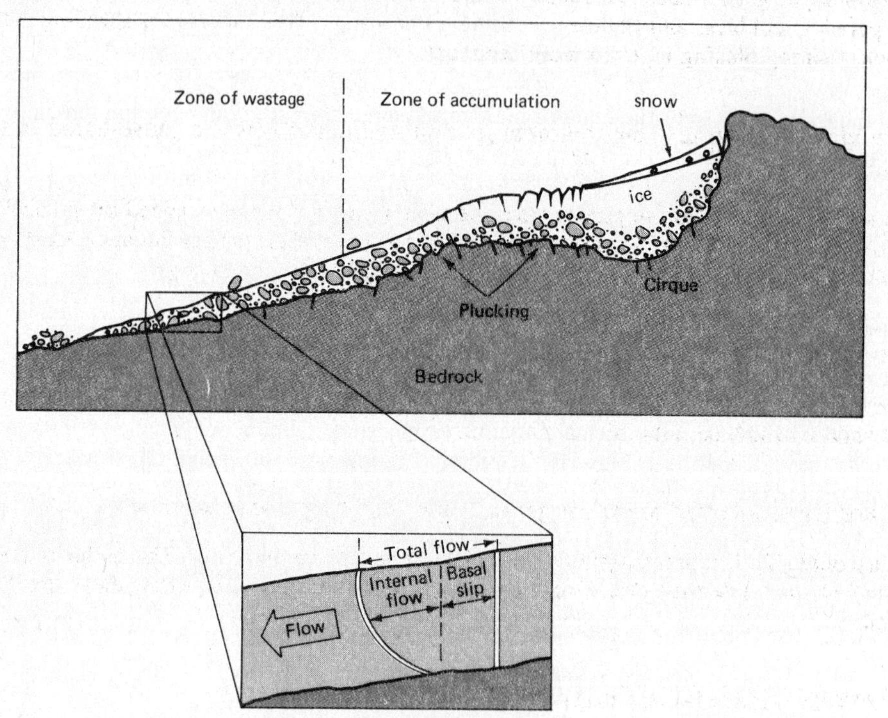

Fig. 10.1 Generalized sketch of a longitudinal view of a valley glacier. Note the snow, ice, and bedrock. The glacier could be hundreds or thousands of meters long. (*From Ojakangas and Darby, 1976.*)

The zone of accumulation is the portion at higher elevations, where more snow is falling each year than melts. The zone of wastage is the portion at lower elevations, where the depletion due to melting is greater than the rate of new ice supply.

10.15 What processes lead to the accumulation of snow and ice on a glacier?

Snowfall, rain that freezes, avalanches of snow from higher elevations, and minor frost forming from the vapor state.

10.16 In the zone of wastage or ablation in the lower reaches of a valley glacier or at latitudes farther from the poles in the case of continental glaciers, what processes cause the ablation?

Melting of ice into water and some sublimation, wherein ice becomes a vapor without passing through the liquid state.

10.17 What relationships between accumulation and ablation, as in Fig. 10.1, will cause a glacier to advance or retreat?

If the net balance is such that accumulation exceeds ablation, the glacier will advance. On the other hand, if ablation exceeds accumulation, the glacier will retreat.

10.18 How do glaciers obtain their loads of sediment?

By picking up loose material, by plucking pieces of bedrock, and/or by grinding fresh rock at the glacier base. Moving ice easily incorporates any loose, weathered material which it passes over. Commonly, the basal ice of a glacier will freeze onto solid bedrock over which it is passing. As it moves again, it can remove jointed blocks or rocks by plucking them out, especially if freeze-thaw (frost-wedging) action by meltwater at the glacier base has dislodged and lifted the blocks a bit. At the bases of glaciers, any silt, sand, pebbles, cobbles, and boulders become tools that act like sandpaper, abrading and gouging the bedrock, thereby picking up even more sediment.

10.19 Valley glaciers enlarge their sediment load in additional ways to those listed in the previous question. How?

Much sediment falls onto the glacier from the higher valley walls or is added by smaller tributary glaciers. Also, the ice not only scours the bedrock at the base of the glacier but also along the sides of the valley-constrained glacier, thereby producing the typical U-shaped valleys.

10.20 What commonly happens to a glacier that flows into a lake or ocean?

It "calves" or gives off chunks of ice, some very large, that float away as icebergs. Also, the water may be warmer than the ice, and this enhances melting.

10.21 Why are icebergs even more dangerous than they might appear to be?

Because only about 10 percent of their mass is exposed above the water. This is due to the difference in density—ice has a density of 0.90 compared to water density of 1.0. Thus, they "float" largely submerged.

10.22 How are icebergs a factor in sedimentation of glacial debris?

As the bergs float away from the glacier front, say, into an ocean, they slowly melt. As this happens, they release their sediment, which varies in size from clay to boulders, and drop it onto the ocean floor. This can occur thousands of miles beyond the front of the calving glaciers.

10.23 Interpret Fig. 10.2 in terms of the velocity of flowing glacial ice.

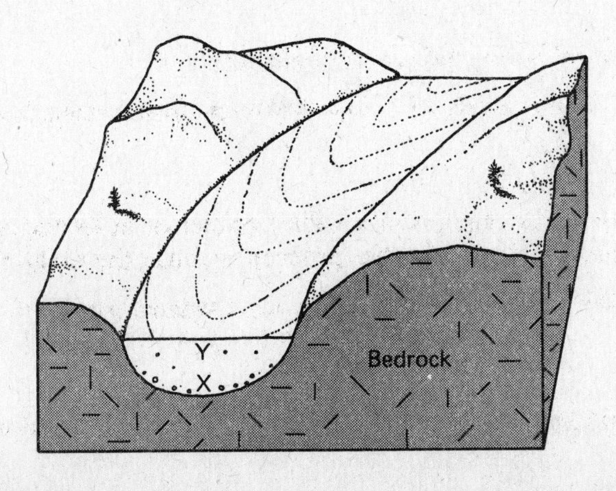

Fig. 10.2 Generalized sketch of a valley glacier.

The lines (crevasses and bands of sediment) show that flow is fastest near the center of the glacier, away from the effects of friction with bedrock on the floor and wall of the glacier. (Note the similarity to stream flow, as in Fig. 9.4 where the water is flowing fastest away from the channel walls.)

10.24 When do valley glaciers have their greatest velocity?

Probably the main factor causing variation in velocity is the temperature. As the temperature increases, meltwater acts as a lubricant and aids sliding, and individual ice crystals deform more easily along glide planes. However, as in a river, many other factors, and especially slope, affect the velocity. (See Problem 9.14.)

10.25 What are crevasses, and where in the glacier of Fig. 10.1 are they located?

Crevasses are large cracks in moving glaciers, and they occur in the upper brittle part of the glacier.

10.26 If you were an Antarctic explorer and fell into a hidden, snow-covered crevass on a glacier that is 200 m (660 ft) thick, would you fall all the way to the base of the glacier?

No. Fortunately, you would probably hit bottom at about 30 m (100 ft), with a thud. (You are indeed "fortunate," for then other members of your team, assuming they are still on top of the glacier, could attempt a rescue.) Below 30 m, the weight of the overlying ice causes the ice to deform in a plastic rather than a brittle way, and crevasses cannot exist. (See Fig. 10.1.)

10.27 See Fig. 10.3. Do the straight lines or striations tell which way the ice that abraded the rock was moving?

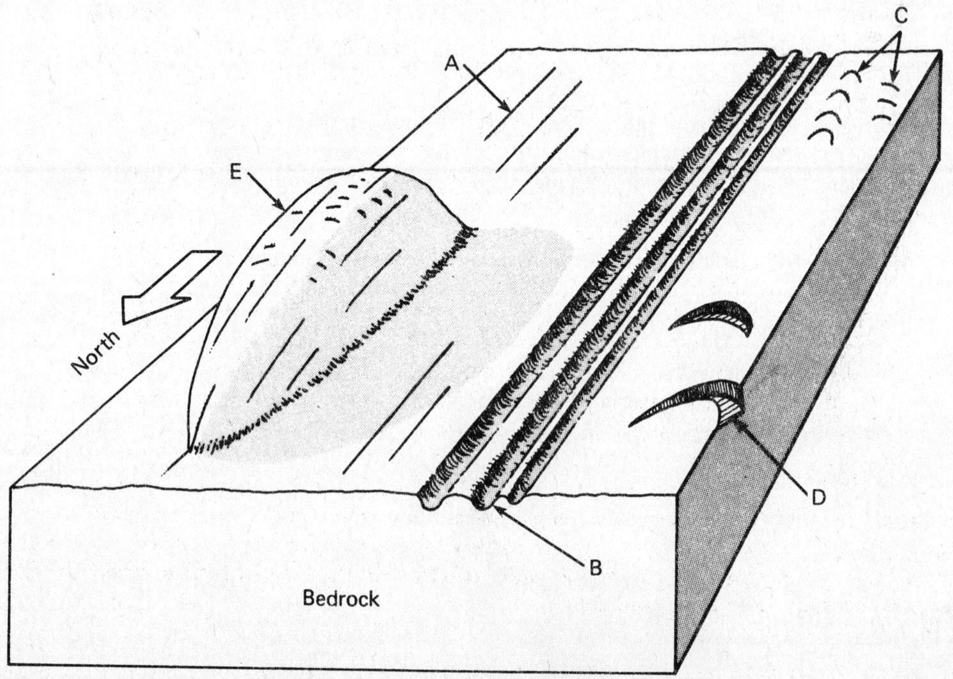

Fig. 10.3 Sketch of a glacially abraded rock outcrop, with striations, grooves, crescentic gouges, and a whaleback. Note north arrow.

No. In this case, you would know only that the ice moved to the north or to the south. Striations and grooves are generally nondirectional current indicators; that is, they may tell you the ice at a given spot may have moved from east to west *or* from west to east. They are not definitive.

10.28 From Fig. 10.3, determine in which direction the glacial ice that abraded the bedrock exposure was moving.

To the north. The crescentic gouge on the right side of the diagram has the steep side of the gouge facing south, or in the up-ice direction, the direction from which the ice came.

10.29 What was removed from the rock exposure of Fig. 10.3 during abrasion by the overriding ice?

Mostly fine particles of rock, probably mostly clay- to sand-sized. Much of it is very fine-grained and is the "rock flour" that makes glacial meltwater streams whitish in color. Much glacially scoured debris is fine-grained. An analogy would be the fine material produced by sand papering (abrading) a piece of wood.

10.30 On Fig. 10.4, identify and define a glacial cirque.

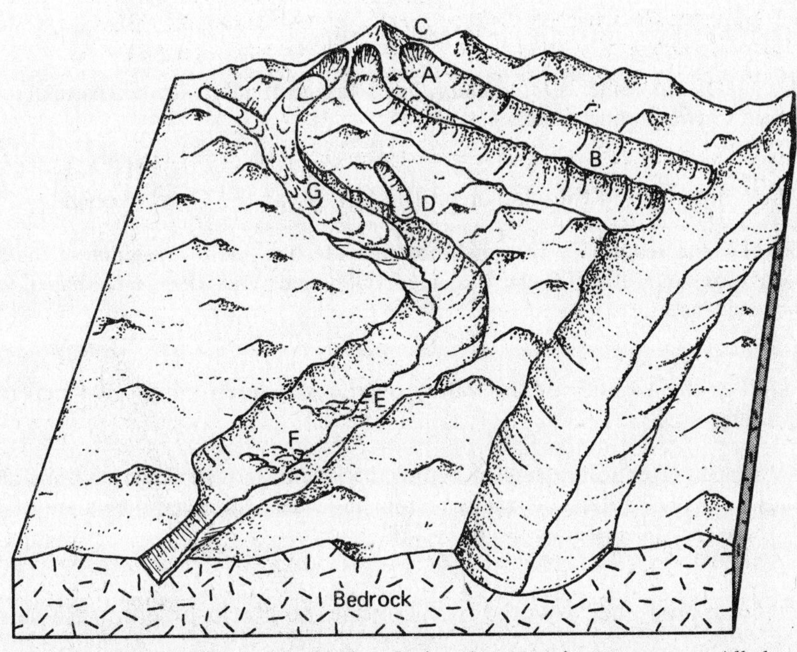

Fig. 10.4 Generalized sketch of a glaciated mountainous terrane. All the valley glaciers have melted away except for the one in the upper end of the valley on the left.

Feature A is a cirque. A *cirque* is a bowl-like feature scoured at the source of a valley glacier by the forming glacier, largely by plucking action as water repeatedly freezes to the bedrock. Therefore, a glacier valley is somewhat lengthened by headward erosion, much as a stream valley is lengthened by headward erosion. (See Problem 9.30.)

10.31 On Fig. 10.4, identify and define a glacial horn.

Feature C is a horn. A *horn* is a mountain peak that has been cut into by cirques on several sides. The Matterhorn in the Swiss Alps is the best-known horn.

10.32 On Fig. 10.4, identify and define an arete.

Feature B is an arete. An *arete* is a sharp knifelike ridge between two parallel glacial valleys.

10.33 On Fig. 10.4, identify and define a hanging valley.

Feature D is a hanging valley. A *hanging valley* is a glacial valley high on the side of a larger glacial valley, formed by a tributary glacier that did not cut as deeply as the main glacier because the base level for the tributary glacier was the top of the ice of the main glacier.

10.34 On Fig. 10.4, identify and define a terminal moraine.

Feature F is a terminal moraine. A *terminal moraine* is an irregular ridge of glacial debris formed at the line of maximum advance of a glacier as it stood and melted awhile before retreating.

10.35 On Fig. 10.4, identify and define a recessional moraine.

Feature E is a recessional moraine. A *recessional moraine* is a ridge of debris formed where a retreating glacier temporarily stood and melted, upstream from its maximum advance.

10.36 On Fig. 10.4, identify and define a lateral moraine.

Feature G is a lateral moraine. A *lateral moraine* is a ridge of debris formed parallel to glacier flow along the line of join of two parallel glaciers.

10.37 On Fig. 10.4, how far did the glacier on the left advance? Give two lines of evidence.

It advanced as far as feature F. Terminal moraines are one line of evidence of the former extent of a valley glacier. The point at which the U-shaped valley ends and where a V-shaped valley begins is the other line of evidence.

10.38 What existed in the vicinity of the valley on the left side of Fig. 10.4 before glaciation, and what is the evidence?

A young, V-shaped river valley existed before the glacier moved down it, transforming it into a U-shaped valley by abrasion along the valley bottom and sides. Note that the valley is still V-shaped at its lower end where the valley glacier did not reach.

10.39 On Fig. 10.5, an area once covered by a continental glacier, identify the terminal moraine.

Feature A is a terminal moraine. (See Problem 10.34.)

10.40 On Fig. 10.5, identify and define a ground moraine.

Feature F is a ground moraine. A *ground moraine* is a broad layer of poorly sorted sediment deposited beneath a moving glacier or beneath a stagnant glacier that melted in place.

10.41 On Fig. 10.5, identify and define an outwash plain.

Feature C is an outwash plain. An *outwash plain* is a gently sloping plain formed of sorted sediment deposited in front of a melting glacier by meltwater from the glacier. With vegetation not yet established in such a location, the meltwater streams are commonly braided streams, as shown by feature C on Fig. 9.5.

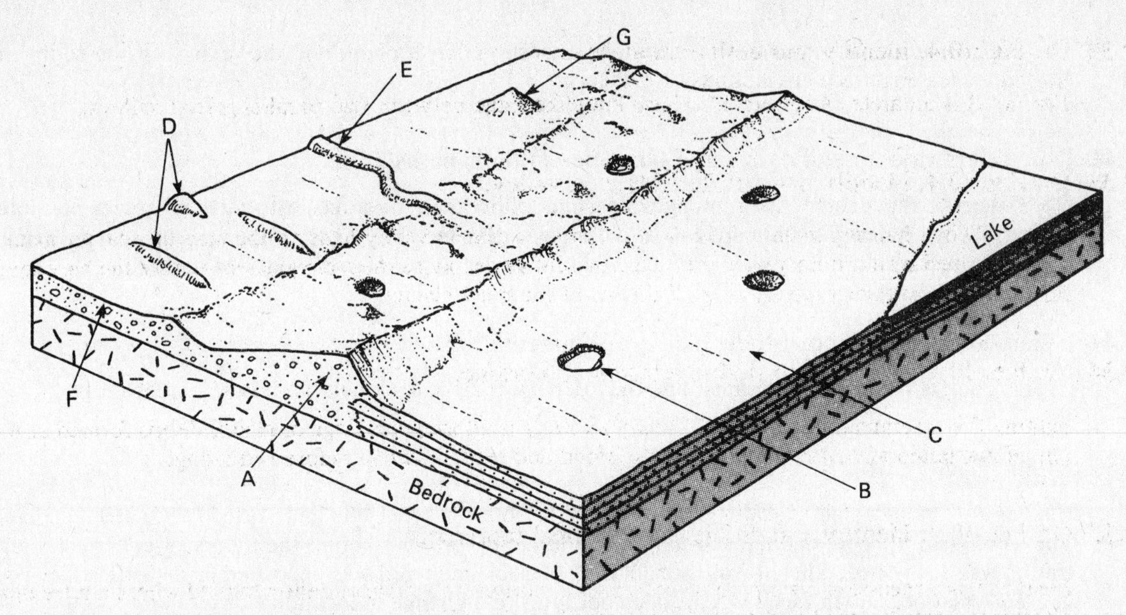

Fig. 10.5 Generalized sketch of an area glaciated by a continental glacier and covered with glacial deposits. Patterns of sediment on the left cross-section of the diagram represent poorly sorted material with some large clasts. The patterns on the right side of the diagram represent finer-grained, bedded sediment.

10.42 On Fig. 10.5, what is feature D? Explain its origin.

Feature D is a *drumlin,* a streamlined hill of glacial till, shaped by the overriding glacier. It is elongated in the direction of glacier movement and it has a steep end in the up-glacier direction. Drumlins commonly occur in swarms of tens or hundreds.

10.43 On Fig. 10.5, what is feature B? Explain its origin.

Feature B is a *kettle lake* (or pond if small), formed where a piece of glacial ice fell off the front of the melting glacier and was buried by sediment, later to melt and form a depression. Later, water filled the depression.

10.44 On Fig. 10.5, what is feature E? Explain its origin.

Feature E is an *esker,* a snakelike sinuous ridge formed by a subglacial meltwater river that flowed in a tunnel in the ice at the base of a glacier.

10.45 Define glacial till, and locate it on Fig. 10.5.

Glacial till is unsorted sediment deposited directly by glacial ice. Moraines consist of till, so on Fig. 10.5, it underlies the upper half (left side) of the area.

10.46 On the basis of Fig. 10.5, how can one distinguish between ice-deposited and meltwater-deposited sediment?

Ice-deposited material is unsorted (i.e., *till*). The sediment deposited by meltwater is sorted into layers of silt, sand, and gravel.

10.47 From Fig. 10.5, describe three lines of evidence that could be used to ascertain the maximum line of advance of a glacier.

The irregular terminal moraine, the relatively smooth outwash plain, and the texture of the sediment (sorted versus unsorted) are all lines of evidence.

10.48 Do glaciers ride up and over large obstacles in their paths?

No. Glaciers behave much like running water, and follow the lowest, easiest routes wherever possible. Hence, valley glaciers in mountains follow old river valleys (see Fig. 10.4). However, the upper portions of a thick ice sheet can flow over obstacles that ice lower in the glacier would go around.

10.49 When do glaciers deposit their loads of sediment?

Again, glaciers behave like streams, and they deposit their loads when there is a decrease in velocity.

10.50 Define glacial drift.

Glacial drift includes all glacially related deposits, both unsorted tills and sorted meltwater deposits. The word is an old one, dating back to the time in the early 1800s before the theory of continental glaciation was developed, when it was thought that all glacial material was deposited from drifting icebergs in a universal ocean. (In this usage, it does not refer to a change in location.)

10.51 Is the front of a continental glacier a big, massive, rather straight ice wall?

No. Because glacial ice behaves like water and moves down topographically low areas such as river valleys whenever it can, the front of a continental glacier commonly consists of several tongues or lobes of ice. Some may have extended hundreds of kilometers beyond others.

10.52 Lakes are commonly the result of glaciation, as in the upper Midwest portion of the United States and in Canada. How are they formed?

Many small lakes form in depressions on irregular moraines. Some are kettle lakes due to melting of buried ice blocks (see Problem 10.43). Others form behind terminal or recessional moraines. Still others may be on scoured bedrock rather than on glacial drift and are thus the result of glacial erosion rather than deposition.

10.53 Did the continental glaciation of North America during the last Great Ice Age enhance or hinder the search for mineral deposits?

Both. In places the bedrock and any mineral deposits in the bedrock are buried beneath tens to hundreds of meters of glacial drift. In other places, as in much of the eastern half of Canada, any weathered-rock material (regolith) was eroded away by the glaciers, exposing many valuable mineral deposits.

10.54 What is a fjord?

A *fjord* is simply a glacially scoured valley that has been inundated by the ocean.

Supplementary Problems

10.55 If accumulation on a glacier exceeds ablation, what will be the result?

The amount of ice will increase, the ice will become thicker, and the glacier will advance or move outward from its former position.

10.56 What would happen if the glaciers of Antarctica were to undergo a rapid increase in velocity and flow into the surrounding ocean?

The ice, both as either a floating glacier (ice shelf) or as icebergs, would melt and raise sea level thereby drowning the world's seaports.

10.57 Is ice a mineral, and is glacier ice a metamorphic rock?

Yes. Recall from Problem 2.21 that a mineral is defined as a naturally occurring, homogeneous, inorganic, crystalline material with a relatively fixed chemical composition. Ice certainly qualifies, but liquid water does not. A rock is an aggregate of a mineral or minerals; thus glacier ice is a monomineralic rock. Recall from Chapter 6 that a metamorphic rock is a preexisting rock that has been changed by temperature and pressure. The original sediment (snow) and sedimentary rock (beds of snow) have reformed and recrystallized to firm and then to massive interlocking ice crystals, and therefore glacier ice is a metamorphic rock. In fact, the ice is commonly banded and the crystals are commonly aligned as in many other metamorphic rocks.

10.58 What is the rock knob on the east side of Fig. 10.3, how did it form, and is it useful?

It is a whaleback or a roche moutonnée and is a streamlined rock knob formed during glacial abrasion. Ideal whalebacks have steep ends in the direction toward which the glacier was moving. This end is steep because jointed rock blocks have been plucked out or lifted out by the glacier when the basal ice froze to the bedrock while the glacier continued to move. The end pointing in the direction from which the ice came is gently tapered.

10.59 What is permafrost?

Permafrost is a term applied to conditions where the ground has remained below freezing temperature for years. Thus, it is permanently frozen soil.

10.60 What forms glacially polished rock?

Glacial polish is the result of abrasion of a bedrock surface by silt and sand tools in a glacier.

10.61 Do all glaciers create grooved, striated, and/or polished bedrock surfaces as in Fig. 10.3?

No. The glacier must be sliding along its base and it must contain rock tools. In order for sliding to occur, the base of the glacier must be at or above the melting point of the ice.

10.62 What is the explanation for a striated bedrock outcrop in a glaciated area, with two distinct sets of striations, one set oriented north-south and the other oriented east-west?

Since striations are formed by tools in a moving glacier and give the general direction of glacier movement, we can interpret the striations as the result of two glaciers or two lobes of a single glacier. Note that we have made a paleocurrent determination, much as we did with cross-beds in sedimentary rocks. However, we do not know whether the north-south striations were made by a glacier moving to the north or to the south, nor do we know whether the east-west striations were made by a glacier moving toward the east or toward the west. Crescentic gouges or whalebacks are necessary for such a determination.

10.63 Many loess deposits are related to glaciation. What are they, and how did they form?

Glacial loess is a deposit of fine, angular wind-blown and wind-deposited silt. It is commonly derived from unvegetated glacial deposits soon after they were deposited by ice or meltwater. (See Problems 11.15 and 11.16.)

10.64 How might one glacial till differ from another?

In the types of rock materials they contain or in the presence or lack of grain orientation. The material depends, or course, upon the types of rock available in the area over which the glacier passed. For example, one many contain an abundance of limestone and shale fragments, and another may be dominated by plutonic and volcanic rock fragments. A till deposited beneath a moving glacier will have oriented clasts, whereas one deposited by a stagnant glacier will not show an orientation.

10.65 Feature G on Fig. 10.5 is a kame. How did it form?

A *kame* is a conical hill of sediment deposited by meltwater in an opening on a stagnant glacier. As the ice melts away, the kame will be exposed in its original conical shape.

10.66 Since kames, as in the previous problem, are found under a glacier, don't they consist of unsorted sediment?

No. Because meltwater was the depositing agent, sorting will have occurred. (Running water sorts!)

10.67 Imagine you are looking at a 40-m-high bank along a river. The bank is being cut into by the river and consists completely of glacial till. How might you determine whether it is the result of a single glaciation or of more than one glaciation?

By studying the till carefully, you might detect some layering, with different rock types present or dominant in certain parts of the cut. Or, buried soil layers, fossil wood layers, or weathered layers might be revealed by a careful study. You would be trying to apply the principle of superposition of Problem 5.10.

10.68 Peat is commonly found in glaciated areas such as Minnesota or Finland. Why?

Peat is partially decayed plant matter that accumulated in glacial lakes. The lakes gradually filled with vegetation. Peat layers can be several meters thick. (If buried beneath thick sediment, the peat would become low-grade coal, with the loss of appreciable water and other volatiles.)

10.69 In Problem 10.22, sediment deposited on the ocean floor by icebergs was discussed. How would one recognize such sediment in the ancient rock record?

The presence of oversized stones in layers of mud and silt, which were deposited in low-energy environments, would suggest that the stones were dropped in from above. (If they were carried in by strong bottom currents, the mud and silt should have been carried away.) These "dropstones" commonly pierce or bow down the underlying sediment.

10.70 What do the arrows in Fig. 10.1 represent?

The velocity of the glacier, with the longest arrow indicating the velocity due to internal flow of the ice and the shorter arrow due to basal slip of the glacier on the bedrock. Note the velocity is slower near its base.

10.71 Where is the cirque on Fig. 10.1?

The bowl-shaped depression at the upper end of the glacier is the cirque. (See Fig. 10.4 again.)

10.72 In Fig. 10.2, is the glacial ice moving faster at point X or point Y, and why?

At point Y, as it is farthest from the friction at the base and the sides of the glacier.

Chapter 11

Shaping the Landscape: Wind

INTRODUCTION

Wind is an active geologic agent in areas of low or seasonal precipitation, and thus wind is most effective in desert areas. However, periods of drought in any area will enhance eolian (wind) geologic activity. A lack of vegetation, or sparse vegetation, is essential to the process.

WIND EROSION AND DEPOSITION

Wind will easily move loose clay, silt, and if the wind is strong, fine sand as well. Clay-sized sediment can be carried by high-level winds all the way around the world. Volcanic eruptions, for example, can put fine ash particles high into the atmosphere where they may circulate for as long as a few years, and perhaps even alter the length of the seasons by reflecting some of the sun's rays. Loess deposits are blankets of wind-blown silt or "rock flour," and are generally associated with the frontal areas of melting glaciers or with deserts.

Wind moves fine- to medium-grained sand, too, but the sand is rarely lifted more than 1 or 2 m above the ground. However, it can be moved hundreds of miles in this manner, with countless short arcuate "jumps." This is the process of saltation. If the sand in a given area is loose and dry, it can be moved from one area to another, forming deflation basins or "blowouts" in erosional areas and sand dune fields in areas of deposition. There are many different shapes and sizes of sand dunes. Gravels, pebbles, and coarser sediment are left behind by the wind, forming lag gravels which then protect the underlying sand from the wind.

Wind-blown sand commonly has several identifying characteristics, including a rather uniform grain size, a high degree of grain roundness, and frosted grain surfaces. Cross-beds formed by wind are commonly very large-scale.

Hard rock can be worn down by wind that is carrying silt and sand as tools of abrasion, but this activity is commonly overestimated. Normal chemical and physical weathering of rock, as described in Chapter 4, first loosen and dislodge grains, which are then free to be moved by running water or by wind.

Solved Problems

11.1 Where is wind erosion most effective?

It is most effective where there is a lack of vegetation and moisture.

11.2 What is transported by wind—clay, silt, sand, or gravel?

The fine-grained particles—clay and silt—are most easily transported by wind. Some sand is also moved, as evidenced by sand dunes. Coarser sediment remains behind.

11.3 What is a *lag gravel?*

The coarse sediment left behind where wind has removed the finer-grain sizes.

11.4 What effect will a lag gravel have on wind erosion?

It will prevent further removal of fine sediment from beneath the gravel layer.

11.5 What is a *desert pavement?*

An area in a desert covered by lag gravels.

11.6 Are sandstorms common?

No. Sand is rarely lifted more than 1 or 2 m off the ground. Most "sandstorms" are really dust storms, composed of clay- and silt-sized particles.

11.7 How is most sand moved by the wind?

It is moved along the surface of the ground, much like bedload in a stream. Some grains are lifted a short distance (as high as 1 or 2 m) and then dropped again, with the impact putting other grains into motion. The motion may be a rolling, sliding, or a "bouncing" (see Problem 11.41).

11.8 What factor most affects the ability of wind to carry sediment of different grain sizes?

The velocity of the wind. The higher the velocity, the larger the particle that can be carried. However, the maximum size that can normally be carried is sand, unless a tornado or hurricane is present. (Then anything goes!)

11.9 Does wind do much abrading of solid rock?

Locally, wind acts as an effective "natural sand-blaster," blowing sand against rocks and thereby abrading them. (Automobile windshields can be frosted in a severe sandstorm.) Usually, however, the wind simply transports away the material that is already loosened by weathering and perhaps already moved some distance by running water.

11.10 What is a ventifact?

A *ventifact* is a stone eroded by the sand-blasting effect of the wind. Smooth, inclined, and polished faces or facets are characteristic (Fig. 11.1).

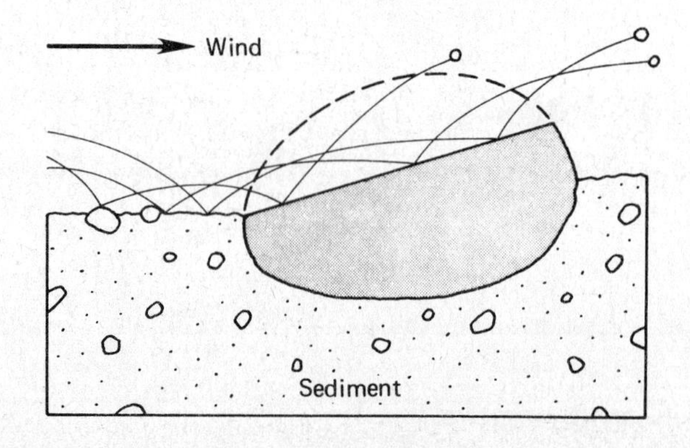

Fig. 11.1 The origin of a ventifact (commonly a flat-sided stone) by the natural sand-blasting of a pebble, cobble, or boulder.

11.11 What is a deflation basin?

A *deflation basin* is a shallow basin eroded by wind in areas of dry, fine-grained sediment. Some deflation basins in desert areas can be tens of kilometers long and tens of meters deep.

11.12 Why is wind abrasion more effective than water abrasion in rounding sand grains?

Because air does not provide a protective cushion around sand grains like water does—it is only 1/800 as dense as water. Thus grains hit each other frequently in strong winds, and corners are chipped off, rounding the sand grains.

11.13 What surface texture of quartz sand grains is commonly attributed to wind action?

A "frosted" appearance that can be caused by small crescent-shaped cracks in the homogeneous quartz; these diffuse the light.

11.14 How far can wind-blown dust be transported?

Very far. Fine volcanic ash, for example, can be thrown several kilometers high by a volcanic eruption and can circle the earth for years. When the volcano Krakatoa in the East Indies exploded in 1883, volcanic ash rose to 130,000 ft (nearly 40,000 m) and circled the earth for years, even making the days slightly darker than usual.

11.15 What is loess, and what is it composed of?

Loess is a term applied to widespread, thick deposits of wind-blown silt. It generally is made up of quartz and feldspar. Calcite is commonly added after deposition.

11.16 Where does loess form?

In two main places, either downwind from ablating (melting) glaciers or downwind from deserts. Examples of glacial loess are found in the Mississippi and Missouri River valleys of the upper Midwest. Desert-related loess exists in northwestern China, probably derived from the Gobi Desert of central Asia.

11.17 The minerals in loess are commonly angular, unweathered, and lack clay minerals. Why?

They are derived from glaciers or deserts, where chemical weathering is not an important process, and therefore clay is not produced. Silt-sized grains do not round easily as they don't have enough mass to strike each other very hard. (The rock flour of glacial meltwater streams may become wind-blown loess.)

11.18 In general, what conditions are necessary for the formation of *sand dunes* (natural accumulations of wind-blown sand)?

Dry sand and a lack of binding vegetation. Note that desert conditions are *not* necessary. For example, dunes commonly occur along many shorelines of lakes and oceans where the upper beaches are ready sources of sand. Other sources of loose sand are floodplains, alluvial fans, and glacial outwash plains.

11.19 Are all sand dunes composed of quartz sand?

No. Virtually any minerals can be wind-blown, although the common lighter ones (quartz and feldspar) are the most common.

11.20 What mineral are the sand dunes of White Sands, New Mexico, composed of?

Gypsum, a very soft mineral. The gypsum sand has not been transported far.

11.21 How large can sand dunes become?

With a source of abundant sand and strong, stable winds, dunes can reach heights of 250 m, as in the Saudi Arabian Desert.

11.22 Are sand dunes stationary?

Some are, if they have been stabilized by vegetation. However, most dunes are actively moving downwind, or migrating. Rates vary from less than a meter a year to tens of meters, and some may migrate much faster.

11.23 What is illustrated by Fig. 11.2?

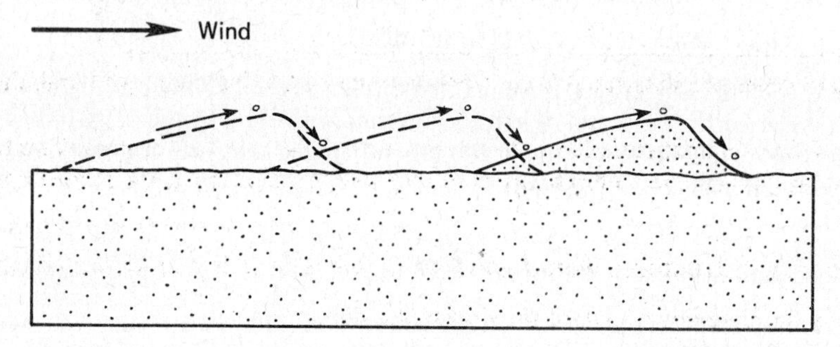

Fig. 11.2 Illustration of the migration of a sand dune from left to right.

Sand dune migration. Sand is blown up the windward side of a dune until it passes over the crest and moves down the leeward side, either by rolling, sliding, or slumping.

11.24 Is the sand in sand dunes well sorted or poorly sorted?

Well sorted. The fine silt and clay are generally carried out of the dune area by the wind, and coarser grains are left behind. The sand itself is better sorted than river or beach sands and is generally fine- to medium-grained sand.

11.25 What are some common shapes of sand dunes? See Fig. 11.3, and name the three types illustrated.

There are three main types of sand dunes, one longitudinal (parallel to the dominant wind direction) and two transverse types (perpendicular to the dominant wind direction). Long, wind-parallel ridges are *longitudinal dunes*. Crescent-shaped dunes with the points pointing downwind are called *barchan dunes,* and crescent-shaped dunes with the points pointing upwind are called *parabolic dunes*. There are other types as well. The type varies with the sand supply, the amount of vegetation present, the wind intensity, and the constancy of the wind. For example, longitudinal dunes generally occur where the sand supply is limited, and transverse dunes (perpendicular to the wind direction) occur where sand is abundant and the wind has a constant direction.

11.26 How would one distinguish between a barchan dune and a parabolic dune?

In barchan dunes, the points and the steep lee side of the dune are downwind. In a parabolic dune, the steep lee side is still downwind (it always is), but the points are upwind.

11.27 Which sedimentary structures are commonly found on dunes?

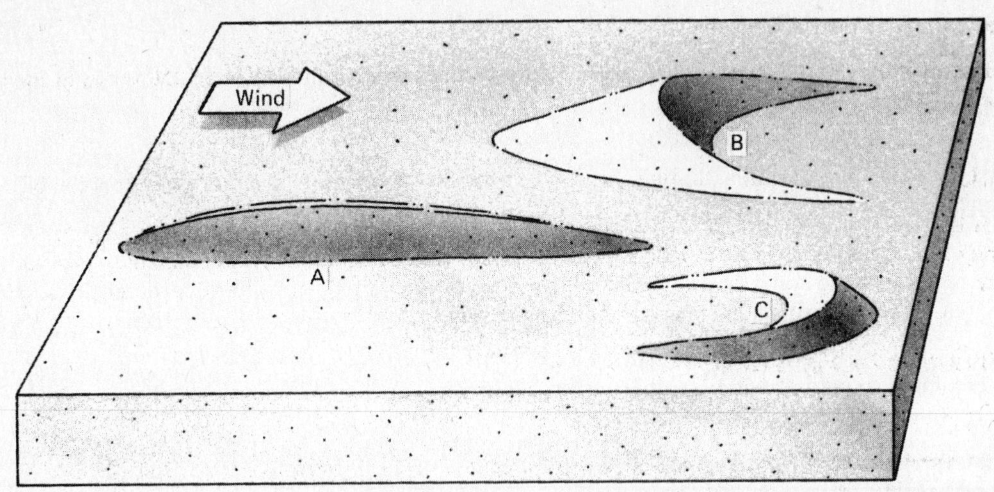

Fig. 11.3 Three dune types are illustrated here. (*a*) Longitudinal. (*b*) Barchan. (*c*) Parabolic.

Ripple marks (the asymmetrical current type, similar to those in water) especially on the windward sides of the dunes, and avalanche (slide) features on the steeper, leeward sides of the dunes.

11.28 What sedimentary structures would you find *within* a dune if you dissected one with a shovel?

Cross-bedding, dipping down toward the downwind direction.

Supplementary Problems

11.29 What are some characteristics of wind-blown sand that could identify it as such in the ancient rock record?

The grains are well rounded, well sorted, and may be frosted if they have had a long history of wind abrasion. The cross-bedding is generally large-scale, with cross-beds several meters high and tens of meters long.

11.30 Have dunes been a source of problems to humans?

Yes, in many places. Cities have been buried by the Gobi Desert of Mongolia. Fertile farmland or grassland can be covered by dunes, as along the edge of the Sahara.

11.31 How can moving sand dunes be stopped?

It is difficult to stop a natural process such as blowing sand. The sand in the source area cannot be hauled away. Vegetation planted on dunes may stabilize the sand. Any obstruction, such as a fence or a hedge, causes a decrease in velocity and the sand drops its load there. (This is the same principle as a snow fence.) The pharaohs of Egypt built great walls to keep sand away from the fertile Nile Valley. Commonly, nature wins the battle.

11.32 Are all deserts covered with sand dunes?

No. Deserts are defined as areas receiving less than 25 mm (10 in) of precipitation per year. Not all such areas have loose sand available for movement by the wind. Many deserts have large areas of solid bedrock. Much of Antarctica is a desert, and there is no sand available there.

11.33 Why is glacial loess important?

It weathers relatively quickly, because of its fine grain size (silt) into a fertile soil, as in the region of the Mississippi and Missouri River valleys.

11.34 Dust storms can occur in arid areas during the dry season. Do they occur in areas with appreciable precipitation?

Ordinarily no, but under special situations wind can move sediment even in the winter. During winter in the agricultural parts of the upper Midwest, dirty "snirt" (snow and dirt) can be blown around. (It may be a snowy but brown Christmas in some agricultural areas.)

11.35 The direct erosional activity of wind is rather minor. However, winds are responsible for many major processes on the earth's surface. Name some.

Wind distributes clouds over land and sea, it redistributes heat from the sun, generates the waves and currents in lakes and oceans, and distributes plant seeds and pollen.

11.36 What conditions led to the "Dust Bowl" of the thirties?

Overcultivation and overgrazing were followed by a severe drought that lasted several years, and the vegetation died. The soil was thus unprotected and blew away, destroying valuable agricultural land.

11.37 You are a geologist. Why would you be interested in dune deposits in the ancient rock record?

A geologist would be able to tell that the area was dry enough for dunes to form, the wind direction could be ascertained by measuring the directions toward which many cross-beds dip, and the composition of the sand would indicate which minerals (and rocks) were exposed to wind erosion.

11.38 You are a geologist, and you find a layer of ventifacts in a sedimentary rock column. So what?

This would tell you that the area was probably semiarid to arid, with enough sand blowing to abrade flat faces on exposed pebbles, cobbles, and boulders.

11.39 Where does the word *eolian,* meaning wind, come from?

From Aeolus, the Greek god of the winds.

11.40 Can you think of a good reason for riding a camel, rather than a jeep, in a windy desert?

Your face is above the elevation of most wind-blown sand. (Your boots may be abraded, but not your face.)

11.41 What is saltation?

Saltation is the process whereby sand grains bounce along in low arcuate jumps in high-velocity wind. As a sand grain descends and hits loose sand, its energy is transferred to another grain which then is lifted into the air.

11.42 Here is a quote: "Every square kilometer of land surface contains some sediment from every other square kilometer of earth's land surface." Explain.

Wind-blown dust is the explanation. The world's air circulation patterns result in dust being blown all over the face of the earth.

11.43 Is wind-blown dust deposited in the oceans?

Obviously so. Some estimate that 20 percent of deep-sea sediment was carried to the ocean by wind.

11.44 What causes wind?

As the sun heats air in a given area, it expands and rises. As this occurs, denser and cooler air moves into the area. As it moves in, wind (moving air) is the result.

11.45 Where are most of the world's deserts?

The large tropical deserts with very low rainfall, such as the Sahara and Kalahari Deserts of Africa and the Great Australian Desert, lie between 10° and 30° latitude both north and south of the equator. These are stable zones of high atmospheric pressure and sinking air that is warmed as it sinks, resulting in very low humidity. Deserts are also located further north, on the leeward sides of mountain ranges or far from the sources of moisture.

11.46 How much of the earth's land surface is desert?

About 19 percent (27.5×10^6 km^2 or 10.62×10^6 mi^2 of earth's total land area of 145×10^6 km^2 or 40.92×10^6 mi^2). An additional 14 percent is semiarid.

11.47 What is a *wadi,* also known as an *arroyo,* a *gulch,* or a *dry wash?*

A valley that becomes a stream channel during the uncommon desert rainfalls. They can very quickly fill with water during a cloudburst and should be avoided during inclement weather. Many people, animals, and automobiles have been lost in this manner.

11.48 On Fig. 11.4, identify (1) lag gravel, (2) lee side of a dune, (3) windward or stose side of a dune, (4) cross-bedding, and (5) deflation basin.

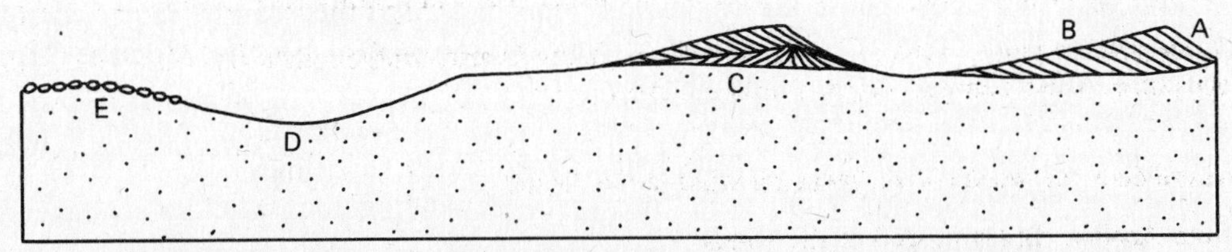

Fig. 11.4 Diagram of an area dominated by wind action. (*a*) Lee side or slip face of a dune. (*b*) Windward or stose side of a dune. (*c*) Cross-bedding within a dune. (*d*) Deflation basin. (*e*) Lag gravel.

(1) E. (2) A. (3) B. (4) A, B, or C. (5) D.

11.49 On Fig. 11.4, is the wind blowing left to right or right to left?

Left to right. The clues are the shapes of the dunes (steep side downwind) and the dips of cross-bedding (cross-beds dip down in the direction toward which the wind is moving).

11.50 Why are some sunsets red?

Dust particles in the air scatter light, and when the sun is low in the sky, the rays are passing through a greater distance of dusty atmosphere. The dust particles scatter the shorter light waves at the blue end of the spectrum more than they (the dust particles) scatter the longer red waves, and more red light passes through the dust to be seen by the viewer. Smoke also contains particles, and can redden a sunset.

Chapter 12

Shaping the Landscape: Shoreline Processes

The ultimate sites of deposition for sediment generated on the continents are the oceans. But once there, the journey of the sediment is not yet over, for the sediment load that rivers deposit at the edges of oceans (or even large lakes) is commonly reworked and redistributed by moving water in the "standing bodies" of water.

Wind-generated waves move sediment to and fro, making it more mature texturally and mineralogically (see Problems 5.17 and 5.19). Wind-generated currents move sediment along the shore. Density-generated currents, especially turbidity currents, move sediment off of the shelves and down the continental slopes to the continental rises at the bases of the slopes where it is resedimented (see Problem 5.18).

Most reworking of sediment occurs in shallow water along shorelines, with the reworking accomplished by waves, tidal currents, and longshore (littoral) currents. The latter currents are caused by waves hitting a shoreline at an angle; water generally moves up the beach face at an angle other than perpendicular, but then runs back down the steepest slope, or about perpendicular to the shoreline. Repetition of this by each wave causes water, and the sediment that it is carrying or dragging, to move, in effect, parallel to the shoreline. This dragging action on the sand abrades and rounds it. The beach, with its nearshore processes, has been referred to as a "river of sand' because so much sand is moved along the shore.

Along very shallow coastlines, barrier beaches can form a short distance offshore, as the waves pile up sand in shallow water. These are somewhat temporary, and they can be modified considerably during storms. Therefore, it is better not to build on the barrier beaches, for the "land" on which you built your house or cabin could disappear.

Many erosional features, including wave-cut cliffs, benches, and stacks, and many depositional features, including beaches, barrier bars, and deltas, occur along shorelines.

Shorelines can be emergent or submergent (drowned), depending upon whether sea level rises or falls relative to the shoreline.

Solved Problems

12.1 What is the main cause of water movement in oceans and lakes?

The drag (i.e., friction) of the wind as it moves over the water surface. It creates both waves and water currents.

12.2 How important are the major surface currents of the oceans (e.g., the Gulf Stream) to sedimentation?

They are surface currents and rarely touch the ocean floor, and therefore carry little sediment except that which drops in from the air or is captured when the weak currents meet the shoreline of a landmass. The major effect of such currents is climatic.

12.3 In waves, what path does water follow?

The water moves in nearly circular orbits, which decrease in size downward until a depth of one-half of the wavelength is reached. (See Figs. 12.1 and 12.2.)

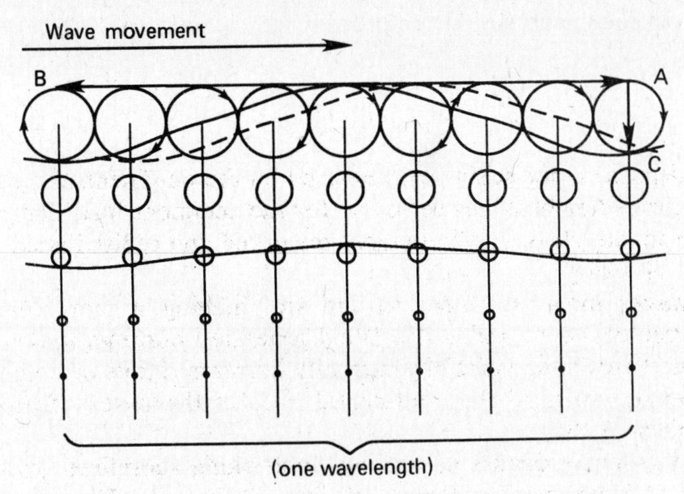

Fig. 12.1 Diagram illustrating the motion of water particles in waves.

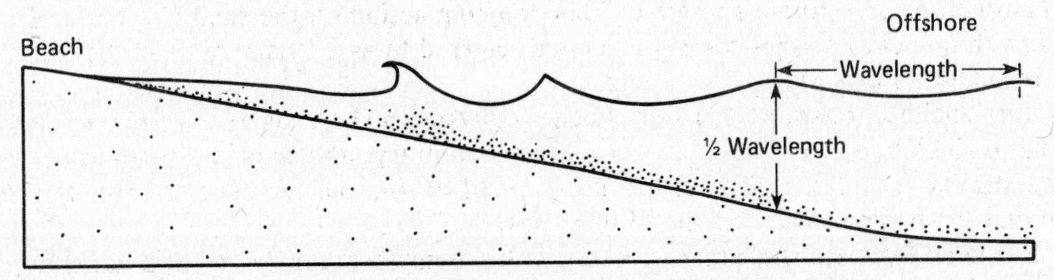

Fig. 12.2 Diagram illustrating the development of surf in shallow water.

12.4 On Fig. 12.1, which line represents the wavelength and which line represents the wave amplitude or height?

Line AB is the wavelength, measured from wave crest to wave crest, and line AC is the amplitude.

12.5 To what water depth are waves effective in moving sediment on the ocean floor or on a lake floor (Fig. 12.2)?

To a depth of about one-half of the wavelength of the waves.

12.6 Assuming the wavelength is (*a*) 2 m on a lake and (*b*) 200 m in an ocean, at what depths will the waves affect bottom sediment?

At a depth of (*a*) 1 m (3.28 ft) in the lake and (*b*) 100 m (328 ft) in the ocean.

12.7 What process is indicated in Fig. 12.3?

Longshore drift or littoral drift (beach drift) that is caused by waves hitting the beach at an angle other than perpendicular. The velocity of this wave-generated longshore current may reach velocities of a few meters per second.

12.8 How effective in moving sand is the process of the previous problem?

Very effective. The beach has been called "a river of sand" because so much sand can be moved parallel to the shoreline for long distances.

12.9 How far out from the beach would the process indicated in Fig. 12.2 be effective?

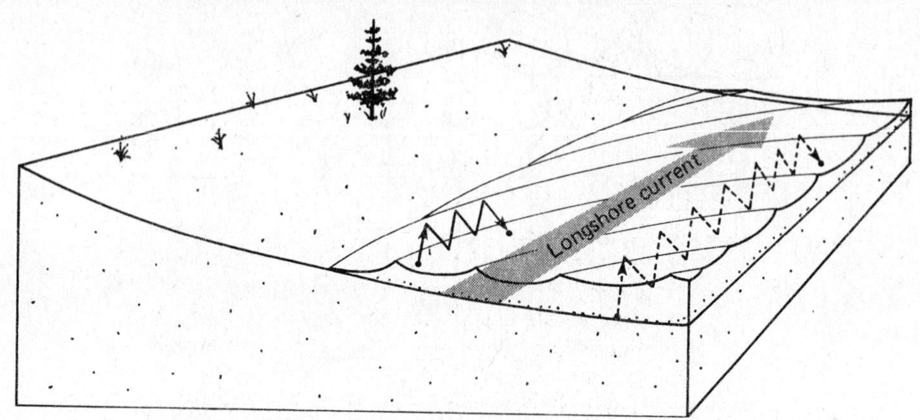

Fig. 12.3 Sketch illustrating the longshore drift of sand caused by waves hitting the beach at an angle. Any individual sand grain is carried up onto the beach at the same angle, but as the water from the wave moves back down toward the water body, it moves perpendicular to the shoreline (down the steepest slope). Thus the sand grain migrates down the beach. Similarly, sand in the shallow water migrates in the same direction (dotted line). Any floating object will also be carried parallel to the shoreline by this "longshore current."

Out to a water depth equal to about one-half of the greatest wavelength, which would of course be a greater depth and a greater distance outward from the shore during a big storm than during a gentle breeze.

12.10 What does Fig. 12.2 illustrate?

Waves dragging on the bottom and "breaking" when water depth is less than one-half the wavelength. When the water depth becomes too shallow, the water particles cannot complete their circular orbits (as in Fig. 12.1) and so they curl over and collapse.

12.11 Why should surfers be interested in Fig. 12.3?

Because it illustrates the origin of the "breakers" that they ride through the surf zone.

12.12 What process is illustrated by Fig. 12.4? Explain it.

Wave refraction is the process. As waves move toward the shoreline, they drag on the bottom (where the water depth is equal to one-half of the wavelength) and are theoretically and ideally turned to a direction perpendicular to the shoreline so that the waves are parallel to the shoreline when they meet it. One part of a long wave hits bottom first, and the wave front swings or pivots toward an orientation more parallel to the shoreline.

12.13 If waves are refracted, as in the previous problem, so that they are parallel to the shoreline, why does the situation illustrated in Fig. 12.3 exist?

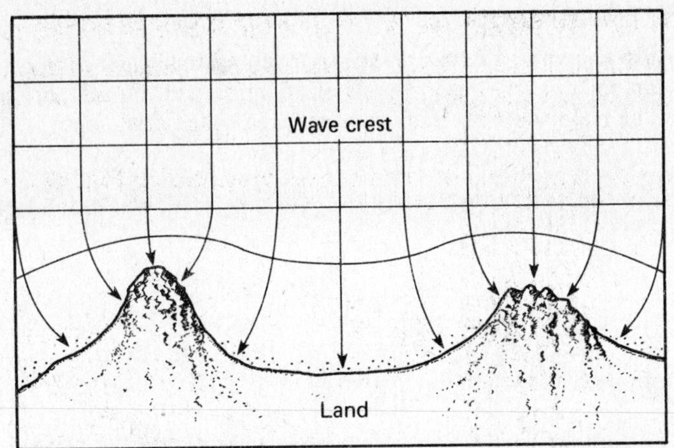

Fig. 12.4 Diagram illustrating wave refraction. Note that offshore, each segment of a wave has the same energy. As the shore is approached, the waves drag on the bottom and tend to hit the shore at right angles. Thus the wave energy of a given segment of wave is dispersed over a greater length of shoreline in bays (i.e., less energy) and deposition occurs. On headlands (points), energy is concentrated and erosion occurs.

Because waves are commonly incompletely refracted before they meet the shoreline, and they meet the shore at some angle other than straight on.

12.14 In Fig. 12.4, where is wave erosion likely to be the strongest and why?

On the headlands (points of land), for the amount of wave energy per given segment of wavelength offshore is concentrated into a shorter portion of the wavelength on points.

12.15 In Fig. 12.4, where is deposition of sediment most likely and why?

In the bays where the velocity of the waves decreases. The energy per given segment of wavelength offshore is spread over a longer portion of the wavelength in the bays.

12.16 Where would be the best place to build your boat dock in Fig. 12.4?

In the bay. It would be destroyed on the points where the wave energy is the greatest.

12.17 What is a beach?

A *beach* is an accumulation of sediment along the edge of a body of water. Sand or gravel beaches are probably best known, but pebble and cobble beaches can exist along shorelines of high energy.

12.18 What is the difference between a bar and a spit?

A *bar* is any loose accumulation of sand or gravel deposited offshore by waves and currents; included are spits, baymouth bars, barrier beaches, and *tombolos* (sand bars connecting islands to the shoreline or to each other).

A *spit* is a sand bar (commonly with a curved hook on the end) that projects into a body of water from the coastline or from a barrier bar. It forms because of currents moving parallel to the shoreline.

12.19 What is a barrier island?

A *barrier island* (also known as a *barrier beach*) is an elongated island composed of sand, paralleling the coastline. Some can be large; the resort hotels of Miami Beach, Florida, are built on such an island. It is separated from the mainland by a shallow water area called a *lagoon*.

12.20 On Fig. 12.5, find the following: wave-cut cliff, beach, wave-cut bench or terrace, stack, raised terrace.

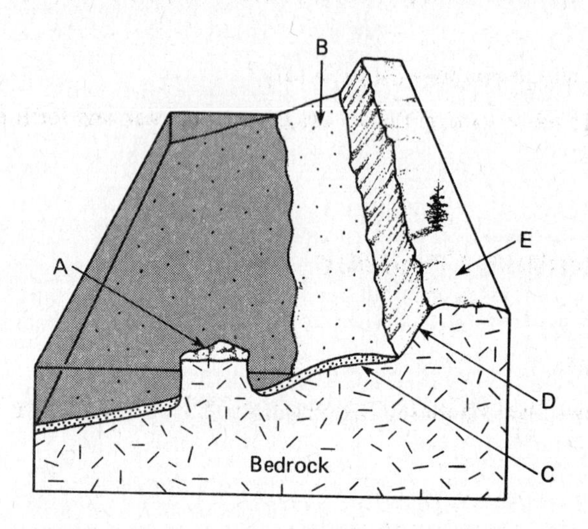

Fig. 12.5 Diagram illustrating shoreline features.

(*a*) Stack. (*b*) Beach. (*c*) Wave-cut bench. (*d*) Wave-cut cliff. (*e*) Raised terrace.

12.21 What is a delta, and why does it form?

A *delta* is a body of river sediment deposited where a river meets a standing body of water (generally the ocean) and deposits its load because of the decrease in velocity at that point. For a delta to form and exist, the river must supply sediment at a more rapid rate than the wave and current action in the ocean can carry it away from that point. Shapes vary; the Nile Delta of Egypt is a rather simple shape, like the Greek letter Δ from which the feature received its name. The Mississippi Delta, on the other hand, is a "birdsfoot" shape and projects seaward a considerable distance because the Mississippi River brings in much more sediment than the waves and currents of the Gulf of Mexico can disperse.

12.22 Name and briefly explain five major types of currents in the ocean.

Major surface currents, such as the California Current or the Gulf Stream, are generated by prevailing winds. *Tidal currents* are generated by the pull of the moon and the sun on the oceanic water surfaces. *Density currents* occur where denser salty waters move beneath fresher waters (as out of the Mediterranean Sea at Gibraltar), and where denser and colder waters move beneath warmer waters (as polar waters moving along the ocean floor toward the equator). A special type of density current is a *turbidity current,* the result of heavy sediment-laden currents moving downslope (see Problem 5.14 and 5.15). *Littoral currents* (longshore currents) are caused by waves meeting the shoreline at an angle other than perpendicular (Problem 12.7).

12.23 Which of the currents named in the answer to the previous problem are important in the transportation of sediment?

The major surface currents and the density currents caused by salinity or temperature differences transport little sediment. Tidal currents are important in moving sediment to and fro along coastal zones and estuaries (river mouths). Turbidity currents have deposited much of the sediment on the continental rises (Problems 1.17 and 5.18). Longshore currents can be very important, transporting sand for hundreds of kilometers.

12.24 How effective is erosion by waves?

Very effective. The cliff and bench of Fig. 12.5 were cut by wave action. As waves hit a cliff, they can hurl large rocks as tools against the rocky cliff, slowly breaking it down and rather quickly rounding the erosional products (pebbles, cobbles, and boulders). The White Cliffs of Dover, England, composed of soft limestone, have retreated several kilometers since the days of Roman occupation about 1600 years ago.

12.25 Does wave action round sand grains, which round with difficulty?

It seems likely that waves in the surf zone round sand grains by dragging them back and forth across other sand grains.

Supplementary Problems

12.26 What is the fetch of the wind on an ocean?

The *fetch* is the distance that the wind blows over the water. The greater the fetch, the larger the waves will be.

12.27 What is a tsunami?

A *tsunami,* also called a tidal wave, is a sea wave generated by an earthquake that caused a sudden movement of a part of the ocean floor.

12.28 The damming of rivers has reduced sand supply to ocean beaches. How can the sand be kept on the beaches, rather than being allowed to move along the shoreline by longshore drift?

This may be an impossible task. Weirs (walls) or jetties or groins are built outward into the water, but these generally serve only as temporary traps—the sand eventually may move around them. This may be useful in one regard—it can provide steady jobs as sand is dredged from reservoirs, ship channels, and bays and is reintroduced onto the beaches.

12.29 How can a drowned (submergent) coast caused by a relative rise in sea level be distinguished from an emergent coast caused by a relative drop in sea level?

A drowned coast is very irregular with numerous bays and estuaries. Much of the East Coast of the United States is clearly drowned. An emergent coast is straighter and commonly has wave-cut cliffs, and deep water is close to the beach. Much of the West Coast of the United States is emergent.

12.30 Why are some coastlines submergent and others emergent?

If the sea level is rising relative to the land, the coast will be drowned. This may be the result of an actual increase in sea level, caused by, for example, the melting of glaciers or by an elevation increase of part of the ocean floor by tectonic processes or by a subsidence of the land. If sea level is falling relative to the land, the coast will be emergent. This may be the result of an actual decrease in sea level caused, for example, by the growth of glaciers on earth's land surface, by the subsidence of a part of the ocean floor, or by the uplifting of the land.

12.31 Are the features of Fig. 12.5 illustrative of emergent or submergent conditions?

Emergent.

12.32 Are shorelines sites of major erosion as well as sites of major deposition?

Yes, the zone of erosive activity along shorelines, while narrow, is very long—there is a great deal of shoreline on the earth's surface. Erosive energy in the narrow zone can be very pronounced, especially along emergent shorelines (Fig. 12.5). Sediment carried off the land by streams accumulates along the shore unless it is carried into deeper water. The zone of deposition is much wider than the zone of erosion, and essentially encompasses the entire continental shelf and slope.

12.33 Southwestern Finland is a low-lying area, with an elevation slightly above sea level. However, its elevation is currently increasing. Why?

This region further illustrates the complexity of processes affecting the relative sea level along coastlines. In this case, sea level has risen in the last 10,000 years because of the melting of earth's glaciers. But this rise has not been equal to the more rapid uplift of the land due to postglacial rebound which followed the melting of the glaciers and which is still going on because of the great mass of ice that depressed the crust.

12.34 Is it right for governmental agencies to try to control the erosion of barrier islands and beaches so as to save seaside homes and cabins?

Human attempts at preserving sand beaches commonly result in their total destruction. One could argue that tax dollars should not be spent here for the benefit of a few people who should not have built or purchased property there in the first place. Building and zoning regulations should be stringent and should be regional in scope rather than local. However, if beaches or barrier islands are already built up, should the forces of nature be allowed to return such areas to their pristine state? Science gives the answer—"let nature take over." However, the answer may not be the best solution.

12.35 What is the relative influence of the sun and the moon in causing ocean tides?

Both exert a gravitational pull on the earth and especially on ocean waters. The sun, although much larger than the moon, is much farther away and therefore exerts only about ⅖ of the attractive force of the moon.

12.36 Why are there two tidal bulges in the oceans, essentially on opposite sides of the earth?

One bulge is formed as ocean waters on the side nearest the moon are attracted toward it by the moon's gravity. On the opposite side of the earth, farthest from the moon, the oceanic waters are attracted least of all, and less than the earth itself, and therefore they form another bulge.

12.37 What are spring and neap tides?

Spring tides are unusually high tides that occur twice a month due to the alignment of the moon and the sun on the same side of the earth, thus exerting a strong (combined) gravitational attraction on the water. A *neap tide* is an unusually low tide, caused twice a month when the moon and the sun are oriented at 90° relative to each other and the earth, and therefore pull at right angles to each other.

12.38 What are flood tides and ebb tides?

Flood tides are tidal waters moving shoreward from deeper waters. *Ebb tides* are the result of the water which has moved in during flood tide, moving back out to the deeper water (i.e., ebbing). Generally the flood and ebb tides move across tidal flats or in estuaries.

12.39 Do lakes have tides?

All bodies of water are indeed influenced by the gravitational pull of the moon and the sun, but the effects are significant and noticeable only on very large bodies of water (i.e., oceans).

Chapter 13

Groundwater

INTRODUCTION

Groundwater is water that is in the ground, as opposed to water actually on the earth's surface (see Fig. 8.1). Of the earth's total supply of water (an estimated 359 quintillion gal or 359×10^{18} gal), about 97 percent is in the oceans, about 2 percent is tied up as glacier ice, much less than 1 percent is in lakes, rivers, and the atmosphere, and about 0.67 percent is in the ground.

Most groundwater is *meteoric water,* water that has soaked into or infiltrated into the ground. A small part of groundwater may be *connate water,* water that was trapped in sediments when the sediment was being deposited. Some may contain traces of *juvenile (magmatic) water,* water that has been brought to the vicinity of the earth's surface from the crust or mantle by magmatic activity, usually volcanism.

POROSITY AND PERMEABILITY

Groundwater is located in pores (spaces) between grains in a sediment or sedimentary rock or in solution cavities in carbonate rocks. If the pores are connected and the sediment or sedimentary rock can transmit water, it is said to be *permeable,* as well as *porous*. A permeable layer of sediment or sedimentary rock from which water may be obtained is referred to as an *aquifer*. In most areas, unconsolidated sediment layers are the aquifers, but, for example, a buried unit of ancient sandstone may also be an aquifer. Well-rounded, well-sorted, uncemented gravel or sand makes the best aquifer because it can have as much as 35 percent porosity (by volume), and the pores are all connected.

WATER TABLE

The *water table* is the boundary between the upper *unsaturated zone* in which the pores are empty or only partially filled with water, and the underlying *saturated zone* in which the pores are full of water (Fig. 13.1). The water table commonly follows the relief of the surface but is more subdued. Groundwater moves slowly, therefore, when water is pumped out of an aquifer. A *cone of depression* may result where pumping is excessive and where there is not sufficient *recharge time* (time to refill the aquifer). Locally, a water table may be *perched* above an impermeable layer, higher than the main water table.

In an *artesian well,* the water level in the well rises above the level of the aquifer. An artesian system requires a permeable layer (the aquifer), impermeable layers above and below, and a *hydrostatic head* or pressure created by a higher elevation of the water elsewhere in the aquifer (Fig. 13.2). Some artesian wells flow constantly and should be capped in order not to waste water from the aquifer.

CAVES

Groundwater is commonly slightly acid because of the presence of carbonic acid—H_2CO_3—formed by reaction of water with CO_2 that is largely derived from the decay of plants and humic acids. (Groundwater is even more acid today because of human activity that causes acid rain with its sulfuric and nitric acids.) Acid groundwaters are important in the weathering of rock, especially in the

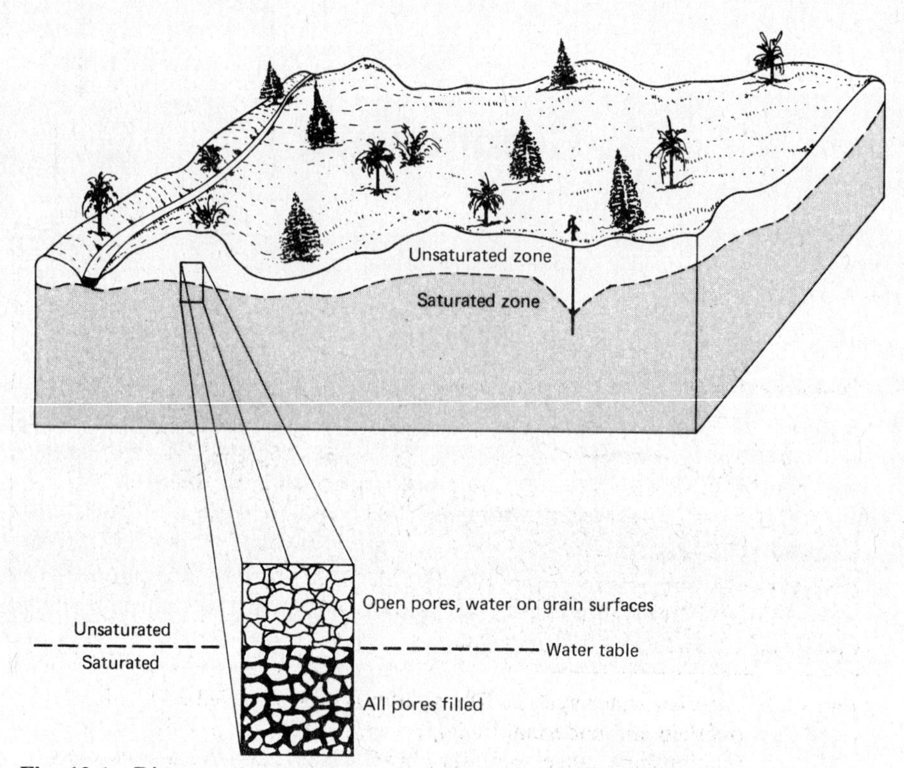

Fig. 13.1 Diagram showing the water table, the unsaturated zone, and the saturated zone. (*From Ojakangas and Darby, 1976.*)

breakdown of the abundant feldspars to form clays (see Chapter 4). In addition, acid waters readily dissolve limestone, commonly creating *caves* (large hollows in the rock) along joints and bedding planes along which the acidic waters move. Some of the dissolved calcite, carried in solution as calcium bicarbonate (see Problem 4.14) may be reprecipitated in the caves as *dripstone* (stalagmites and stalactites) as the water loses its CO_2. Cave networks commonly underlie areas of *karst topography* which commonly have irregular and rugged hills that have resulted from the solution of limestone or dolomite. Karst areas contain *sinkholes* which are surficial depressions caused by limestone solution, commonly accompanied by collapse into underground cave networks.

WATER USAGE

Groundwater in general is in short supply. Most population centers have water shortages, as water is needed for agriculture and industry as well as for personal consumption. About one-half of the water is used for irrigation; in California, the figure is 87 percent, and about half of that is lost to evaporation and transpiration (refer back to Fig. 8.1). It has been estimated that to produce 1 lb of rice requires 200 gal of water, 1 lb of wheat requires 60 gal, 1 corn plant requires 50 gal, and 1 gal of milk requires 932 gal! In Idaho, with extensive irrigation and hydropower in the state, the water usage is 22,000 gal per day per person, whereas in Vermont the usage is only 250 gal per day per person.

The pumping of great amounts of water out of aquifers can lead to problems other than shortages. The land within a 65-mi radius of downtown Houston has subsided as much as 3 m because of the removal of groundwater. Coastal cities may have the additional problem of seawater entering the aquifer as fresh water is removed.

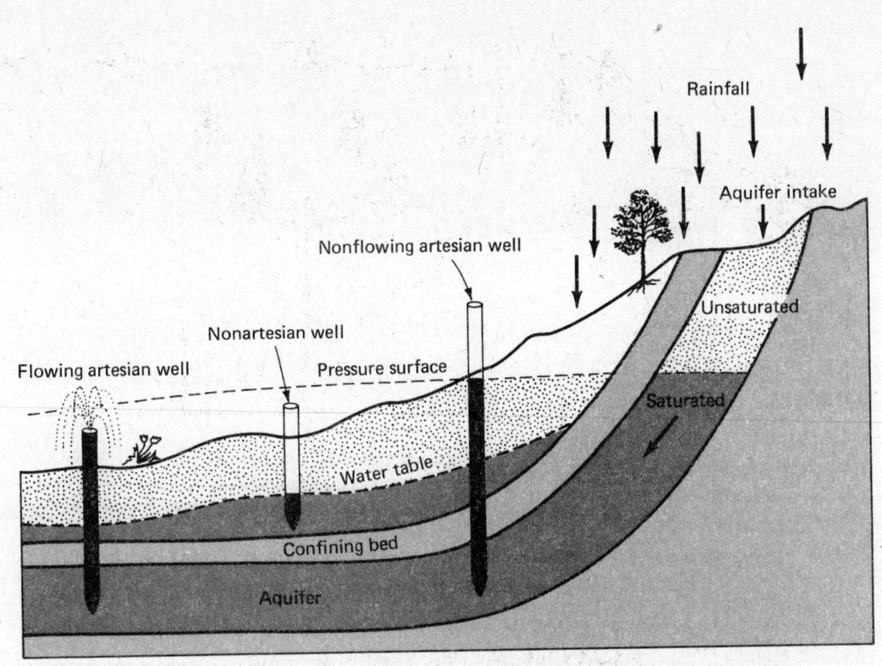

Fig. 13.2 Artesian water system. The aquifer is a permeable bed and is overlain and underlain by impermeable (confining) beds. These relationships cause a pressure head to exist, and the water stands higher near the recharge area. When a well is drilled into the aquifer, the pressure forces the water to rise above the aquifer. If the pressure is high enough, the well may be a flowing artesian well. The water tends to rise to the same level as the water table in the aquifer, as in a U-shaped tube, but friction prevents it from rising as high. (*From Ojakangas and Darby, 1976.*)

Solved Problems

13.1 Is groundwater abundant?

No. It makes up only 0.67 percent of the earth's water, and part of this has already been contaminated by pollution. Groundwater is a very scarce and very valuable commodity.

13.2 Why don't people in semiarid or arid areas just drill deep wells to obtain water?

This may be possible in some areas, but nearly all groundwater is meteoric water that infiltrates after rain falls or after snow melts. So groundwater is scarce in semiarid or arid areas.

13.3 What is a water table?

The *water table* is the surface above the zone of saturation that separates it from the overlying unsaturated zone. It is a planar feature that generally follows the surface of the land but with a more subdued relief.

13.4 What is a perched water table (Fig. 13.3)?

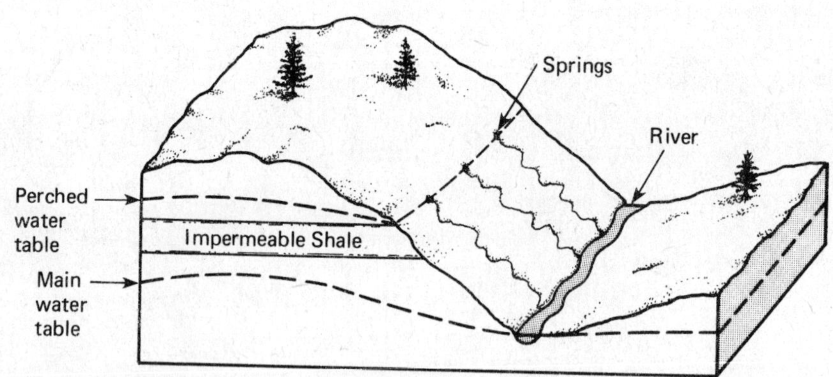

Fig. 13.3 A perched (local) water table.

A *perched water table* is a local water table that is at a higher elevation than the main general water table. It exists largely because of the presence of an impermeable layer such as shale beneath it.

13.5 What is the difference between porosity and permeability?

Porosity is a measure of the relative amount of pore space or voids in a sediment or in a rock, expressed as the percent of voids to the total rock volume. *Permeability* is the ability of a sediment or a rock to transmit water (i.e., to allow water to pass) through the connected pores. A material can be porous and yet be impermeable, as e.g., pumice or moderately cemented sandstone.

13.6 Does a shale have porosity?

Yes, most shales are porous, with small spaces between the clay and silt grains. The amount may be 15 percent by volume.

13.7 Since a shale has porosity, shouldn't it be a good aquifer?

It is not a good aquifer because the pores are so small that surface tension holds the water in the pores, resulting in a low permeability.

13.8 Would a vesicular basalt (see Fig. 13.4) be a good aquifer?

It has porosity, a necessary attribute, but if the vesicles are not connected, it will not have permeability.

13.9 Can a granite or a gabbro be an aquifer?

Yes, but the porosity and the permeability would be due to joints in the igneous rocks. Meteoric water can fill a network of joints and provide a large reservoir of water.

13.10 Are "underground pools" and "underground rivers" sources of groundwater?

Only in some areas with abundant caves. Theoretically, a water well driller could intersect a cave or an underground pool. However, the "underground pools and rivers" commonly referred to by people as the sources of their water are not actual hollows in the rock but are porous and permeable sand or gravel beds.

13.11 Farmer Jones was having a water well drilled. As soon as the driller hit water, Farmer Jones ordered the driller to stop. Was he indeed frugal and wise?

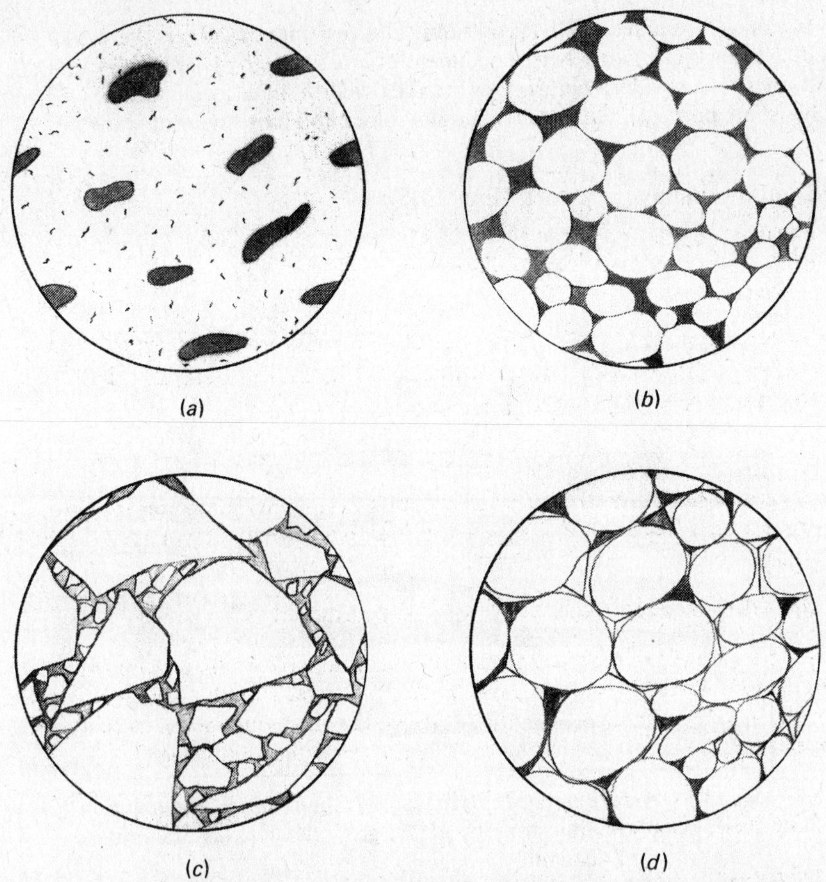

Fig. 13.4 Sketches of microscopic views of rock textures. Dark areas are pore spaces. (*a*) Vesicular basalt. (*b*) Well-rounded, well-sorted sand. (*c*) Angular, poorly sorted sediment. (*d*) Well-rounded, well-sorted, well-cemented sandstone.

Frugal yes, wise no. Because the water table varies with the amount of rainfall, his well may be dry in a dry year. He should have allowed the driller to drill deeper, to assure a more reliable supply.

13.12 What characteristics are necessary for an artesian system (Fig. 13.2)?

An aquifer, impermeable layers above and below, and a hydrostatic head.

13.13 Should Farmer Smith let his flowing artesian well flow?

No. The aquifer may be depleted by the constant flow. If recharge capability is slow, his flowing well may cease to flow. Many wells which were flowing wells decades ago are not flowing any longer. It is a blatant waste of a valuable resource.

13.14 What is a *cone of depression?* Find one on Fig. 13.1.

It is a cone-shaped, downward-pointing depression of the water table due to the rapid removal of water through a well in an aquifer that has a slow rate of recharge.

13.15 Which of the sediments and rocks of Fig. 13.4 is the best aquifer and which is the worst? Why?

The best is *b*, a well-rounded, well-sorted sand. The vesicular basalt *a* is the worst if the vesicles are not connected. The angular, poorly sorted sediment of *c* has less porosity than *b*, and even if the pores are connected, there is a smaller volume of pores. The well-rounded, well-sorted, well-cemented sandstone *d* may have some porosity, but the cement may block the movement of water.

13.16 See the sketch of a cave network (Fig. 13.5).

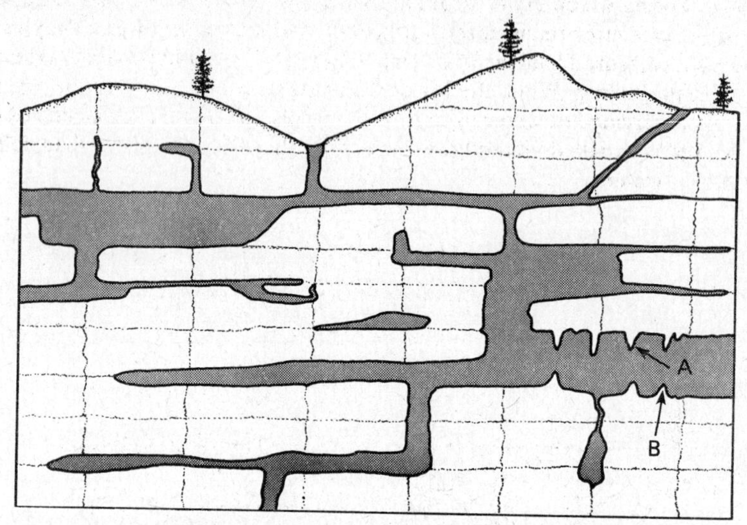

Fig. 13.5 A cave network in limestone, the result of the solution of the limestone by acidic groundwater. (*a*) Stalactites. (*b*) Stalagmites.

(*a*) How did it form?

By the solution of the limestone by acidic groundwater.

(*b*) Which are the stalagmites and which are stalactites?

Stalagmites (*b*) point upward, whereas *stalactites* (*a*) hang downward from the top of the cave. How can you keep these similar words straight? One device is to think of the *g* in the word *stalagmite* representing the ground (i.e., they form on the ground) and the *c* in *stalactite* representing the ceiling (i.e., they form on the ceiling).

(*c*) How are stalagmites and stalactites formed?

As groundwater seeps into caves which were formed earlier by solution, it partially evaporates, and part of the dissolved calcium bicarbonate is precipitated as calcite. As the drop falls to the floor of the cave, it evaporates further.

13.17 What is connate water, and where is it usually found?

Connate water is water trapped in the sediment during sedimentation and is commonly found in deep oil wells. It is usually salty, both because of its marine origin and because of solution of trapped salts by the connate water.

13.18 What are springs? Where on Fig. 13.3 might a spring be found?

Springs are places where a local water table intersects the surface. Springs can also occur along faults or shear zones. On Fig. 13.3, springs could occur anywhere along the dashed line that depicts the intersection of the water table with the surface of the ground.

13.19 What causes hot springs?

These exist in areas where groundwater comes into contact with rocks that are still hot due to geologically recent igneous activity or in contact with hot gases from such areas. Some water may be heated by frictional heat of major fault zones.

13.20 What is a geyser, and how does it function? (See Fig. 13.6.)

A *geyser* is a hot spring which emits steam and/or hot water with great force. Superheated water upon reaching the lower-pressure area near the surface turns to steam. Hot water deep in a fracture system or a conduit is under pressure from the overlying water, and it can become superheated to temperatures above 100°C without boiling. When the lower water suddenly becomes steam, the resultant volume increase forces the overlying hot water to rise. As it moves up and reaches a zone of less pressure, it also becomes steam and is expelled violently. Furthermore, hot water is lighter than cold water and tends to rise through colder water.

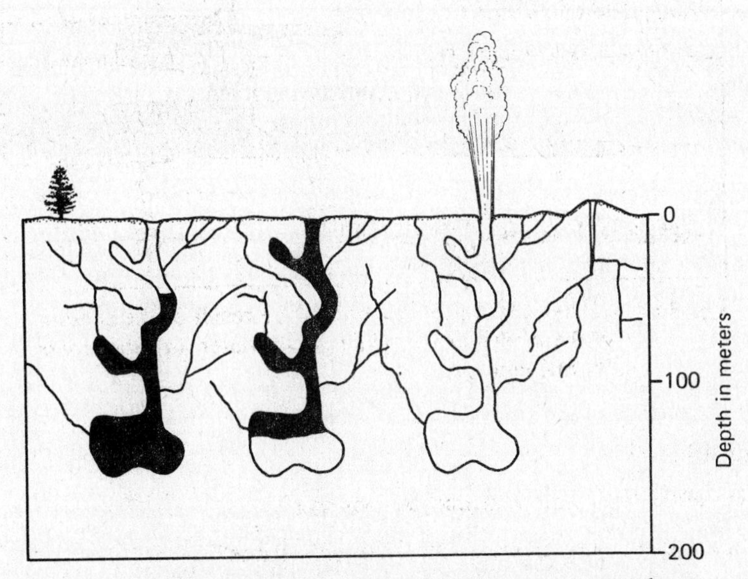

Fig. 13.6 The formation of a geyser. (*a*) Fracture system has filled with meteoric water. (*b*) The boiling point at the bottom is higher than 100°C because of the weight of the overlying water which raises the boiling point; part of the lower water has turned to steam which raises the water above it in the fracture system. (*c*) As the water rises into regions of less pressure, it all boils and turns to steam and erupts suddenly as a geyser.

13.21 Name an area that contains many hot springs and geysers.

New Zealand, Iceland, and Yellowstone National Park in Wyoming are the three largest areas. Many other areas contain one or a few hot springs and geysers.

13.22 What is a sinkhole?

A *sinkhole* is a depression at the surface in an area underlain by carbonate rocks and is a product of both solution and collapse into a cave network.

13.23 What is karst topography?

Karst topography is a topography developed on areas underlain by carbonate rocks that have undergone considerable solution, resulting in sinkholes and underlying cave networks. Such topographies can be very rugged.

13.24 Where in the United States can karst topography be found?

Kentucky, Tennessee, and Florida are three of the best areas for karst development. In several other states, small karst areas are present. Puerto Rico also has karst topography.

Supplementary Problems

13.25 Why is groundwater important in the transformation of sand into sandstone?

It may carry in the cementing material (usually silica, calcite, or clay) in solution. It may also dissolve previously deposited cement and even sand grains (usually feldspar grains or rock fragments), thereby creating a "secondary porosity" (as opposed to a primary depositional porosity) and a potential petroleum reservoir.

13.26 Some areas of karst topography lack surface streams. Why?

All the water sinks into caves and smaller pore spaces by moving downward through sinkholes or joints in the rock. Any streams in such an area are literally underground streams.

13.27 What is hydrology?

Hydrology is the science of the study of water, including groundwater.

13.28 Is groundwater virtually inexhaustible?

No! It can be used up if it is taken out faster than it can be replaced (recharged) into the ground. In west Texas and New Mexico, the water table in a major aquifer has dropped 30 m (100 ft) in the last century. It would take thousands of years of recharging to fill it to its former level.

13.29 How can humans help recharge a depleted aquifer?

By pumping water back into the groundwater system through recharge wells and by constructing reservoirs to enhance seepage before the water can run off in rivers.

13.30 What is "hard water"?

Hard water is groundwater that contains dissolved calcium carbonate (as calcium bicarbonate—see Problem 4.14) and magnesium. Hard water is undesirable, as it makes a scum when soap is added to the water. Therefore, water softeners are used, utilizing processes that remove the undesirable components. However, many water softeners replace the calcium and magnesium with sodium that is also undesirable.

13.31 What are the white and gray deposits around the hot springs and geysers of Yellowstone Park?

Travertine ($CaCO_3$) and sinter or geyserite (SiO_2) are deposited as the hot spring water comes to the surface and cools, precipitating much of its dissolved load.

13.32 How do "water-witchers" ("deviners" or "dousers") find water?

There is no scientific explanation of how a forked peach, willow, or hazel branch (or a bent wire coathanger) can indicate the presence of water. Most witchers, perhaps even without realizing it, know the pattern of water wells in the area in which they work, and probably subconsciously incorporate these data into their analyses. Groundwater is present in most areas, but it can be too deep, too little, or too salty.

13.33 What is the rate of movement of groundwater?

Usually only a few centimeters per day. In some situations, it may move 100 cm per day or 0.5 cm per day. In the rare case of groundwater being present in underground caverns, it can move much faster.

13.34 What is Darcy's law?

In 1856, Henry Darcy of France found that the discharge of groundwater is directly proportional to the hydraulic slope and cross-sectional area of an aquifer that is transmitting the water. It is expressed as $Q = VA$ and $V = KI$. Therefore, $Q = KIA$, where Q = discharge, K = hydraulic conductivity, I = change in elevation divided by the distance traveled, A = cross-sectional area of the aquifer, and V = velocity of flow.

13.35 Using Darcy's law, what is the expected flow velocity V of water for clear sand 100 ft thick in a 1 mile wide valley? The sand has a K value of 500 ft per day and a water table slope of 5 ft/mi. The porosity of the sand is 20%. Note that 1 mile equals 5280 ft.

First, solve for Q:

$$\begin{aligned} Q &= KIA \\ &= (500 \text{ ft/day})\left(\frac{5 \text{ ft}}{5280 \text{ ft}}\right)(100 \text{ ft} \times 5280 \text{ ft}) \\ &= 250{,}000 \text{ ft}^3\text{/day} \end{aligned}$$

Now solve for the average velocity of flow:

Since $Q = VA$, then $V = \dfrac{Q}{A}$

Because the actual flow is only through the area of voids (porosity) in the cross-section, then the velocity of flow through the sand is:

$$\begin{aligned} V = \frac{Q}{A \times \text{porosity}} &= \frac{250{,}000 \text{ ft}^3\text{/day}}{100 \text{ ft} \times 5280 \text{ ft} \times 20\%} \\ &= \frac{250{,}000 \text{ ft}^3}{105{,}600 \text{ ft}^3} \\ &= 2.36 \text{ ft/day} \end{aligned}$$

13.36 How is the depth to the water table measured?

It is the depth at which water stands in a well.

13.37 What happens to the water table during a drought?

Because there is little or nor recharge of the aquifer, the water table will drop as water is removed via wells, springs, or base flow.

Chapter 14

Earthquakes

INTRODUCTION

An earthquake can be defined as a shaking of a part of the earth's surface because of the passage of a series of shock waves through the underlying rock. Most earthquakes are caused by sudden movement along faults, but some are due to movement of magma in magma chambers. Rocks under stress can be bent, and then when the stress is suddenly relieved by movement, the rocks may spring back to their original prestress form, but with a net displacement across the fault. These characteristics are summarized by the elastic-rebound theory of earthquakes. Studies along the San Andreas fault, the biggest earthquake laboratory in the world, verify this (Fig. 14.1).

Fig. 14.1 The San Andreas fault zone and some related faults in California. (*From Ojakangas and Darby, 1976.*)

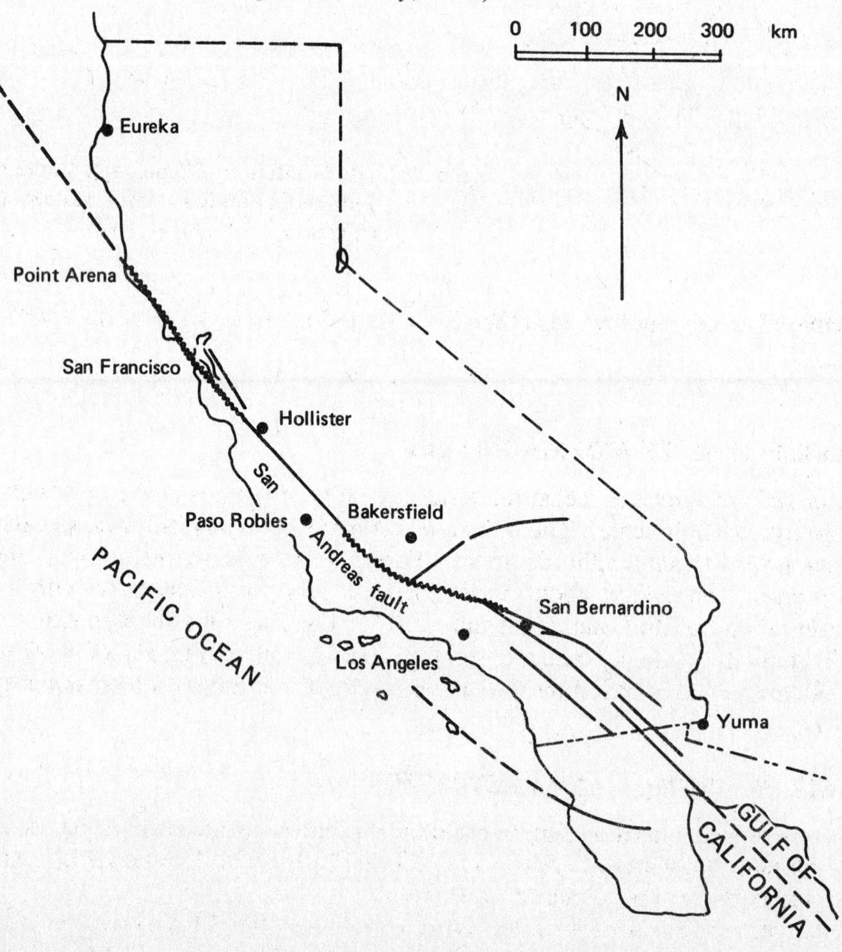

EARTHQUAKE WAVES AND LOCATION

There are three main types of waves generated during an earthquake—the P waves (primary or push-pull waves), the S waves (secondary or shear waves), and L waves or long waves. The P and S waves are body waves and go through the earth's crust, whereas L waves travel near the surface.

The P wave is about twice as fast as the S wave, and this lag in arrival of S waves from the *epicenter* (the point on the earth's surface above the actual earthquake location or *focus*) allows the distance from a seismograph to an earthquake to be calculated. If an earthquake is recorded on three seismographs in different areas, circles can be drawn around each seismograph, and the actual location is the common point of intersection of the three circles (Fig. 14.2).

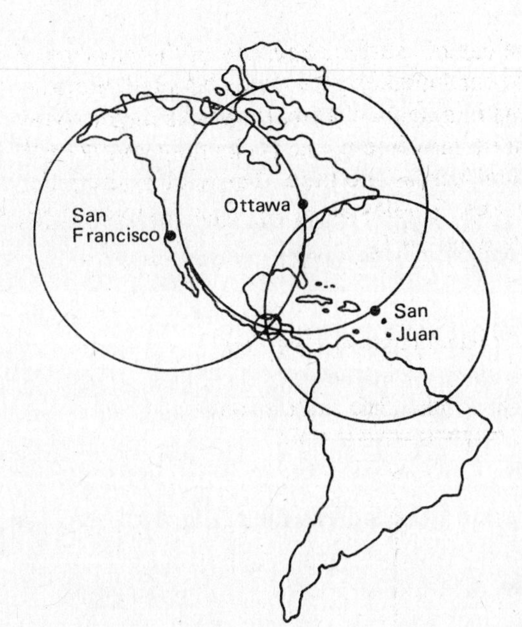

Fig. 14.2 Three different seismograph stations with the same earthquake plotted at each station. (*From Ojakangas and Darby, 1976.*)

Most earthquakes are shallow, less than 8 km deep, but some have been recorded at a depth of about 700 km.

EARTHQUAKE INTENSITY AND MAGNITUDE

The severity of an earthquake is measured either by an intensity scale of I to XII or by a magnitude scale of 0 to 10 (the Richter scale). The latter, in common use, is logarithmic, so each whole number of the scale has a vibrational amplitude about 10 times as great as the next lower whole number and an increase in energy release of about 30 times. In general, an earthquake with a magnitude of 6 causes moderate damage, and one of magnitude 7 on the Richter scale can cause severe damage. The Mercalli intensity scale is based upon the compilation of reports of shaking and damage, whereas the Richter scale is based on the actual recorded vibration on a seismograph.

EARTHQUAKE PREDICTION AND PREVENTION

Seismologists are working hard to discover means of predicting earthquakes. These include the measurements of foreshocks, water depth in wells, tilting of the ground, magnetism, radon in wells, and electrical conductivity. So far, successes are few.

Earthquake prevention is even more difficult than earthquake prediction. Prevention is obviously not yet possible, and may never be possible. However, it is known that faults can be lubricated with water to cause slippage, and it has been suggested that major strain along a fault zone might be relieved in this manner. Proper building construction can reduce earthquake damage, but it is even better to delineate particularly hazardous areas and to avoid building in such areas.

Solved Problems

14.1 What is an earthquake?

An *earthquake* can be defined as a shaking of a certain part of the ground and the underlying rocky crust, caused by the passing of a series of shock waves through the crust.

14.2 What is the cause of most earthquakes?

Most are caused by the sudden movement along faults in the earth's crust. These may be new faults formed as a rock fractures, or more commonly, the movement may be along preexisting faults.

14.3 Name another less important cause of earthquakes.

The movement of magma in chambers beneath volcanoes causes numerous small earthquakes, most so small that they are not felt but are recorded on instruments. Commonly they are a warning of a major volcanic eruption.

14.4 Explain why there are numerous earthquakes along the San Andreas fault (Fig. 14.1).

The two blocks of the earth's crust on opposite sides of the fault are under pressure that causes the western block to move northward relative to the eastern block. However, the rocks on opposite sides of the fault are touching, and the friction is difficult to overcome. When the pressure buildup exceeds the friction, sudden movement occurs and the crust at that locality is shaken by the seismic (shock) waves.

14.5 What is the elastic-rebound theory of earthquakes?

The *elastic-rebound theory of earthquakes* states that the crust behaves as an elastic medium; that is, rock can be deformed or bent, but will return to its original shape after the stresses are removed. So rocks under pressure, as along the San Andreas fault, will bend (i.e., accumulate strain energy) until the pressure becomes greater than the friction and movement occurs. After the sudden movement, the rocks have been relieved of the strain and will have returned (rebounded) to their original shapes, but with a net movement (displacement) along the fault.

14.6 How do earth scientists know that the San Andreas fault will move again and cause another earthquake?

Detailed surveys across various segments of the fault show that the rocks on opposite sides of the fault, but not right at the fault itself, are moving at rates of as much as 2 cm per year in opposite directions. Elastic strain is accumulating, and when it exceeds the frictional forces along the fault face of that segment of the fault, movement will occur. In addition, we have a long recorded history of earthquakes along the fault—why should they stop now?

14.7 When and where will the next movement occur along the San Andreas fault?

We don't know precisely when or where, for the San Andreas is really a fault zone with countless faults and fractures. Depending on the accumulated strain, movement could occur along any of the fractures.

However, the detailed surveys indicate the general areas in which the stresses are accumulating most rapidly (i.e., where the fault is "locked"). Nevertheless, we cannot yet predict precisely when or precisely where the pressure will exceed the friction.

14.8 There are three main types of seismic or earthquake waves generated by an earthquake. Describe each.

The P and S waves are body waves that travel through the earth. The *P wave or primary wave* (or compressional wave) can be thought of as a "push-pull" wave, with each minute particle of rock alternately compressed and decompressed in the direction that the wave is traveling. In an *S wave or secondary wave* (or transverse wave), the particles are oscillating vertically as the wave passes through the rock, much as the path that a rope with one end tied to a post will assume as the other end is moved up and down. That is, the movement is perpendicular to the direction that the wave is traveling. The *L wave or long wave* (also known as a surface wave) is a complex wave with orbital motion that causes major oscillation, and thus is the wave that causes earthquake damage as it travels near the earth's surface.

14.9 What is a seismograph?

A seismograph is an instrument that records earth movement (earthquakes).

14.10 Describe, using Fig. 14.3, how a seismograph works.

Fig. 14.3 A seismograph. (*From Ojakangas and Darby, 1976.*)

Part of a seismograph is attached to solid rock of the earth's crust. The other part, with a recorder (a pen or a light beam), is constructed so as to not move when the earth beneath it vibrates. The difference in the vibration is recorded on the drum, which is turning and shaking. The product is a seismogram (see Fig. 14.4).

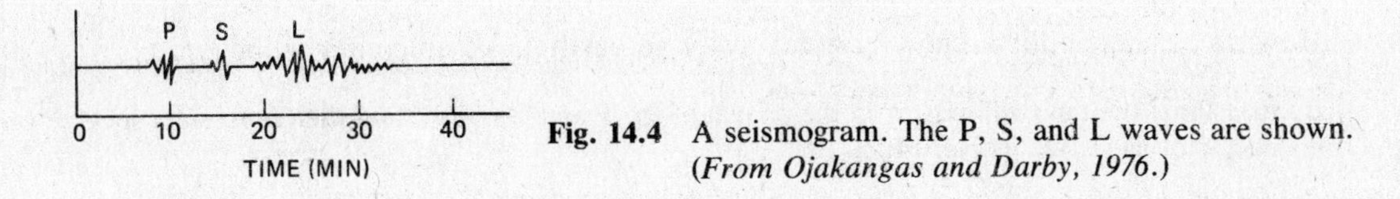

Fig. 14.4 A seismogram. The P, S, and L waves are shown. (*From Ojakangas and Darby, 1976.*)

14.11 What is the epicenter of an earthquake?

The *epicenter* is the point on the earth's surface directly above the point at which the earthquake occurred.

14.12 What is the focus of an earthquake?

The *focus* is the actual spot within the crust where the earthquake occurs.

14.13 On the seismogram of Fig. 14.4, identify the P, S, and L waves.

The first peaks on the left are caused by the P wave, the middle high peaks are caused by the S wave, and the large vibrations on the right are the L waves.

14.14 How fast do earthquake waves travel?

P waves travel from 5 to 13.8 km/s (kilometers per second) and are faster with increasing depth. S waves travel about half as fast (3.2 to 7.3 km/s). The L waves travel at 4.0 to 4.5 km/s.

14.15 What are the relative speeds of the earthquake waves as they pass through the earth?

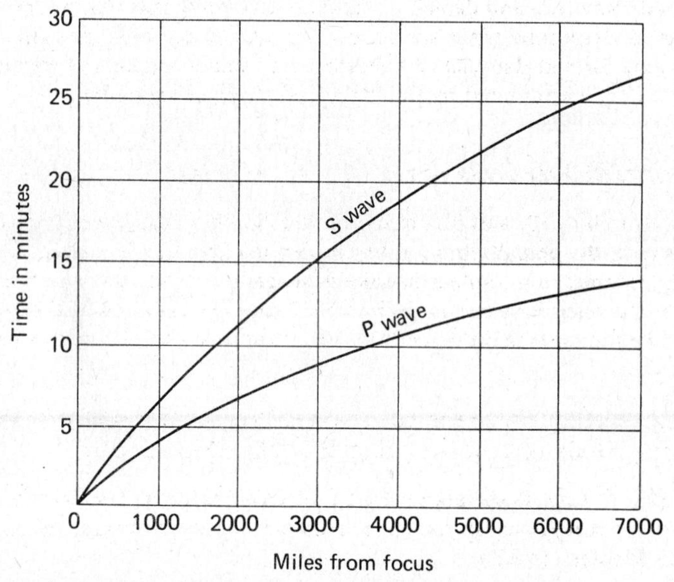

Fig. 14.5 Travel-time curves for the P and S waves.

From Fig. 14.5, it can be seen that the P wave arrives first and is the fastest, the S wave arrives next, and the L wave last.

14.16 How do earth scientists know how far away an earthquake epicenter is located?

By measuring the time differences in the arrival of the P and S waves, and then utilizing a special time-distance plot as shown in Fig. 14.5.

14.17 By utilizing Fig. 14.5, determine the distance to an earthquake epicenter if the time lag is (*a*) 7 min and (*b*) 10 min.

(*a*) If the lag is 7 min, the earthquake is about 4800 km (3000 mi) away. The P wave has arrived in 8 min and the S wave in 15 min. (*b*) If the lag is 10 min, the earthquake is about 8300 km (5200 mi) away. The P wave has arrived in 12 min and the S wave in 22 min.

14.18 (*a*) How is the epicenter of an earthquake located?

By the three-circle method, as shown in Fig. 14.2. The distance to an earthquake is calculated for three different seismographs. The distance to an earthquake is known, but the direction to the earthquake is not given by a single seismograph. If the distances from each of three seismographs are plotted on a single map, they will intersect at the epicenter of the earthquake.

(*b*) Could two seismographs be used to locate an earthquake?

No. Note that any two circles will intersect at two points, and the earthquake could be at either point. At least three circles are essential.

14.19 What is the *Mercalli earthquake intensity* scale?

A scale going from I to XII that measures the effects of an earthquake on people and people-made structures. People are questioned as to what they experienced or saw in their area. The scale is somewhat subjective, as people may not always recall the event in detail, either unintentionally or intentionally. For example, an earthquake of intensity VI is felt by everyone, people run outdoors, heavy furniture is moved, some plaster may fall, and damage is slight. One of intensity XI causes masonry buildings and bridges to collapse, underground pipes are put out of commission, rails are quite bent, landslides occur, and related cracks are formed. Isoseismal lines on a map connect points of equal disturbance; the highest intensity area will be surrounded by the highest-value isoseismal line.

14.20 Briefly describe the *Richter scale of earthquake magnitude*.

It is a scale going from 0 to 10, and it is based on the amplitude of waves recorded on a seismograph. Thus it is a measure of the actual vibration (magnified in order to see it well), and hence the energy of the quake. It is logarithmic; e.g., an earthquake of magnitude 4 has 10 times the vibrational magnitude and about 30 times the energy of one of magnitude 3, and an earthquake of magnitude 5 has 100 times (10 times 10) the vibrational magnitude and 900 times (30 times 30) the energy of a level-3 earthquake.

14.21 You read about an earthquake in the newspaper. Your companion asks you if this earthquake with a Richter magnitude of 7 was a "biggie." What do you say?

Yes, it was a "biggie." Any quake above 6 on the scale can cause considerable damage. A 7 is very strong and can be very dangerous. If you can feel an earthquake, its magnitude is at least 2½. Some are larger than 7. (See Problem 14.32.)

14.22 What are aftershocks?

Characteristically, a large earthquake is followed by numerous (sometimes hundreds) of smaller earthquakes called *aftershocks*. Some can be very strong. They are due to readjusting movements along the fault where the focus was located. Similarly, minor earthquakes can precede a major earthquake.

14.23 How deep are earthquakes?

Earthquakes are classified as shallow, intermediate, and deep. Most are shallow (85 percent) and occur within 60 km (36 mi) of the surface, some occur at depths of 60 to 300 km (36 to 180 mi), and only a few are deeper than 700 km (420 mi).

14.24 What is a tsunami?

A *tsunami* is a seismic sea wave, originating from a large earthquake that occurs under the ocean. The energy of the earthquake is transferred to the water as well as to the earth. When such a wave reaches shallow water, it can rise to heights of many meters, and where the shallow water is confined in a narrow channel, it can create a wave more than 15 m (49.2 ft) high. The tsunami that followed the Good Friday 1964 earthquake in Alaska had speeds of 640 km/h (400 mi/h) on its way to points such as California and Hawaii.

14.25 Which parts of the United States are most active seismically?

The West, and three small areas in the East, as shown on Fig. 14.6.

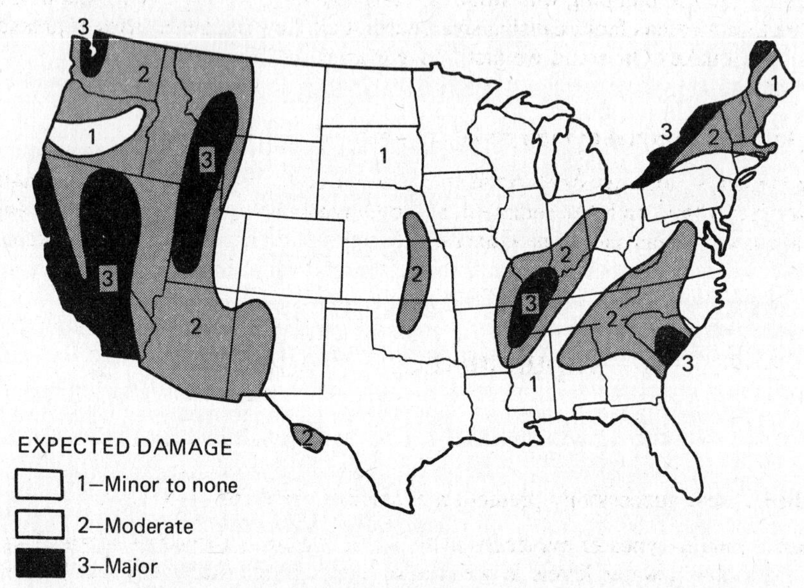

Fig. 14.6 Map of the United States showing earthquake danger. Areas of greatest danger are shown in black, areas of moderate danger are shown in dark gray, areas of minor to no danger are not shaded. (*After Environmental Science Services Administration, 1969.*)

14.26 When and where was the biggest earthquake in the United States?

December 16, 1811, at New Madrid, Missouri (240 km, or 150 mi, south of St. Louis near the southeastern tip of the state) on the Mississippi River. Although there were not instruments in those days to measure the earthquake magnitude, the intensity was 12 on the intensity scale. The Mississippi River flowed backward (that is, it started flowing in what was normally the upstream direction, but only for a short time) in that area. A depression 240 km (150 mi) long was formed. Reelfoot and St. Francis Lakes are two remnants of the depression; each is 32 to 48 km (20 to 30 mi) long and 6.4 to 8 km (4 to 5 mi) wide. About 2.6 million km^2 (1 million mi^2) of the United States shook, and it was felt from Canada to the Gulf and from the Rockies to the Atlantic. It even caused church bells in Boston to ring. Few people lived in that part of Missouri, which was still the wild frontier, and only six persons were killed. Over the next 3 months there were 1800 aftershocks.

14.27 Does the New Madrid area show on Fig. 14.6?

Yes. It is the area of major earthquake potential south of the Great Lakes and north of the Gulf of Mexico.

14.28 Are all earthquakes harmful?

No, small ones may not be, and humans purposely generate small earthquakes with seismic vibrators or explosions in order to bounce the shock waves off buried rock layers, thereby determining the structure of buried rock layers. This technique is of great value for locating petroleum traps, such as buried anticlines. (It takes longer for the waves to reach the surface after bouncing, or reflecting, off the lower sides of an anticline than for those that are bounced off the higher top of the anticline.)

14.29 Can earthquakes be prevented?

We do not know how to prevent them. However, we do have some ideas. For example, we know that water pumped down into an oil field in Colorado acted as a lubricant along faults, thereby causing daily earthquakes. When the pumping was stopped, earthquakes gradually diminished in number. Could we do this to the San Andreas fault, causing small earthquakes and small releases of pressure, thus heading off a major earthquake? Or would we just "trigger a bigger"?

14.30 How can future earthquake damage be minimized?

By making buildings more resistant to the stresses caused by the shaking of an earthquake, by building on solid rock rather than on loose sediment, and by having disaster plans in place. In addition, all people in earthquake-prone areas should be educated as to what to do when an earthquake occurs.

Supplementary Problems

14.31 How did the Chinese successfully predict an earthquake in 1976?

By synthesizing many types of evidence, including many small recording instruments in villages (thus noting the preshocks), water levels in wells (they may change drastically just before an earthquake), electric currents in the ground, variations in ground tilt, and even the anomalous behavior of animals (bees left the hive, birds left the nests, cattle were edgy, and so on).

14.32 What was the Richter magnitude of the most severe earthquake on earth since seismographs were installed?

8.6 in Columbia, South America (1906) and in Assam, India, in 1950. The energy release during such an earthquake is about twice as much as the energy given off by all the coal and oil produced annually in the world.

14.33 What was the most deadly recent earthquake?

In July 1976, an earthquake in northeastern China killed 655,237 people and injured 779,000. Twenty square miles of the city of T'angshan were leveled. Ironically, the earthquake that they successfully predicted occurred in the same year; this serves to illustrate that people cannot yet successfully predict when earthquakes will occur. An earthquake in 1556 in central China killed 830,000 people.

14.34 Are there earthquakes in Japan?

Yes! Tokyo has been devastated about once per century for the last 2000 years.

14.35 What is the best predictor of earthquakes?

Foreshocks, but unfortunately they are not present before every quake.

14.36 During earthquakes, do faults literally "swallow up" people, animals, and dwellings?

Not really. However, the tremors commonly cause landslides. When a landslide starts, the upper end may open as a great gash, and then close as the slump proceeds, thus very rarely "swallowing." (This evidently happened to a cow in the San Francisco earthquake of 1906.) The faults which caused the quake do not open up.

14.37 Can land actually be raised or lowered as the result of a single earthquake?

Yes. About 88,000 km^2 (34,000 mi^2) of land was raised or lowered as a result of the Good Friday 1964 earthquake in Alaska. Some land went down as much as 1.6 m (5.4 ft), and other land went up as much as 12.9 m (42.3 ft).

14.38 What factors control the speed of earthquake waves, and how is their velocity calculated?

The equations for the velocities of the P and S waves are as follows:

$$V_P = \left(\frac{B + 1.33G}{\mathrm{d}}\right)^{(0.5)}$$

$$V_S = \left(\frac{G}{\mathrm{d}}\right)^{(0.5)}$$

where d = rock density, G = measure of the rigidity of the rock, and B = measure of the incompressibility of the rock.

14.39 Based on the equations of the previous problem, which wave is the faster, the P or the S?

The P wave. Note that the numerator ($B + 1.33G$) in the formula for the velocity of the P wave is larger than the numerator G in the formula for the velocity of the S wave. Since the denominators are the same d, the P wave has the higher velocity.

14.40 What is the difference between the *intensity* of an earthquake and the *magnitude* of an earthquake?

The intensity (measured on the Mercalli scale) is a measure of the local disturbance and/or destruction caused by an earthquake and is greatest at the epicenter. The magnitude (measured on the Richter scale) is a measure of the amplitude of an earthquake, based upon measurements from seismograms; it can be used to determine the amount of energy released during an earthquake.

14.41 San Francisco and Los Angeles are on opposite sides of the San Andreas fault (Fig. 14.1). Are they moving toward each other or away from each other? How soon will they be side by side, assuming a movement of 2 cm per year?

They are moving toward each other. The western block which contains Los Angeles is moving northward relative to the eastern block which contains most of the San Francisco Bay area. The two cities are now about 600 km (360 mi) apart.

$$600 \text{ km} = 6 \times 10^2 \text{ km} = 6 \times 10^5 \text{ m} = 6 \times 10^7 \text{ cm}$$

Then, 6×10^7 cm divided by the annual rate of movement 2 cm gives 3×10^7 years, or 30,000,000 years. (Perhaps both cities can go into deficit spending now, in anticipation of the great savings that will result when they can have only one city government?)

14.42 From Fig. 14.6, what can be said about the nature of the crust beneath the area of the New Madrid earthquake (central United States), the two areas on the East Coast (Boston and Charleston), and much of the West, especially California?

Faults are present beneath all those areas, and there *will be* earthquakes in the future!

Chapter 15

Earth's Interior

INTRODUCTION

Drill holes have penetrated only about 9000 m (30,000 ft) into the earth's crust, and the deepest mines are not that deep. Thus, humans have barely scratched the surface. Our knowledge of the earth's interior is based on indirect observations from earthquake (seismic) waves and, to a lesser extent, from meteorites.

WAVE REFLECTION AND REFRACTION

Seismic waves, the P and S waves of the previous chapter, are either reflected off rock layers of different density and seismic properties or are refracted (bent) by those layers as they pass through. The boundaries (surfaces) between the layers of different velocities act as reflecting surfaces and are called *discontinuities*. Major discontinuities separate the interior of the earth into three zones known as the crust, mantle, and core.

At distances of 11,000 to 16,000 km from an earthquake (on the almost spherical earth, 103 to 143° of arc from the epicenter of the earthquake), P and S waves do not exit at the earth's surface. That is, they are "blacked out," forming a "shadow zone." The reason the S waves do not exit is that they cannot penetrate into the core, whereas the P waves do not exit between 103 to 143° because they are refracted (bent) away from that zone.

EARTH'S CRUST

The crust is separated from the underlying higher-velocity mantle by the *Mohorovicic discontinuity,* commonly known as the *Moho*. The crust consists mainly of two types—continental and oceanic. The continental crust consists of two layers; seismic waves passing through the upper layer indicate that it is granitic, whereas the lower layer is basaltic, but the boundary between these two layers is usually quite indistinct. The crust under many mountain ranges is considerably thicker than under adjacent areas, showing that the mountains have "roots." The crust is also thicker under plateaus than it is under plains which have lower elevations.

Oceanic crust in the deep ocean basins generally is made of a thin layer (several hundred meters) of oceanic deep sea sediment, a layer of denser sedimentary rocks and lava flows about 800 m thick, and a basaltic layer between 4 and 6 km thick. The oceanic crust thus totals 5 to 8 km in thickness. However, the sediment thickness is thousands of meters in the continental rises at the bases of the slopes (see Problems 1.17 and 1.18), adjacent to the continents.

EARTH'S MANTLE

The mantle consists of rock through which sound waves move at a higher velocity than they do in crustal rocks. The upper mantle, along with the earth's crust, constitutes a strong and brittle layer called the *lithosphere*. Beneath this is a 100-km-thick, low-velocity "plastic" zone called the *asthenosphere*. Its existence is probably related to water in the minerals or to partial melting of the rock material, or to both.

Stony meteorites composed largely of olivine and pyroxene arrive on earth from elsewhere in the solar system (mostly from the asteroid belt between Mars and Jupiter) and are thought to be representative of the composition of mantle rocks.

EARTH'S CORE

The earth's core consists of two zones, a liquid outer layer and a solid inner layer. The evidence for the liquid layer is the failure of S waves to penetrate this layer and the slowing of P waves. The P waves again speed up nearer the center of the earth, and this is the evidence that the inner core is solid.

Seismic wave velocities, and models generated to explain the density of the earth, indicate that the core is mainly composed of iron and nickel. Iron-nickel meteorites are thought to be representative of the composition of the earth's core.

EARTH'S MAGNETISM

The earth acts as a giant magnet. This magnetism cannot be the result of a giant bar magnet located in the earth's interior, because the interior is so hot that a magnet made of iron would lose its magnetic properties; this happens at the Curie temperature for iron of 578°C. The strength of the magnetic field varies slightly with time, as does the direction of the field. In fact, the field has reversed directions numerous times, with the North Pole becoming the South Pole and vice versa. All these characteristics suggest that the earth's magnetism is generated in the liquid outer core, with fluid motion acting as a generator.

Solved Problems

15.1 What happens to seismic waves as they move through the earth's interior?

They are both refracted (bent) and reflected as they meet zones of different material within the earth. For simplicity, Fig. 15.1 shows the refracted waves but not the reflected waves.

15.2 What are discontinuities?

Discontinuities are the boundaries between zones of different material within the earth, as in Fig. 15.1.

15.3 How was the earth model shown in Fig. 15.1 developed?

It is based on discontinuities between the zones of different seismic velocities.

15.4 What are the differences in the types of materials through which the P and S waves can move?

P waves can move through a solid, liquid, or gas, but S waves can travel only through a solid.

15.5 What is the evidence that the outer core is liquid?

S waves cannot penetrate the liquid outer core, and the velocity of P waves is decreased considerably (see Fig. 15.1).

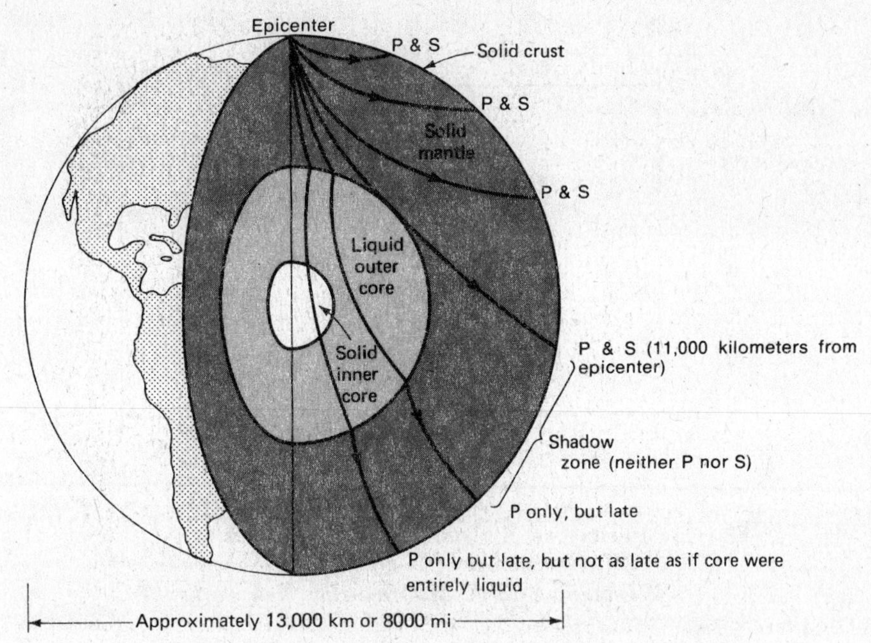

Fig. 15.1 Model of the earth's interior, largely based on the behavior of earthquake waves at depth. Also shown are a hypothetical earthquake and the movement of waves. (*From Ojakangas and Darby, 1976.*)

15.6 What causes the shadow zone of Fig. 15.1?

The *shadow zone* is a zone on the earth's surface, 11,000 to 16,000 km (or 103 to 143°) from an earthquake epicenter, in which P and S waves do not arrive from the earth's interior. This is because the P waves are refracted, or bent, into the core and because the S waves cannot penetrate the liquid outer core.

15.7 What is the evidence that the inner core is solid?

There is a discontinuity at a depth of 5120 km (3179 mi), at which the velocity of the P waves increases.

15.8 How thick are the crust, mantle, and core? (See Figs. 1.5, 15.1 and 15.2.)

The crust is generally between 5 and 40 km thick, the mantle is about 2900 km thick, and the core has a radius of 3450 km.

15.9 What is the Moho?

The *Moho* is the Mohorovicic discontinuity between the crust and the mantle, named after the Yugoslavian scientist who discovered it. It is also called the M discontinuity. It is located at a depth of 5 to 40 km.

15.10 How are the seismic velocities of Fig. 15.2 related to rock density?

The denser the rock, the higher the velocity. However, other factors affect the velocity even more. (See the next problem.)

15.11 What factors beside rock density affect the seismic velocity?

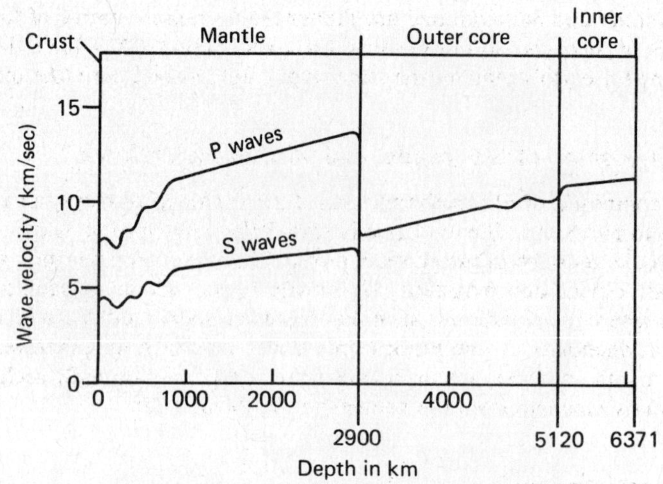

Fig. 15.2 Velocities of P and S waves within the earth. (*After Allison et al., Geology, The Science of a Changing Earth, McGraw-Hill, New York, 1974.*)

The rigidity of the rock and the incompressibility of the rock. (See Problem 14.38.)

15.12 Based on the relative seismic velocities shown in Fig. 15.2, what can be said about the density of the mantle with increasing depth?

It becomes denser. The weight of the column of overlying rock increases with depth, making the rock material more compressed and denser. However, Problem 15.11 shows that the answer is, in reality, much more complicated.

15.13 What can be said of the density of the outer core relative to the lower mantle, and why is this so?

From Fig. 15.2, it can be seen that the seismic velocity decreases considerably at the mantle-core boundary, suggesting a density contrast. However, this is primarily because the outer core is a liquid rather than a solid. (The density of the outer core is actually much greater than the density of the lower mantle.)

15.14 There are two zones in the outer part of the earth that show up on Fig. 15.2 as small wiggles in the velocity curves because they are quite thin. These two zones together include the crust (5 to 40 km thick), a high-velocity layer of the uppermost mantle that is as thick as 50 km, and beneath the high-velocity layer a 100-km-thick, low-velocity layer. What are the two zones that include these three "layers" called?

The crust plus the high-velocity layer is the *lithosphere*. The low-velocity layer is the *asthenosphere*.

15.15 Why is the asthenosphere a low-velocity zone?

This "plastic" layer is explained as being caused by partial melting of rock material or by a higher water content. Either would cause a velocity decrease in both P and S waves, as shown in Fig. 15.2. Above the asthenosphere, the temperature is too low for rock material to be plastic, and below it the pressure is too high for the rock to behave plastically.

15.16 How do meteorites give us insight into the nature of earth's interior?

Meteorites, most of which are thought to come from the asteroid belt between Mars and Jupiter, are of two main types (stony and metallic) and are thought to be representative of the materials of our solar system. Materials of this type do not occur in the earth's crust. They have the proper densities to fit model calculations for earth's mantle (the stony type) and its solid core (the metallic iron-nickel type).

15.17 What is the composition of the mantle, and what is the evidence?

It is very likely composed of ultramafic rocks that are made up primarily of ferromagnesian minerals such as olivine and pyroxene. The evidence is varied. Because seismic waves travel at high speeds in the mantle, the rocks must be dense types composed of dense minerals such as those already named. Meteorites contain olivine and pyroxene. Ultramafic rocks such as eclogite are found on the earth's surface in places where the situation is such that pieces of mantle rocks can be carried upward by earth movements. Also, diamonds can be formed only under very high temperatures and pressures such as are found in the mantle, and they occur in kimberlite, a type of ultramafic rock found in pipelike intrusions that apparently emanate from the mantle (see Problem 7.38).

15.18 Is the mantle homogeneous?

No. The velocity differences shown in Fig. 15.2 indicate that the upper mantle consists of zones of different velocities and hence of different materials.

15.19 What evidence from Fig. 15.2 would suggest that the inner core might be solid rather than liquid?

The higher velocity of the P wave and the marked discontinuity at the inner-outer core boundary.

15.20 What is the composition of the core, and what is the evidence for that composition?

The core is thought to consist of iron with some nickel, a composition similar to the metallic meteorites. Minor sulfur and silica may also be present. Modeling of the densities of the core also fits an iron-nickel core.

15.21 Why is the earth a magnet?

Because movement within the liquid outer core generates electric current.

15.22 What is the nature of the crust, and what is the evidence?

The crust is granitic on the continents and basaltic in the ocean basins. There is some evidence that the basalt layer continues beneath the granitic continental crust, as in Fig. 15.3.

Supplementary Problems

15.23 What are "sial" and "sima"?

Sial is another word for the granitic continental crust (from SIlica and ALumina), and sima is synonymous with the basaltic oceanic crust (from SIlica and MAfic).

15.24 Of what importance are sediments and sedimentary rocks in the earth's crust?

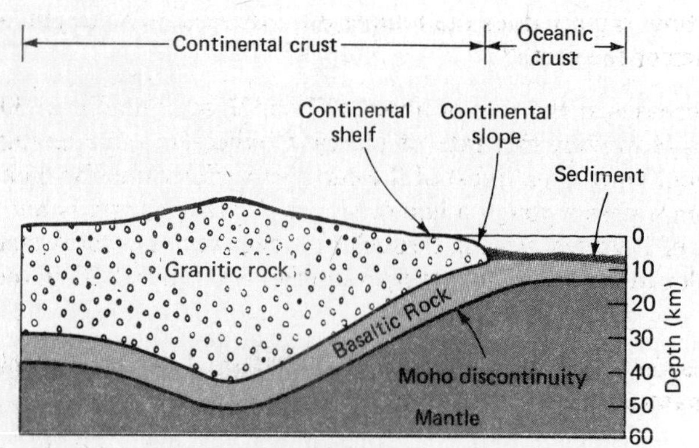

Fig. 15.3 Diagrammatic model of continental and oceanic crust. (*From Ojakangas and Darby, 1976.*)

The oceanic crust is covered by a thin layer of several hundred meters of sediment. About 75 percent of the earth's land surface is covered with sedimentary rock varying in thickness from a few meters to several kilometers. While widespread, the layer forms less than 5 percent of the earth's crust by volume.

15.25 See Fig. 15.3. What is the downward bulge in the crust beneath the mountain range called, and how do scientists know that it is there?

It is called the *root* and consists of granitic rock. The velocities and reflections of seismic waves and lower gravity readings reveal its presence.

15.26 Where is the thickest crust on earth, and how thick is it?

Beneath the Himalayas and the Tibetan Plateau. It is 70 km thick.

15.27 In general, where is earth's crust the thinnest, and how thin is it?

In the deep ocean basins, away from the continental margins, it is 5 to 8 km thick.

15.28 What is the density of the earth, what are the densities of the rocks at the earth's surface, and what is the significance of these numbers?

The earth's density is 5.52, and rocks at the surface have densities of about 2.7 (granite) to 3.0 (basalt). This relationship means that the earth must have a denser interior in order to have an overall average density of 5.52. (See Problem 1.33.)

15.29 The earth behaves as a giant magnet (see Fig. 1.6). Why? A clue is present in Fig. 15.1.

The magnetic field is largely due to electromagnetic currents generated by convection in the fluid outer core. In a bit of circular reasoning, the presence of the earth's magnetic field, which seems to be deep-seated and unrelated to variations in the magnetic properties of rock types in the crust, can be used as support for a fluid outer core.

15.30 What is the evidence that the temperature of the earth increases with depth?

Deep mines and deep oil wells display higher temperatures at depth. Magma comes to the surface as lava and it must have melted at depth. In addition, the outer core appears to be molten based on seismic properties, and this requires a high temperature in the interior.

15.31 In the upper crust, how rapidly does the temperature increase with depth, and what, then, is the temperature at the center of the earth?

The temperature increases at the rate of about 1°C per 33 m of depth, or 30°C per kilometer of depth. (See Problem 1.40.) However, this rate of increase indicates a temperature at the earth's center of 128,000°C, which would mean that most of the earth's interior would be molten. The passage of seismic S waves, which cannot pass through a liquid, shows that the mantle is not liquid, but solid. Also, the increase in velocity of P waves passing through the inner core indicates that it is solid. Other pressure and temperature calculations indicate a temperature of 4000 to 5000°C in the earth's core.

15.32 The foci of a few earthquakes are as deep as 700 km, or 420 mi (see Problem 14.23). What does this indicate about the nature of the earth's interior at that depth?

The fact that most earthquakes are the product of movement along faults indicates that the earth material at that depth is solid and brittle enough to fracture and move. A liquid does not crack, and it would not have the property of elasticity. (See Problem 14.5.)

15.33 Why do earthquakes not occur at depths of more than 700 km?

Probably because the pressures at greater depth are so high that movement along faults is impossible. Faults are most abundant near the earth's surface where the pressure caused by the weight of the overlying rocks is relatively low.

15.34 What is the theoretical reason that the earth is zoned as in Fig. 15.1, with the heaviest material at the center and the lightest at the surface?

The earth may have had a cold origin, forming from the gravitational attraction of smaller bodies that were in orbit around the sun. As they came together, their radioactive elements gave off heat and the angular momentum generated heat as well, causing the earth to melt. In simple terms, as it cooled and crystallized under the influence of gravity, the heavy material moved downward to form the mantle and the denser core, and the light material moved upward to form the crust.

15.35 Why can't the earth's magnetism be the result of a permanently magnetized gigantic bar magnet in its interior?

Because ferromagnetic materials lose their magnetism at temperatures higher than 578°C, and most other materials lose it at less than 800°C. The earth's interior is too hot for permanently magnetized material to exist. The temperature in the core is estimated at 4000 to 5000°C. Besides, the pressure of the weight of the mantle would melt iron at the mantle-core boundary at a temperature of about 3700°C.

15.36 What are the relative volumes of the earth's crust, mantle, and core, and how can this be determined from Fig. 15.1?

This can be determined by calculating the volume of three spheres, one with a radius of about 3500 km (the core), one with a radius of 6360 km (the radius of the core and the mantle combined), and one with a radius of 6400 km (the total earth), and then subtracting. Therefore, using the formula for the volume of a sphere:

$$V = \frac{4}{3}r^3$$

the three volumes can be calculated as follows:

$$= \frac{4}{3}(3500\text{ km})^3 = 17.959 \times 10^{10}$$

$$= \frac{4}{3}(6360\text{ km})^3 = 10.771 \times 10^{11}$$

$$= \frac{4}{3}(6400 \text{ km})^3 = 10.981 \times 10^{11}$$

The volume of the core is thus 17.959×10^{10} km^3, the volume of the mantle is 8.975×10^{10} km^3 (10.771×10^{11} km^3 − 17.959×10^{10} km^3), and the volume of the crust is 0.21×10^{11} km^3 (10.981×10^{11} km^3 − 10.771×10^{11} km^3). In percentages, these constitute 1.9, 81.7, and 16.4 percent for the volumes of the crust, mantle, and core, respectively. (Since the crust of the ocean basins has a thickness of 4-6 km, this is a maximum percentage for the volume of the crust.)

15.37 Is the strength of the earth's magnetic field constant?

No. It has apparently lost about 5 to 6 percent of its strength in the past 100 years.

15.38 What is the significance of this decrease in the strength of the earth's magnetic field?

In 2000 years or so, it could have a strength or intensity close to zero. Navigation by compass would be a problem. And since the earth's magnetic field prevents cosmic rays from bombarding the earth, its loss could theoretically mean a massive bombardment which could alter genes and affect evolution of life forms on earth.

15.39 Why is the earth giving off heat?

While part of the heat may be residual heat from the days of earth's formation (rock is a good insulator and thereby retards the release of heat from the earth's interior) or generated in the core by crystallization or crystal settling, most is probably generated by radioactivity in the earth's crust.

15.40 Why do earth scientists think that the radioactive minerals of the earth are concentrated in the crust rather than in the mantle or the core?

Because granitic rocks which make up the bulk of the earth's granitic crust are 5 to 20 times as radioactive as the mafic minerals which make up the inner zones of the earth. The radioactivity comes largely from radioactive potassium, uranium, and thorium.

15.41 What is eclogite?

Eclogite is a rock that is rarely present on the earth's surface. It is thought to come from the mantle and consists of garnet and an unusual pyroxene. Eclogite does occur in tectonic zones on the earth's surface after being moved upward from the mantle as isolated blocks.

Chapter 16

Deformation and Mountain Building

INTRODUCTION

There are four main types of geologic structures—folds, faults, joints, and unconformities. The first three are largely direct results of deformation of the earth's crust. *Unconformities* are buried surfaces of erosion, generally the product of uplift, erosion, and then deposition of younger sediments upon the surfaces. Because uplift, commonly followed by subsidence, is involved, unconformities are also related to deformation of the earth's crust.

FOLDS

There are two main types of folds. *Anticlines* are fold structures in which the rock layers are folded upward, whereas *synclines* are downfolds. Commonly, anticlines and synclines occur side by side. Folds are the result of compression. Rocks fold when the pressures are applied slowly, so that the rocks may react in a ductile ("plastic") rather than a brittle manner. They also fold at great depths where both the temperatures and pressures are high.

FAULTS

Faults are fractures in the earth's crust along which movement has occurred. There are two main types of movement along faults—in a vertical sense (dip-slip movement) and in a horizontal sense (strike-slip movement). Most faults, called *oblique-slip faults,* have a net movement that is the resolution of a movement which has both horizontal and vertical components.

If a fault is inclined, rather than horizontal, the top block is called the *hanging wall,* and the bottom block is called the *footwall.* These are both old mining terms derived from inclined mine openings, in which the rock above was called the hanging wall and the floor upon which the miners stood was called the footwall. Based on the movement of the hanging wall *relative to* the footwall, the type of dip-slip fault and the stresses which caused the movement can be determined. If the hanging wall moved down relative to the footwall, the fault is a *normal fault,* caused by extensional (tensional) forces. Conversely, if the hanging wall moved up relative to the footwall, the fault is a *reverse fault,* caused by compressional forces. A fault along which the movement was horizontal is called a *strike-slip fault* or a *tear fault*; this type of fault is caused by a force-couple, two forces acting in opposite horizontal directions.

JOINTS

Joints are fractures (cracks) in rocks, along which movement has *not* occurred. They are usually present in parallel sets of great number, and are the result of either extensional or compressional forces. The study of joint patterns can yield information about the stress field that caused them. Continued or additional stresses may result in movement along some joints, making them faults. Another type of fracture has a different origin: columnar joints in igneous lava flows, dikes, or sills are the result of cooling and contraction as the magma crystallized, for solid igneous rock occupies less space than does liquid magma. Another type of joint, subhorizontal, is due to unloading of overlying rock by erosion, and the rock that was deeply buried expands in a direction perpendicular to the earth's surface (see Problem 4.2).

UNCONFORMITIES

Unconformities are buried surfaces of erosion or, theoretically, surfaces of nondeposition as well. An unconformity is thus a record of an erosional event in the history of an area. Sir James Hutton, the "father of geology," identified unconformities and realized that they represented "a succession of worlds." There are three major types of unconformities. An *angular unconformity* is a surface separating overlying rock from underlying, more steeply dipping rock. A *disconformity* is an erosional surface separating beds above and below which are parallel to each other. A *nonconformity* is an erosional surface separating overlying sedimentary rock from underlying igneous or metamorphic rock.

MOUNTAIN BUILDING

There are four main types of mountains. *Volcanic mountains* are large composite volcanic cones [see Fig. 3.5(*a*)] or thick accumulations of volcanic rock. *Fault-block mountains* are the products of movements of the earth's crust along faults, with considerable uplift of the rock on one side of a fault. The Tetons of Wyoming are a classic example, and the Basin and Range Province of southwestern United States contains hundreds of such mountain ranges. *Domal mountains* are broad, high uplifts such as the Black Hills of South Dakota. *Fold-belt mountains* (orogenic mountains) are large mountain ranges which are the products of folding, faulting, intrusion, and metamorphism of thick sequences of sedimentary rocks. The major mountain ranges of the world are of this latter type, and the processes will be further explained in Chapter 17 on plate tectonics.

Solved Problems

16.1 (*a*) When geologists see steeply dipping layers of sedimentary rock, as in Fig. 16.1, they assume that they were tilted sometime *after* they were deposited. What is the basis for this assumption?

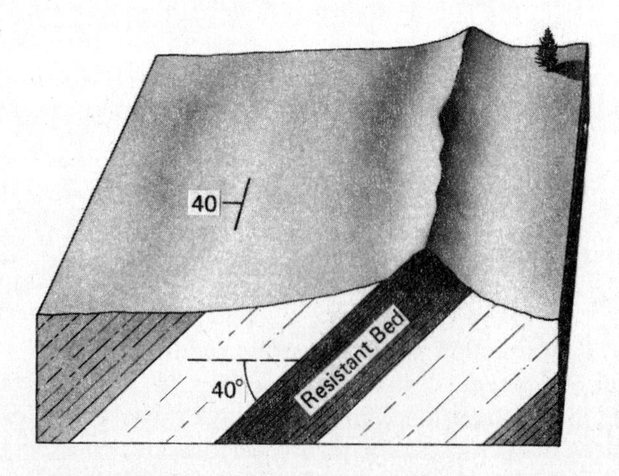

.1 Block diagram of tilted beds with strike-dip symbol to show the orientation of the beds. Such symbols are used on geologic maps. The long line portion of the symbol gives the strike or trend of the bed, and the shorter line, which is always perpendicular to the longer line, gives the direction toward which the bed dips. The amount of dip in degrees is given by the number next to the short line.

Observation of modern sedimentary processes shows that sediment is deposited in horizontal or near-horizontal layers or beds. This is the *principle of original horizontality*. The *principle of uniformitari-*

anism, or "the present is the key to the past," is also being applied here. (Gentle dips, however, may be original, as in the case of alluvial fan deposition at the base of a mountain range.) Tilted beds are generally the result of the erosion of folded rocks, with the more resistant layers standing as ridges.

(*b*) What is the strike and dip of the beds on the diagram?

They strike north-south and dip 40° to the west.

16.2 Name and briefly describe the two main types of folds shown in Fig. 16.2.

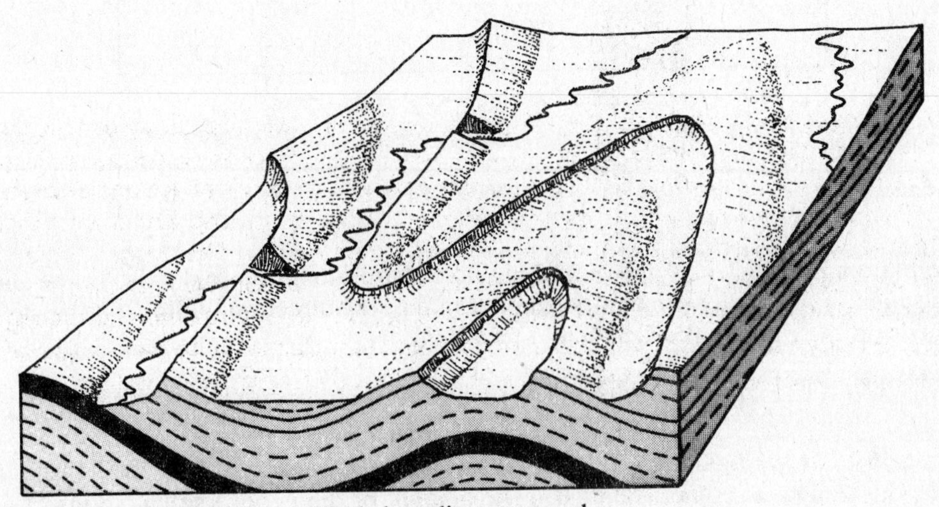

Fig. 16.2 Block diagram of folds in sedimentary rocks.

Anticlines are structures in which the beds are folded upward, and *synclines* are structures in which the beds are folded downward.

16.3 What kind of fold is shown in Fig. 16.3?

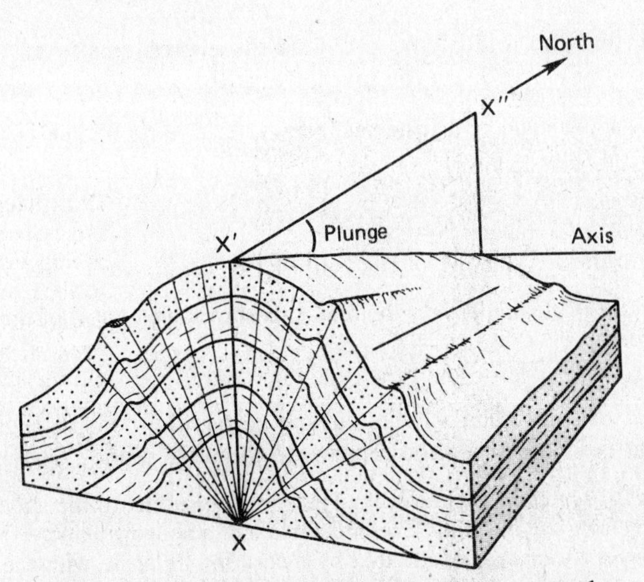

Fig. 16.3 Block diagram of a fold in sedimentary rocks. Note smaller folds on the flanks.

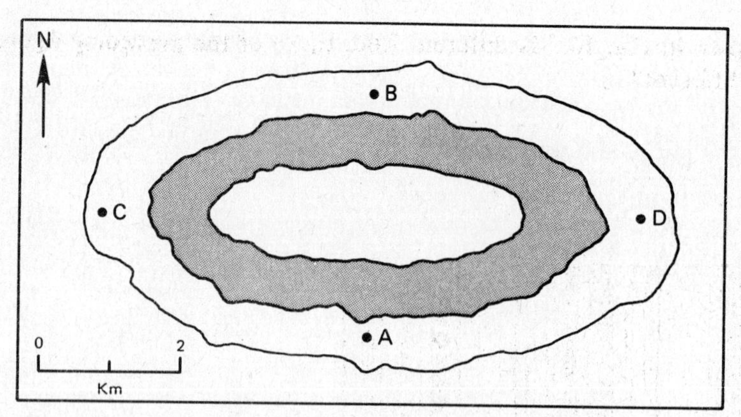

Fig. 16.4 Geologic map showing a fold. Note the scale.

An anticline.

16.4 The fold of Fig. 16.3 was drawn looking northward along the fold axis, in the direction of elongation of the fold. Estimate the amount of dip, the direction of dip, and the strike (trend) of beds on the right flank (right side) of the fold.

The beds dip about 45° to the east and strike about north-south.

16.5 Figure 16.4 is a geologic map of a fold of the same type as that of Fig. 16.3. Place the proper strike-dip symbol at points A and B. (The amount of dip is not known, so insert a question mark instead of a number.)

At point A the symbol should be oriented as follows: $\overset{?}{\perp}$

At point B, the symbol should be as follows: $\underset{?}{\top}$

16.6 You now know that the fold of the previous three problems is an anticline. Place proper strike-dip symbols at C and D on the map of Fig. 16.4.

At point C, the symbol should be oriented as follows: ? ⊣

At point D, it should be as follows: ⊢ ?

16.7 Define and locate the axial plane of the fold of Fig. 16.3.

The *axial plane* is the plane that bisects the fold. It is shown on the end of the fold as line x-x′. In 3-D, it is the plane given by x-x′-x″.

16.8 Note the vertical to subvertical cracks cutting across the beds in Fig. 16.3. What are these cracks, and what is their orientation relative to the axial plane? Of what value are they?

These represent cleavage planes in the rock. They are subparallel to the axial plane but spread out or flare out a bit away from the fold axis. The intersection of a bedding plane with a cleavage plane results in a line, which gives the plunge of the fold by measuring its angle with the horizontal and gives its orientation with respect to north. Therefore, if even a small part of the fold is exposed, much can be learned about the fold. (See Fig. 6.2 to refresh your memory on the origin of cleavage.)

16.9 The fold shown in Fig. 16.5 is different from those of the preceding figures. In what major aspect is it different?

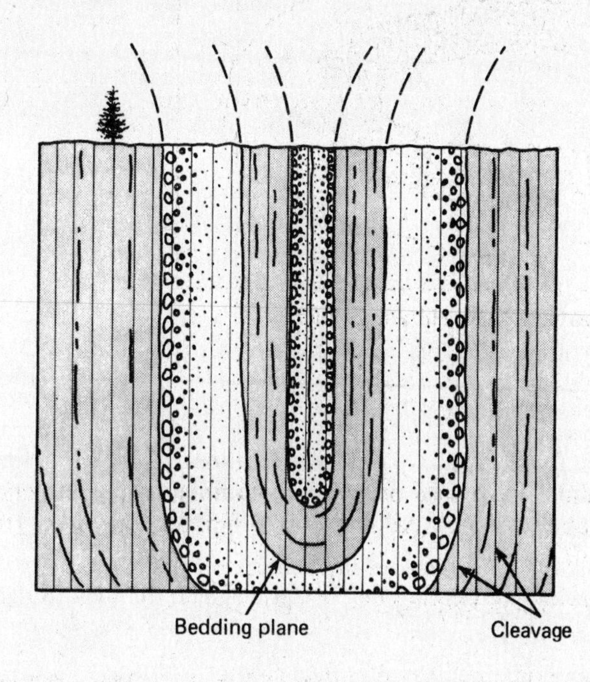

Fig. 16.5 Cross-section showing vertical graded beds and cleavage.

This is an *isoclinal fold* (meaning "the same inclination"). Note that the two flanks of the fold are parallel to each other, whereas in previous figures, the folds are broad and open with the two flanks dipping in opposite directions. Note also the cleavage lines and that the top of the fold has been eroded off.

16.10 From Fig. 16.5, formulate a rule on the relationships of cleavage and bedding in isoclinal folds.

Cleavage and bedding are parallel to each other on the flanks of folds, but the cleavage crosses the bedding at high angles on the crests and troughs of such folds.

16.11 (*a*) How can geologists tell whether the fold of Fig. 16.5 is an anticline or a syncline? (Realize that the dashed part of the sketch represents rock that has been eroded away.)

By observing the graded beds [see Fig. 5.5(*a*) and Problem 5.12], the original tops of the beds can be determined, for the coarser grains are at the bases of graded beds. Or, if the relative ages of the beds were known, e.g., on the basis of fossils in the beds, then the original "top" or "up" direction could be determined.

(*b*) Which type of fold is depicted here?

A syncline. You might complete the sketch by drawing in the closure of the fold at the bottom of the figure by connecting the beds with looping, curved lines.

16.12 (*a*) How can geologists tell whether the faulted fold of Fig. 16.6 is an anticline or a syncline?

By observing the cross-beds. Note that one bed contains curved cross-beds which, when deposited, are concave upward.

(*b*) Which type of fold is shown in the figure?

An anticline.

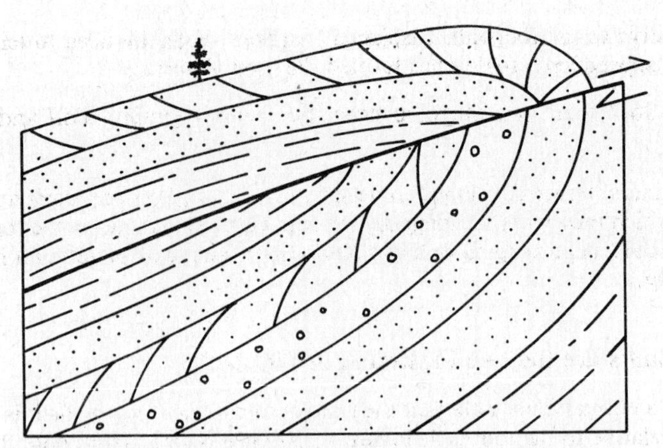

Fig. 16.6 Cross-section of a faulted fold containing a sandstone bed that is cross-bedded.

16.13 The anticline of Fig. 16.6 has an unusual orientation, compared to those of previous figures. What is the name of this type of fold, and why is its orientation significant?

This is a *recumbent fold,* in which the axial plane makes a low angle with the horizontal, rather than a high angle as in previous figures. Note that the lower flank of the fold is overturned (i.e., it is upside down). This is an indication of strong horizontal forces directed, in this diagram, from left to right.

16.14 Are folds the result of crustal shortening or crustal stretching?

Shortening, the result of compressional forces, rather than stretching, which is the result of extensional forces.

16.15 Are both synclines and anticlines the result of compression?

Yes. Recall that they commonly form together, as in Fig. 16.2.

16.16 (*a*) How can normal and reverse faults, as in Fig. 16.7, the two types of faults with vertical movement along their dipping planar surfaces (i.e., dip-slip faults), be distinguished from each other?

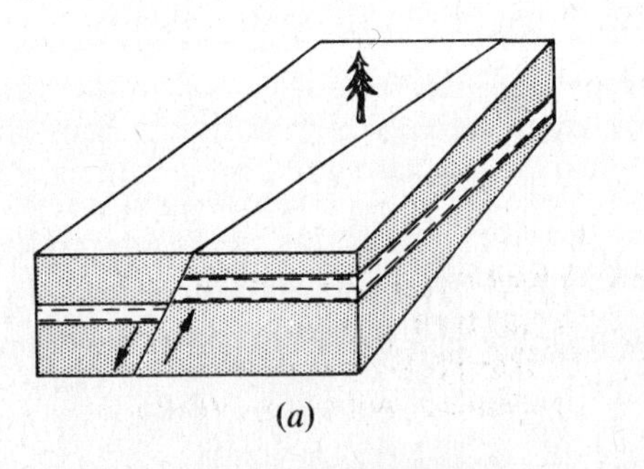

(*a*)

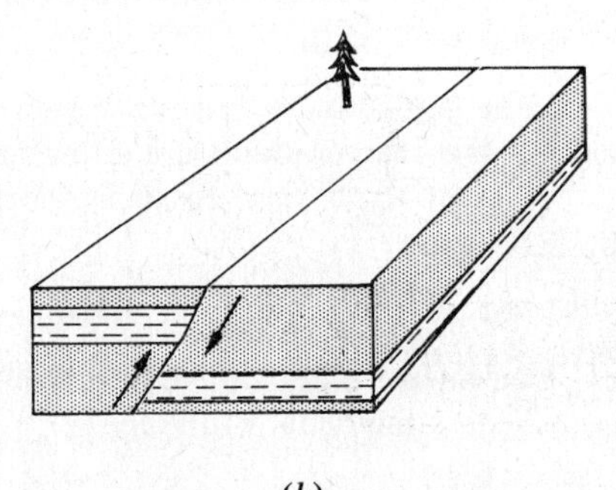

(*b*)

Fig. 16.7 Block diagrams showing the two main types of dip-slip faults. On each diagram, the dashed layer represents a marker bed that is used to determine the sense of relative movement of the two sides of the fault. (*From Ojakangas and Darby, 1976.*)

In a *normal fault,* the side "above" or on top of the dipping fault plane (the hanging wall) has moved down relative to the side below the dipping fault plane (the footwall). In a *reverse fault,* the hanging wall has gone up relative to the footwall. Displaced "marker" beds are used to tell the direction of movement. Note that subsequent erosion has leveled the ground surface.

(*b*) From Fig. 16.7, can you surmise how the terms *hanging wall* and *footwall* came to be used?

These are old miners' terms developed to designate, when mining ore along a fault, the side of the inclined opening which they were standing on (the footwall) versus that on the topside which was "hanging" over their heads. (The hanging wall was obviously a source of anxiety, for if rock slabs were to fall, it could be deadly.)

16.17 What kind of faults are the two illustrated in Fig. 16.7?

Figure 16.7(*a*) is a normal fault; note that the rock on the left side of the fault is the hanging wall, and it has gone down relative to the footwall. Figure 16.7(*b*) is a reverse fault; note that the hanging wall has gone up relative to the footwall.

16.18 (*a*) Is the fault of Fig. 16.7(*a*) the result of crustal shortening (compression) or crustal lengthening (extension)?

Normal faults are the result of extensional forces. The extension has caused the hanging wall to "fall" relative to the footwall, under the influence of gravity. (Note the directions of movement given by the arrows.)

(*b*) Is the fault in Fig. 16.7(*b*) the result of crustal shortening (compression) or crustal lengthening (extension)?

Shortening, the result of compression which caused the hanging wall to be pushed up over the footwall.

16.19 Which type of fault, normal or reverse, should be more common in an area of folded rock?

Reverse faults, as both folds and reverse faults are the products of compressional forces. However, the forces acting upon a given area can change, so an area once under compression could later be subjected to extensional forces. (And, once compressional forces are removed, the relaxation in effect means that extensional forces have taken over.)

16.20 Name and describe the origin of the third major type of fault, illustrated in Fig. 16.8.

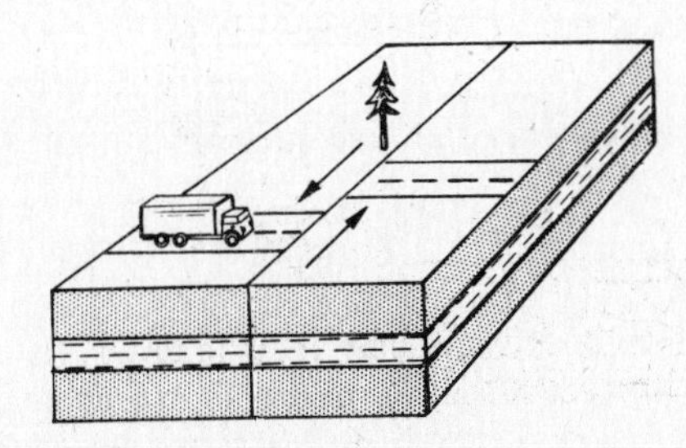

Fig. 16.8 Block diagram showing a third major type of fault, a strike-slip fault. (*From Ojakangas and Darby, 1976.*)

This is a *strike-slip fault*, also known as a "tear fault." It is the result of horizontal movement parallel to the trend (or the strike) of the fault, rather than the result of vertical or dip-slip movement along the plane of the fault.

16.21 Is the fault of Fig. 16.8 the result of compression or extension?

Horizontal movement along a fault is the result of a special case of compressive forces called a *force couple*. The forces are acting toward each other but are in adjacent structural blocks; this results in one block sliding past the other.

16.22 Which type of fault is the San Andreas fault of Fig. 14.1?

The San Andreas is a strike-slip (tear) fault with tens and perhaps hundreds of kilometers of horizontal movement along it. The San Andreas fault is actually a fault *zone* rather than a single fault, and movement may occur along any of the many fault planes in the zone. Therefore, the unpredictability makes the fault even more dangerous.

16.23 On Fig. 14.1, how should arrows be drawn to show the relative horizontal movement of the two sides of the San Andreas fault (i.e., one arrow on each side of the fault)?

The two arrows should be oriented as follows: The arrow on the western block should be pointing north, and the arrow on the eastern block should be pointing south. The reasons for this movement will be explained in the next chapter on plate tectonics.

16.24 In Fig. 16.9(*d*), nearly horizontal beds rest upon steeply dipping beds. What type of a geologic structure is this, and what was the step-by-step history which resulted in this structure?

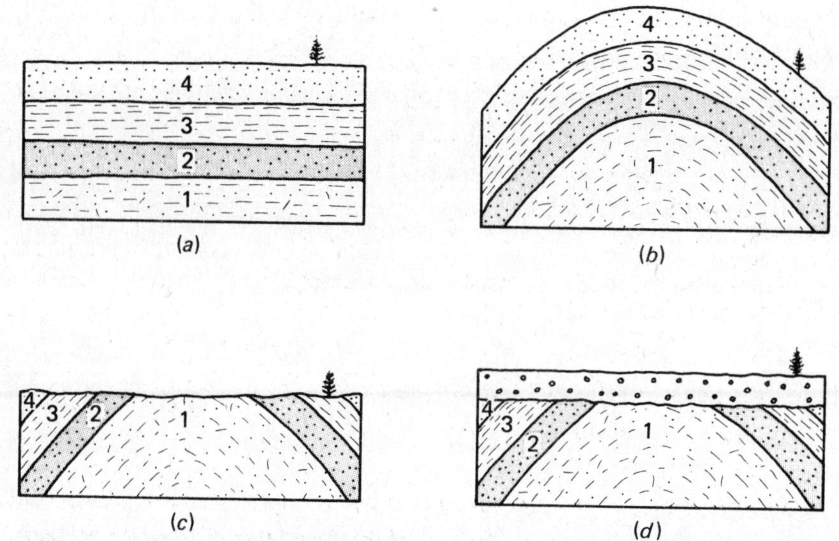

Fig. 16.9 Four-stage diagram showing the development of an unconformity as in (*d*).

The structure is an *angular unconformity*, or buried surface of erosion. The steps in order, from oldest to youngest, are shown in the four parts of the diagram, in order from (*a*) to (*d*). In (*a*), deposition occurred. In (*b*), deformation occurred. In (*c*), uplift and erosion occurred. In (*d*), more sediments were deposited upon the erosional surface of (*c*). A slightly more detailed sequence of events is: deposition of sediment, lithification into sedimentary rock, uplift, tilting, erosion, probably subsidence, deposition of the horizontal sediment, lithification of the sediment into sedimentary rock, and finally, uplift and erosion.

16.25 If the steeply dipping beds of Fig. 16.9 are 150 million years old and the horizontal beds are 55 million years old, when did the uplift and erosion occur?

Sometime between 150 million years ago and 55 million years ago. Additional information is necessary to be more precise.

16.26 What is the origin of joints, sets of widely spaced cracks in rocks with no movement along them?

There are several types of joints. Most are probably due to compressional or extensional forces exerted in the area after the rocks were formed. Others are parallel to an erosional surface and are the result of the upward expansion of rock after removal of the overlying rock. In igneous rocks, some joints are the result of contraction due to crystallization of magma, because the solid rock occupies less space than the liquid magma; columnar joints in lava flows, dikes, and sills have such an origin (see Fig. 3.17).

16.27 There are four main types of mountains. Name them, give a brief origin for each, and an example of each.

Volcanic: Edifices of volcanic rocks built up by continued extrusion (commonly explosive) from a vent, from vents, or from fissures. Examples: Mt. St. Helens or Mt. Ranier in Washington state, Mt. Fujiyama in Japan, Mt. Etna in Sicily, Mt. Vesuvius in Italy, and Mt. Mayon in the Philippines. (See Fig. 3.5.)

Fault-block: Considerable vertical movement along large faults, usually normal faults. Example: Grand Teton Mountains. (See Fig. 16.10.)

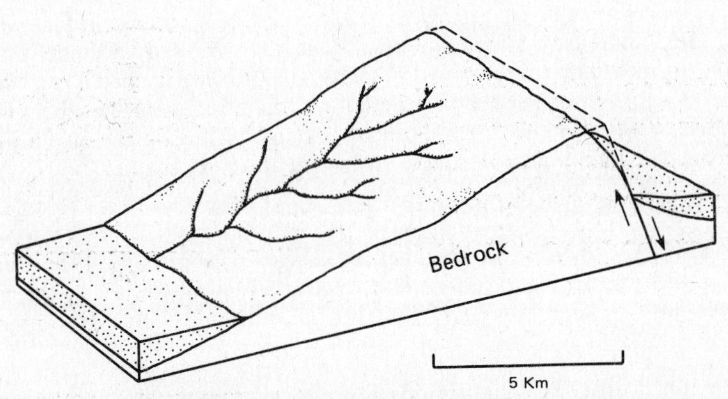

Fig. 16.10 Block diagram of a fault-block mountain range. Note the scale.

Domal: Upward pressures in an area causing considerable uplift. Example: Black Hills, South Dakota.

Fold-belt: Extreme pressures causing major large-scale folding, commonly accompanied by reverse faulting, metamorphism, intrusion, and/or volcanism. This type occurs in regions of very thick sedimentary rock sequences. Examples: The Alps and Appalachians. Most of the world's major mountain ranges are variations of this type.

16.28 *Thrust faults* (*overthrust faults* if they are subhorizontal as in Fig. 16.11) commonly show that thick upper blocks have moved tens of kilometers across the underlying blocks. Note the scale. Forces great enough to push a block up and over another block without totally breaking or crushing the upper block are difficult to explain. How do geologists think this could have happened?

Gravity sliding of an upper block (from a higher elevation) over a lower block is theoretically possible. Even then, it is difficult to explain because of the friction that should be present along the surface between the blocks. Water and wet shale along the fault plane have both been proposed as lubricants. However, the best answer appears to be gravity sliding with the upper block "buoyed up" by water

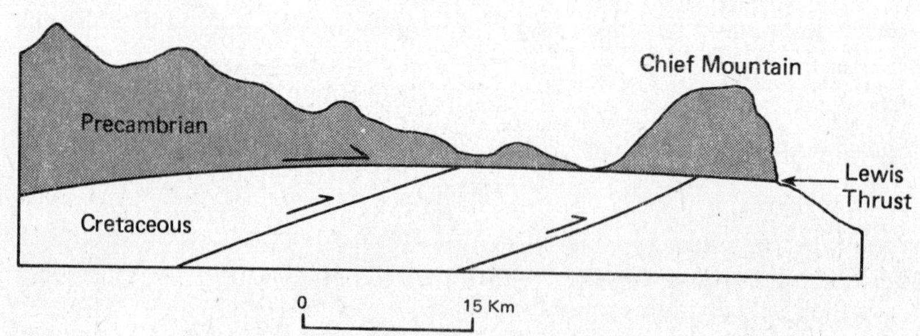

Fig. 16.11 Cross-section of a major zone of thrust-faulting in Montana. Movement along the Lewis Thrust has moved old Precambrian rocks onto younger Cretaceous. Total movement is estimated at more than 50 km. (*After Alt, D., and Hyndman, D. W., Roadside Geology of Montana, Mountain Press Publishing Co., Missoula, 1986.*)

compressed along the fault zone, creating a "pore pressure." With such buoyancy, it is theoretically easy to slide even huge blocks down slight slopes, and even up onto adjacent slight slopes, after rupture has occurred along a weak zone such as a shale bed.

16.29 (*a*) What are geosynclines, and where and how do they form?

According to geosynclinal theory, *geosynclines* are zones of weakness in the earth's crust, tens of kilometers wide and hundreds of kilometers long, that have subsided and received extraordinarily thick sequences of sediment (10,000 m or more). They have commonly formed along the edges of continents. For the last century, it has been debated whether the thick sediment was there because of the subsidence of the crust or whether the crust subsided because of the weight of the thick pile of sediment. Modern plate tectonic theory provides an answer—see the next chapter.

(*b*) Locate two geosynclines on Fig. 16.12.

The Cordilleran and Appalachian geosynclines are labeled on Fig. 16.12. (The Ouachita Geosyncline is an extension of the Appalachian Geosyncline. Note their locations along the margins of the North American continent.

16.30 What relationship do fold-belt mountains have to geosynclines?

The thick piles of sedimentary rock in fold-belt mountain ranges were believed to be deposited in geosynclines. For example, the Appalachian Mountains rose out of the Appalachian geosyncline and the Rocky Mountains rose out of the Cordilleran geosyncline (see Fig. 16.12).

16.31 Which types of forces—compressional or extensional—were dominant during the "building" of fold-belt mountains?

Compressional, as is suggested by the very name "fold-belt." Reverse faults and low-angle thrust faults are common in fold-belt mountains.

16.32 Relatively thick piles of sediment, but not as thick as in geosynclines, have accumulated in basins which formed in the continental interiors, as in Fig. 16.12. How do geosynclines, basins, and synclines differ from each other?

Basins and geosynclines subsided before and/or during sedimentation and received more sediment than adjacent areas. An example of a basin is the Michigan Basin of Fig. 16.12. Synclines were developed by folding after sedimentation and generally after lithification of the sediment. Most synclines are small structures measuring a few to tens of kilometers long, whereas geosynclines and basins are much larger. Synclines (and anticlines) are commonly present within basins and geosynclines.

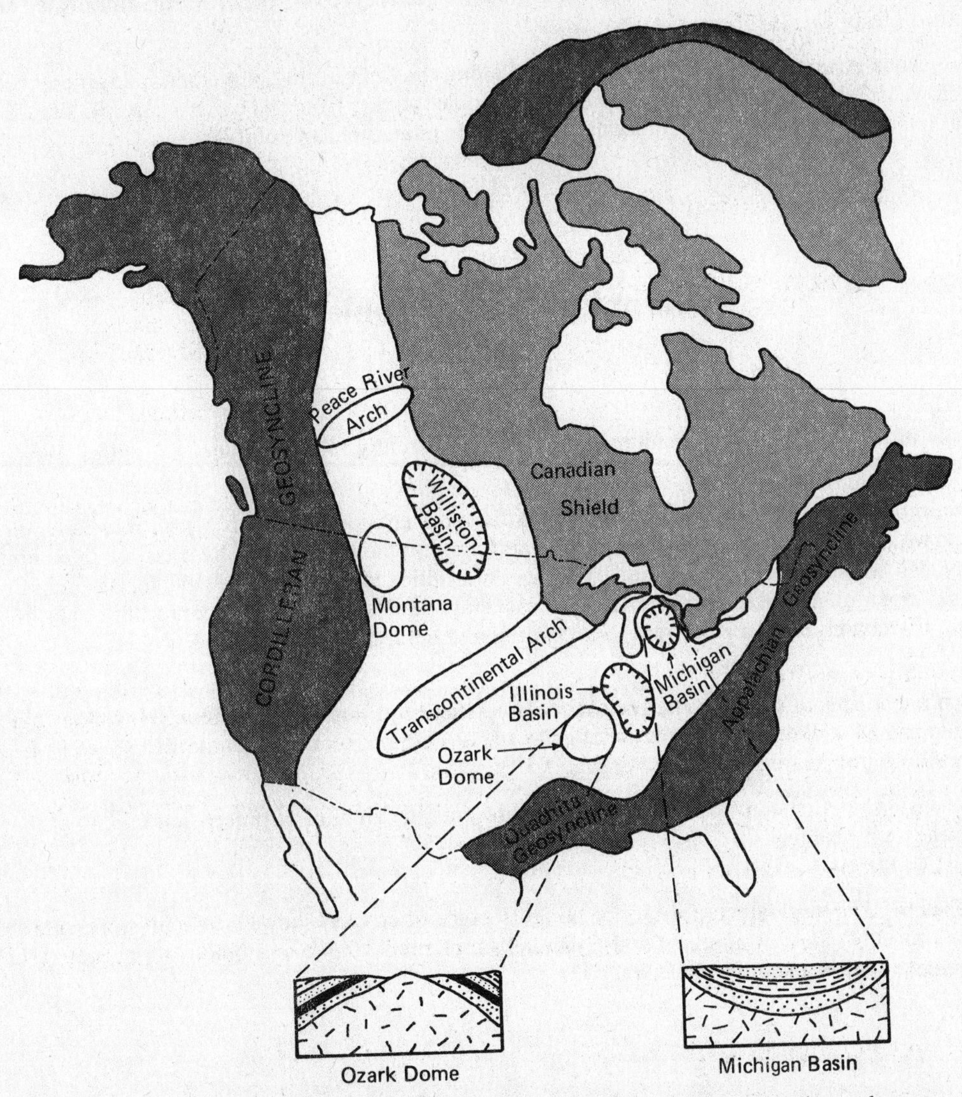

Fig. 16.12 Generalized map of North America, showing major geosynclines, domes, and basins. The Michigan Basin and the Ozark Dome are shown in greater detail. (*After Clark, T. H., and Stearn, C. W., Geological Evolution of North America, Ronald Press, 1968.*)

16.33 Contrast domes and anticlines.

Domes exist in areas that were rising or were even emergent during sedimentation, thus receiving less sediment than adjacent areas. The Ozark Dome is a well-studied example (Fig. 16.12). Anticlines are much smaller structures and can form on the flanks of major domes or in basins. (An anticline that is not markedly elongated is sometimes referred to as a "dome"; this is a totally different connotation than just described and admittedly is confusing.)

16.34 Which type of fold—anticline or syncline—is commonly a petroleum trap, and why?

Anticlines are commonly petroleum traps. Assume the presence of porous and permeable reservoir beds that allow the passage of water, oil, and gas. Natural gas has a lower density than oil, and oil has a lower density than water, and thus both gas and oil rise to the top of water and are trapped at the crests of anticlines (See Figs. 7.3 and 7.8.)

16.35 Why are economic minerals, such as gold in quartz veins, or uranium minerals, sometimes found along faults?

The space generated by the crushing of rock along the fault during movement allows entry of mineralized water, either warm waters from below or cooler waters from the earth's surface. Where the chemical conditions are right, ore minerals may be precipitated out of solution.

Supplementary Problems

16.36 Some of the world's major uranium deposits are apparently related to unconformities. Why might this be so?

Unconformities, because they are buried surfaces of erosion, commonly overlie weathered zones of rock which may be porous and permeable and are commonly overlain by basal conglomerates which may also be porous and permeable. Thus, unconformities may be passageways for uranium-bearing waters. An additional factor may be that the rocks above and below the unconformity may be impermeable, thereby constituting barriers to ascending and descending mineralized waters.

16.37 Each major unconformity in the rock record represents an interval of erosion. How does the angle between the beds above and below an angular unconformity, such as the one in Fig. 16.9, relate to the length (duration) of erosion?

There is no relationship. The angle merely represents the amount of preerosion folding at that locality. See the next problem.

16.38 In Fig. 16.13, a single unconformity or buried surface of erosion shows a high-angle relationship at one locality, a low angle at another, and a parallelism of rocks above and below at another. What are the different types of unconformities called?

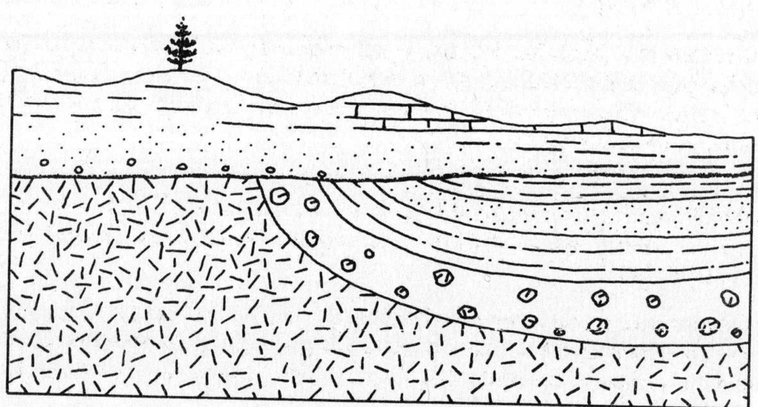

Fig. 16.13 Cross-section of a major unconformity with different relationships of rock units along the unconformity.

Where there is an angular relationship, it is called an *angular unconformity*. Where the beds above and below are parallel, it is a *disconformity*. Where the sedimentary rocks overlie igneous or metamorphic rocks, it is called a *nonconformity*.

16.39 How can one determine that the igneous rock beneath the horizontal sedimentary rocks on the left side of Fig. 16.13 has not intruded, as a magma, up against the overlying sedimentary rock?

There are several lines of evidence. The overlying rock contains rounded fragments of the igneous rock (they have been weathered and transported), there is no contact metamorphism of the sedimentary rock adjacent to the igneous rock, the crystal size of the igneous rock is not smaller near the sedimentary rocks where it would have cooled more quickly, and the contact can be traced laterally into the angular unconformity.

16.40 Assume that the rock units on the map of Fig. 16.4 consist of soft, highly weathered rocks and that it is impossible to measure the attitudes of the beds. However, fossils are present and indicate the relative ages of the rock units. (*a*) If the structure is an anticline, where on the map would the oldest fossils occur? (*b*) The youngest? (*c*) What if the structure were a syncline? (*d*) What underlying principle was utilized in answering these three questions?

(*a*) Near the center of the structure. (*b*) At the outer edges. (*c*) Youngest in the center and oldest at the outer edges. (*d*) The principle of superposition (i.e., the youngest rocks were deposited on top of the older rocks).

16.41 Assume that an area with jointed rocks was later subjected to extensional or compressional forces. Would the joints play any part in localizing the planes of movement?

Yes. As the joints are planes of weakness, movement will most likely occur along some existing joints, unless the deformational forces are so oriented that the forces cause new fractures (joints) to form with different orientations than the preexisting sets.

16.42 Which type of mountain range is the Sierra Nevada of California, considered by many to be the grandest, most majestic mountain range in the conterminous United States?

The Sierra Nevada is a composite-type mountain range. It has a big normal fault along its eastern side, with perhaps as much as 8000 m of vertical movement along it, and could be called a fault-block mountain range. It also consists largely of a huge composite batholith with minor remnants of the original volcanic-sedimentary country rock into which the magmas intruded, and thus might be called a fold-belt mountain range as well.

16.43 What is an oblique-slip fault?

An *oblique-slip fault* is a fault that has had a net movement in both a vertical (dip-slip) sense and a horizontal (strike-slip) sense. (The normal, reverse, and strike-slip faults you have already studied are "end-member" types.) Even our classic strike-slip fault, the San Andreas, has had some vertical movement along some of its segments.

16.44 Define graben and horst, and referring to Fig. 16.14, indicate which fault blocks are grabens and which are horsts. Are they likely to be the products of compressional or extensional forces? Are the faults normal or reverse faults?

The two lower blocks are *grabens*, or down-faulted blocks. The central block is a *horst*, for it has moved up *relative to* the adjacent grabens. (Note that the horst may actually have remained stationary, but the *relative* sense of movement is upward.) The forces would have been extensional, causing normal faults to form.

16.45 How much crustal extension, in meters, has occurred in the area of the faulting shown in Fig. 16.14?

Note that if the grabens are "raised" so that the marker bed of each block is at the same elevation, empty spaces remain. Measurement of the distance across the four spaces will give the amount of crustal extension. In this case, it is about 350 m. (Measure the distance A–B, plus the three similar segments.)

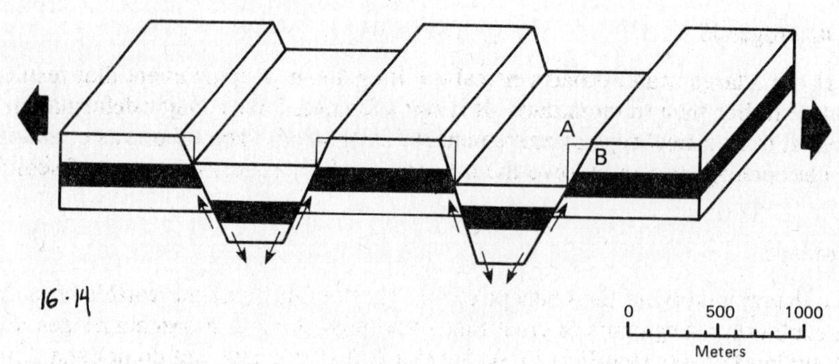

Fig. 16.14 Block diagram of horsts and grabens. Note the scale.

16.46 (*a*) What is a nappe? (*b*) What type of fault is illustrated in Fig. 16.6?

(*a*) A *nappe* is a large recumbent or overturned fold, as in Fig. 16.6. Nappes are characteristic of Alpine-type folding (i.e.. in the Alps of Europe). (*b*) The fault is a thrust fault, as in Fig. 16.11. Nappes and thrust faults commonly are found together in highly folded mountain ranges such as the Alps.

16.47 Assume that the top half of the fold of Fig. 16.6 were eroded off. How would a geologist know that the structure is an eroded nappe with one side overturned?

The cross-beds in the lower part of the fold are convex upward, rather than downward, so the bed has been overturned. Or, if the beds were graded, the coarse grains would be at the top of the bed rather than at the bottom, as originally deposited. Cross-beds and graded beds are thus indicators of the tops of the beds (i.e., ''top indicators'').

16.48 The folds shown in Fig. 16.2 are disappearing or ''plunging'' beneath the surface. What is the plunge of a fold, and how would you describe the plunge of the anticline and the syncline?

The *plunge* of a fold is the angle that axis makes with a horizontal plane. (It is named on Fig. 16.3.) The folds of Fig. 16.2 are broad open folds with vertical axial planes. The syncline is plunging about 20° to the north, as is the anticline.

16.49 What are the small folds on the flanks of the anticline of Fig. 16.3 called, and what is their significance?

They are *drag folds,* smaller versions of the larger fold, and were formed during the same folding event. Commonly a large fold might not be exposed, but drag folds, which are commonly measured in tens of centimeters rather than tens or hundreds of meters, may be completely exposed in small outcrops. Geologists can observe the small fold and determine the direction and amount of plunge of the fold, and thereby obtain an understanding of the deformation in that area or region on the basis of that one outcrop. (This assumes, of course, that the fold is representative.)

16.50 What is the term applied to the ridge formed by the steeply dipping resistant bed of Fig. 16.1?

Hogback. It is the result of the differential erosion of soft and resistant beds.

16.51 In some regions, differential erosion of gently dipping rock units forms broad, relatively low ridges. What is such a feature called?

A *cuesta.*

16.52 What is an orogeny?

An *orogeny* is a mountain-building event or tectonic process that includes folding, faulting, intrusion, and metamorphism.

16.53 What is an epeirogeny?

An *epeirogeny* is a large-scale, broad vertical uplift or down-warping event that results in plateaus, basins, or domes, rather than in mountains. It is not accompanied by major deformation, metamorphism, and intrusion. It is a vertical type of movement rather than a horizontal one as commonly occurs during orogenies. Yet epeirogeny can involve the movement of thousands of meters of sediment.

16.54 What is isostasy?

Isostasy is a theory involving the concept of different segments of the earth's crust being in "equilibrium" with each other. Segments of crust that stand high, such as mountain ranges, also project lower, resulting in mountain roots (see Fig. 15.3). Segments that ride low and do not stand in high relief, such as the oceanic basaltic crust, do not project to great depths and hence do not have roots.

16.55 What is isostatic rebound?

Isostatic rebound is the uplifting of a segment of the earth's crust in response to the removal of overlying material. Postglacial rebound is well-documented. The elimination of the water of a large lake due to climatic change may also result in uplift or isostatic rebound. Theoretically, erosion of a mountain range will result in a continued uplift during and after erosion, so as to eliminate the root of the mountain range.

16.56 Is rock deformation brittle or ductile, or both?

Both. Rocks under pressure near the earth's surface may break, or behave as a brittle material, whereas rocks under pressure at depth may deform as a ductile (i.e., a "plastic") substance. Faults are products of brittle deformation, whereas folds are products of ductile deformation.

16.57 Name three factors other than pressure which increase the ductility (plasticity) of a rock.

These are the rate of application of stress, the temperature, and the amount of intergranular fluids present. The rapid application of stress will cause brittle deformation, whereas a slow application may result in ductile deformation and thus folding. The temperature usually increases with increasing depth; a rock at a high temperature may deform in a ductile manner, whereas the same rock at the earth's surface where it is cool may deform in a brittle manner. The presence of intergranular fluids, usually water, results in a weaker rock that will deform in a more plastic manner.

16.58 Do anticlines form hills and do synclines form valleys?

This is commonly so, but erosion of rock units of differing hardness can result in the opposite situation. For example, a rock unit such as a silica-cemented quartz sandstone in the core of a syncline may be more resistant to erosion than the units on the adjacent anticline. After a long period of erosion, the downfolded rocks of the syncline may exist as a hill rather than as a valley.

16.59 What is the hinge of a fold?

The *hinge* of a fold is a line connecting points of maximum curvature of a fold. On Fig. 16.3, line x–x′ is the hinge.

Chapter 17

Plate Tectonics

INTRODUCTION

Way back in the 1600s, when the first crude maps of the world were made, the "match" between the coastlines of South America and Africa was noted by perceptive and imaginative persons (Fig. 17.1).

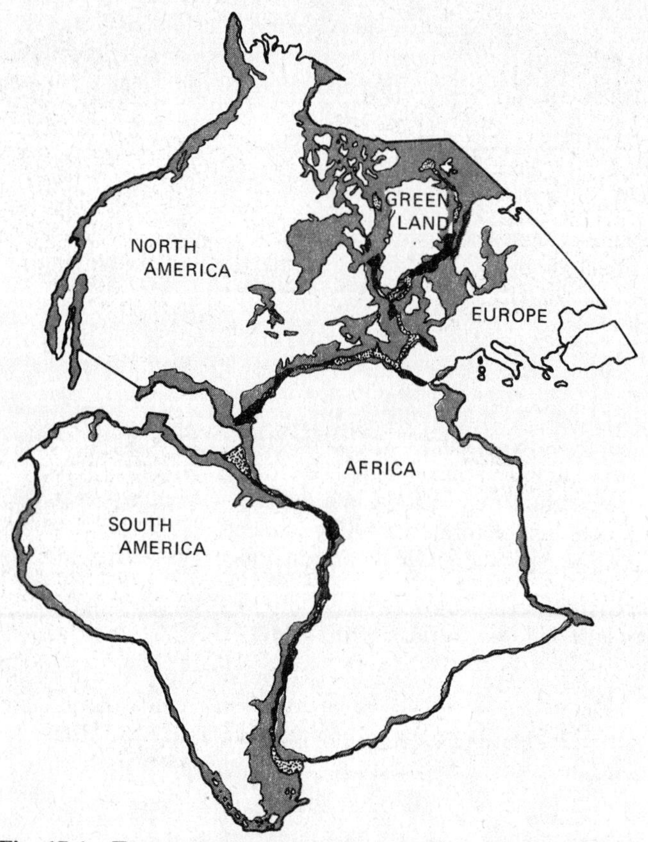

Fig. 17.1 Four continents illustrating how the coastlines match. This is especially true of South America and Africa. (*From Ojakangas and Darby, 1976.*)

However, it wasn't until the first part of the twentieth century that the grand scheme of "continental drift" was developed, largely by Alfred Wegener of Germany. He envisaged a supercontinent that he named Pangea ("all the earth"), which eventually fragmented with landmasses drifting apart to form the present continents (Fig. 17.2). The theory had its proponents, especially in South Africa and South America, but in general its popularity died with Wegener in 1930. Its main shortcoming was the lack of a driving mechanism that could literally move continents. In addition, the database supporting the theory was not a strong one.

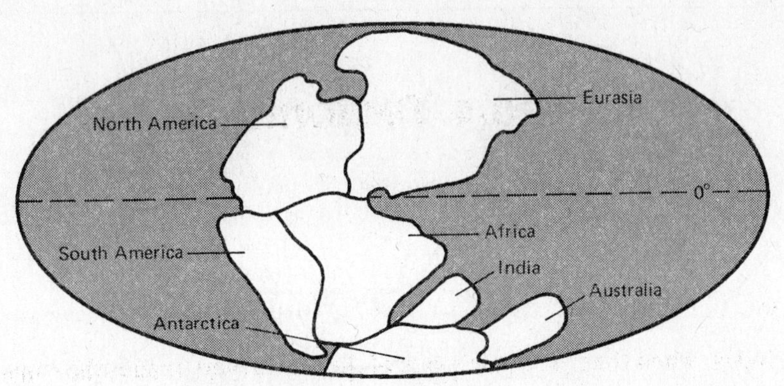

Permian-Triassic time: 225 millions years ago

(*a*)

Jurassic-Cretaceous time: 135 million years ago

(*b*)

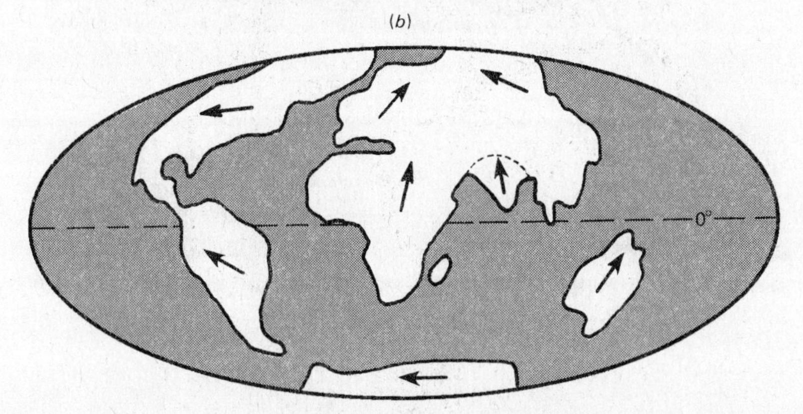

Today: arrows show relative movement directions

(*c*)

Fig. 17.2 Pangea, as hypothesized by Wegener. In (*a*), Pangea is intact. In (*b*), it has split into several individual continents. In (*c*), today's configuration is shown, with arrows indicating present movement.

The early 1960s saw a rousing rebirth of the concept. Many new scientific data that both revitalized and supported the idea led to a more complete concept, "plate tectonic theory." According to this theory, the continents are indeed moving relative to each other, but as rather passive passengers on larger "plates," or pieces, of the earth's outer zone, or *lithosphere* (Fig. 17.3). *Tectonics* is a fancy word for the study of earth's major structural features and the processes that formed them.

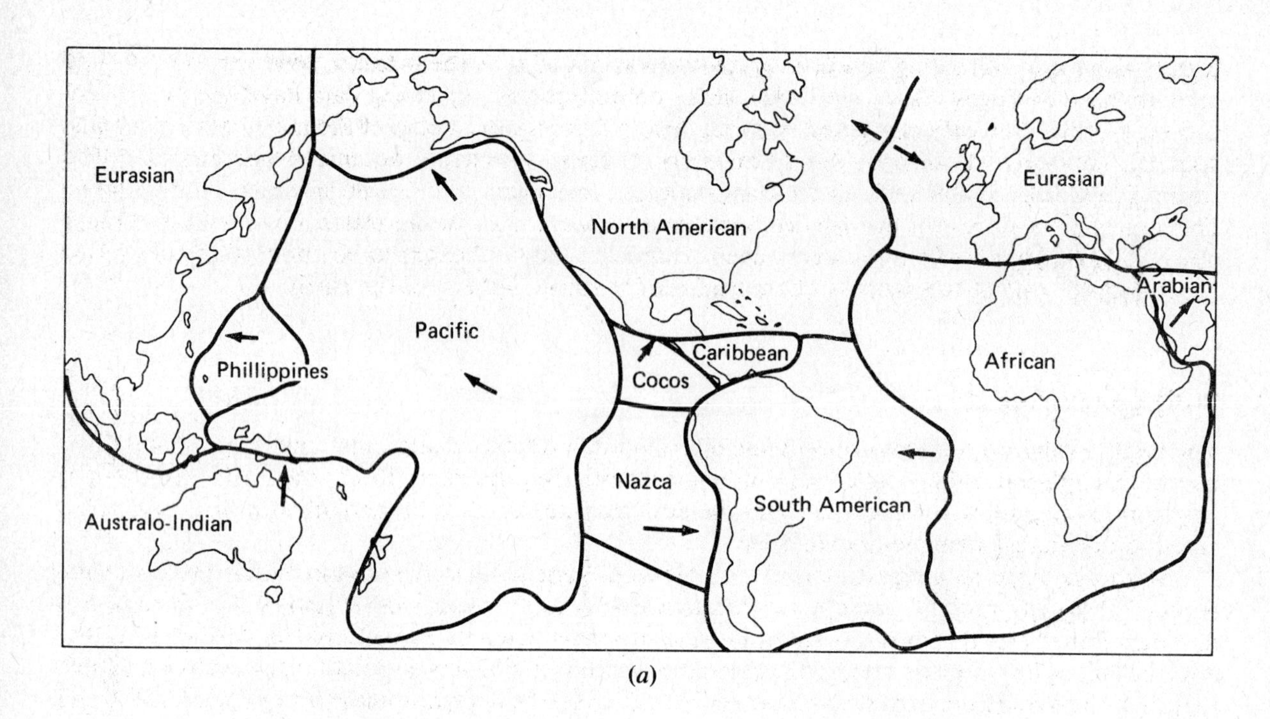

(*a*)

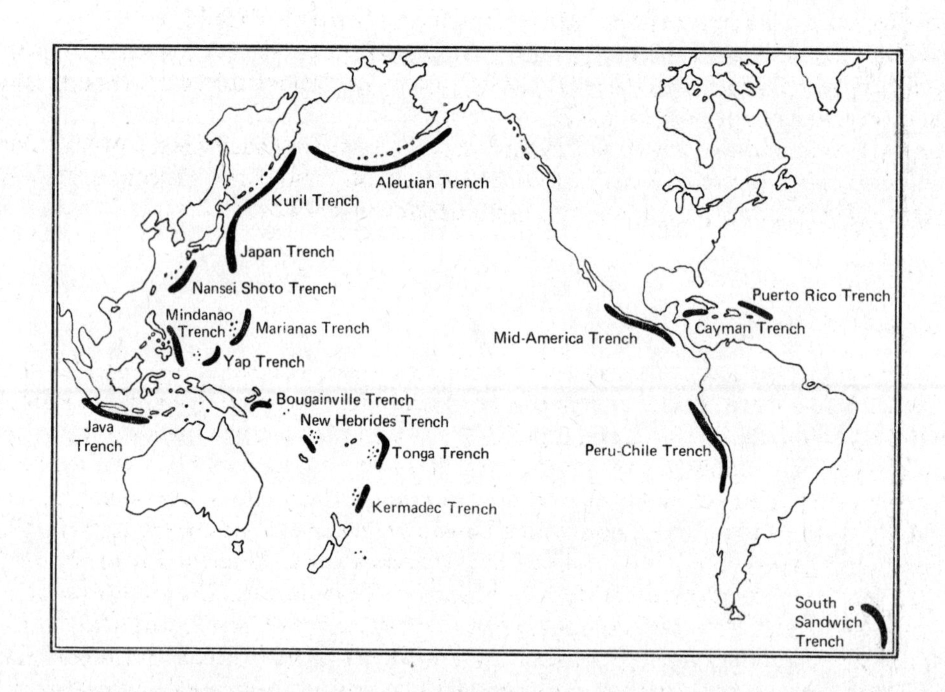

(*b*)

Fig. 17.3 Maps of world. (*a*) The lithospheric plates. (*After Carpenter, J. R., and Astwood, P. M., Plate Tectonics for Introductory Geology, Kendall/Hunt, 1983.*) (*b*) The trenches. (*From Ojakangas and Darby, 1976.*)

The new information that led to this theory included knowledge of the features on the ocean floor (island arcs, ridges, trenches, abyssal plains, and other features), heat flow data, earthquake data, and paleomagnetic ("old magnetism") studies which showed that within the ocean floor basalts there are alternating bands of magnetized rocks of opposite polarity, with the same pattern on both sides of ridges. Since the 1960s, many other types of supporting data have been collected, and today plate tectonics is a grand unifying theory that explains many of the major geological phenomena, including the distribution of mountain ranges, volcanoes, oceanic ridges and trenches, earthquakes, large faults such as the San Andreas, and many other earth processes and features. The largest problem is still the driving mechanism—that is, *why* do the plates move relative to each other? The cause must be a relatively deep-seated one and is thought to be convection cells in the upper mantle, part of the process of heat transfer from the interior of the earth.

PLATE MOVEMENT

The earth's rigid lithosphere—approximately the outer 100 km of the crust and upper mantle that overlies the plastic asthenosphere of the upper mantle (see Problems 15.14 and 15.15)—consists of six major plates and several smaller ones that are in motion relative to each other at slow rates measured at only a few centimeters per year.

Some plates are moving away from each other (divergent motion), some are moving toward each other (convergent motion), and others are moving sideways past each other (strike-slip or transform motion) (Fig. 17.4). If two plates are moving away from each other, with new crust formed in the zone between, doesn't this mean that either the earth is getting larger or that elsewhere on the surface plates are being destroyed or consumed? Only a very few earth scientists support the idea of an expanding earth; most believe that there are zones of *subduction* where two converging plates meet, with one moving downward beneath the other and melting at depth (Fig. 17.5).

Earthquakes occur along all three types of plate boundaries (Fig. 3.2), volcanoes occur along both divergent and convergent boundaries (Fig. 3.2), and major folded and volcanic mountain ranges form along convergent boundaries (Fig. 1.2).

Convergent plate boundaries can be classified into three types, based upon whether the leading edges of the converging plates consist of oceanic or continental crust. Ocean-ocean collisions, ocean-continent collisions, and continent-continent collisions all occur.

HOT-SPOTS AND RIFTING

It has been suggested that a continent such as Pangea splits when a series of hot-spots form at depth, perhaps in the asthenosphere, each causing the overlying plate to bulge upward until three cracks form at about 120° to each other (a triple junction). If two of the cracks from each hot-spot were roughly aligned and if all continued to split, the resultant weak zone could become a wider crack or rift. With continuing thermal activity at depth, the rift could evolve into a divergent plate boundary such as the Mid-Atlantic Ridge, and a new ocean basin would exist in the spreading zone (Figs. 17.6 and 17.7). The ocean basin would be floored with new basaltic crust that would constitute the newest parts of the two spreading plates. The third crack over each hot-spot would become a "failed arm" that would remain as a narrow elongate zone containing basalt and sedimentary rocks.

The new ocean basin could become smaller if a subduction zone were to form along the edge of the basin. Oceanic crust is heavier than continental crust so it is subducted and partially melted, forming a chain of volcanoes. As the oceanic crust is consumed, the continental portions of the two plates would eventually collide and the ocean basin would no longer be present. Such a history of plate movement encompassing the opening and closing of an ocean basin has been named a *Wilson cycle,* after J. Tuzo Wilson of Canada who first recognized the process. The Red Sea is an example of an ocean basin in the process of formation (Fig. 17.7).

There are also individual hot-spots. One has been hypothesized as being present beneath the Pacific plate, having led to the development of the Emperor Seamount chain (including the newest islands in the chain, the Hawaiian Islands) as the plate has moved northwestward over the periodically active hot-spot.

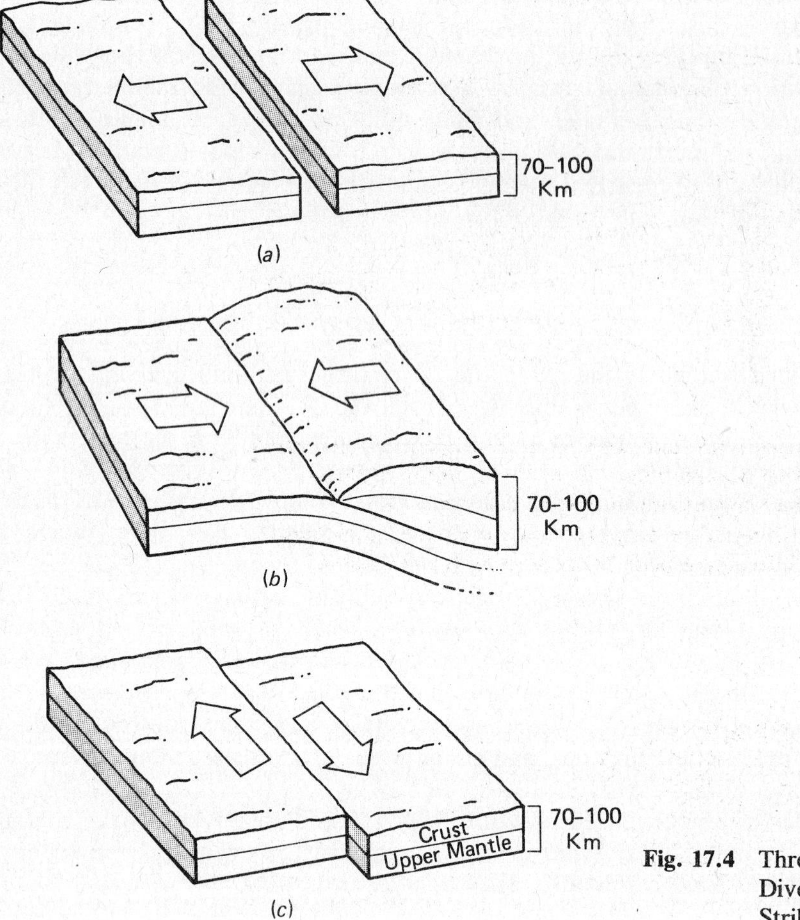

Fig. 17.4 Three types of plate boundaries. (*a*) Divergent. (*b*) Convergent. (*c*) Strike-slip or transform.

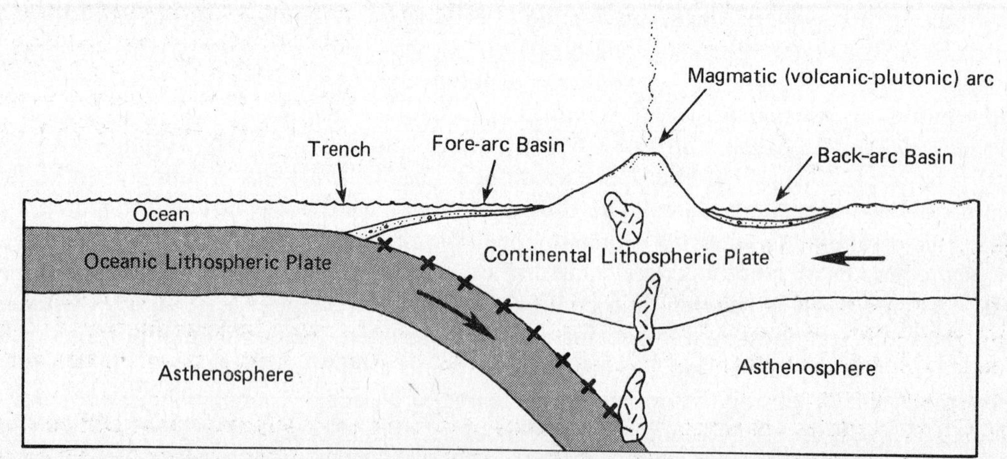

Fig. 17.5 Cross-sectional diagram illustrating the subduction of an oceanic plate beneath another plate which could include a continent or which could be another oceanic plate. X's denote location of earthquake foci, the zone of earthquakes that is known as the Benioff zone. Associated sites of sediment accumulation are named. Cross-hatched areas beneath volcano represent igneous plutons. Dotted areas represent sediment.

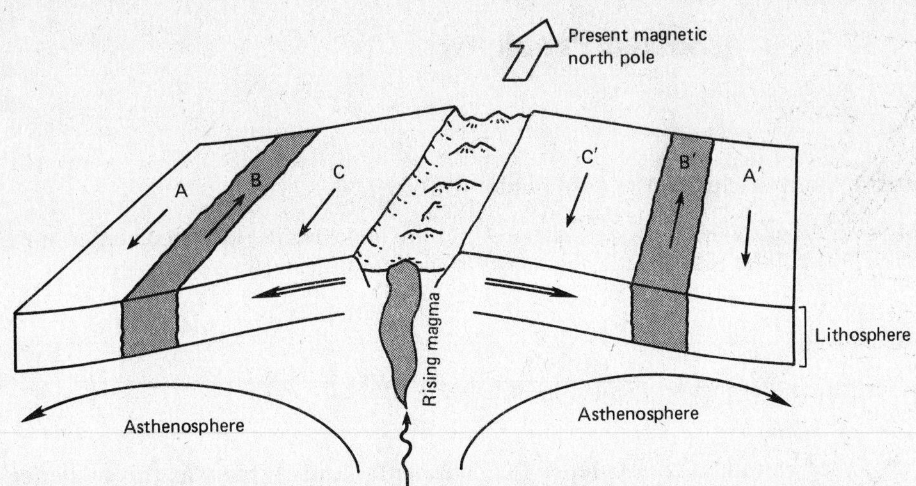

Fig. 17.6 Block diagram representing a few alternating bands of oppositely polarized basalts on both sides of an oceanic ridge. The arrows on top of the diagram point toward the magnetic north pole as determined by paleomagnetic measurements in the stripes. Such a magnetic striping has been measured along all divergent plate boundaries (spread centers).

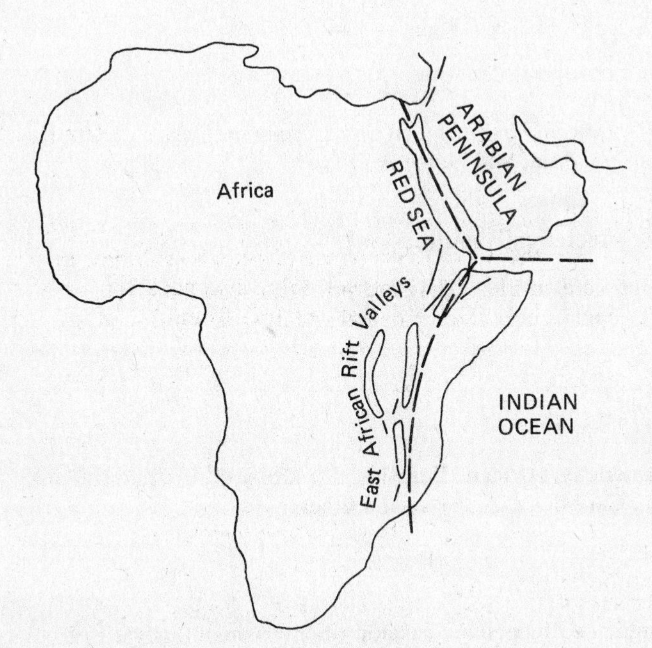

Fig. 17.7 Map of Arabian Peninsula, Africa, Red Sea, Gulf of Aden, and the East African Rift Valleys. See Problem 17.49 for interpretation. Heavy dashed lines represent hypothetical arms of triple junction.

CONTINENTAL ACCRETION

As heavier oceanic crust is subducted beneath lighter continental crust, as is currently going on along the west coast of South America, it partially melts, and new magmas form and rise to be emplaced as volcanic and plutonic rocks along the edge of the continental mass. In this way, continents can grow larger.

In addition, as plates converge, various pieces of oceanic or continental crust collide with continents and are welded against continental margins to become parts of the continent. These exotic or "suspect" terranes, among the latest findings in the grand plate tectonics scheme, include many terranes or blocks of rock along the west coast of North America, and make up nearly all of Alaska. As many as 200 separate additions to the western edge of the continent have been suggested. Other continents include similar exotic terranes.

Solved Problems

17.1 What was probably the first clue to the continental drift idea?

The east coast of South America and the west coast of Africa are quite parallel to each other, as in Fig. 17.1. The fit is even better if the bases of the continental slopes are fitted against each other.

17.2 What is Pangea?

This is the name for Alfred Wegener's supercontinent (see Fig. 17.2).

17.3 In the early 1900s, Wegener amassed evidence for continental drift. What was this evidence?

It included: (*a*) shapes of continents; (*b*) late Paleozoic (roughly 250-million-year-old) glacial deposits on the southern continents (seemingly part of a major ancient glaciation of a larger landmass); (*c*) similar geological formations (rock, structures, mountains) of similar age on opposite sides of the Atlantic; and (*d*) similar fossils, including plants such as the *Glossopteris* flora and animals such as the reptile *Lystrosaurus*, now found on several continents but best explained as having spread over a single large landmass.

17.4 What are Gondwanaland and Laurasia?

The southern and northern parts of Pangea. *Gondwanaland* included South America, Africa, Australia, Antarctica, and India. *Laurasia* included Europe, Asia, and North America.

17.5 Why is the concept of "continental drift" technically outmoded?

Because the continents are not drifting as separate entities. They are passengers on larger plates, as shown in Fig. 17.3. That is, some major plates consist of both continental and oceanic crust, and others consist only of oceanic crust.

17.6 How many lithospheric plates are there? (See Fig. 17.3.)

Six major plates (North American, South American, African, Eurasian, Pacific, and Australo-Indian), plus numerous smaller plates (Nazca, Cocos, Caribbean, Arabian, and others).

17.7 Describe the vertical zonation of a lithospheric plate.

The upper portion consists of the crust, (granitic continents and basaltic ocean basins); this varies from about 5 to 40 km in thickness. The lower portion consists of the rigid upper mantle which is about 65 km thick. Thus plates are 70 to 100 km thick.

17.8 Figure 17.4 illustrates the three main types of plate boundaries. Name and describe each.

(*a*) *Divergent (spreading) boundaries* along which two plates are moving away from each other.
(*b*) *Convergent boundaries* along which two plates are moving toward each other.
(*c*) *Strike-slip or transform boundaries* along which two plates are sliding past each other.

17.9 Give examples of each of the three main types of plate boundaries illustrated in Fig. 17.4.

(*a*) *Divergent:* Mid-Atlantic ridge; East Pacific Rise.
(*b*) *Convergent:* West coast of South America; Japan, Philippines.

(*c*) *Strike-slip:* San Andreas Fault, California; Denali Fault, Alaska; major faults in New Zealand and the Philippines.

17.10 Name three major scientific developments that provided the basis for plate tectonic theory.

Perhaps the most important are the physiography of the ocean basins, paleomagnetism, and heat flow studies. (Gravity studies of the ocean floor and radiometric dating of oceanic rocks are others.)

17.11 Where on earth is new oceanic crust being generated?

Beneath all the divergent plate boundaries. The three most important are the Mid-Atlantic Ridge, the East Pacific rise, and the Mid-Indian Ocean Ridge. The magma rises from the upper mantle and is emplaced primarily as lava flows along the divergent boundaries (spread centers).

17.12 What happens to oceanic crust (basalt) at a converging plate boundary?

It is subducted because it is heavier than continental crust. If two oceanic plates are converging, one or the other will be subducted. See Fig. 17.5.

17.13 See Fig. 17.5. What happens to the subducted plate?

It is partially melted at depth and rises as magma.

17.14 What feature on the sea floor marks a subduction zone?

A trench. (See Figs. 17.3 and 17.5.) Two examples of trenches are the Marianas Trench in the western Pacific and the Peru-Chile Trench off the western coast of South America. Trenches are the deepest parts of ocean basins.

17.15 What other feature is always paired with a trench?

A magmatic (volcanic-plutonic) arc, either an island arc (e.g., Japan or the Philippines) or a continental arc (e.g., the Andes of South America). Topographically, these are mountain chains.

17.16 What is the origin of the Hawaiian Islands (Fig. 3.16)?

The Hawaiian Islands are the result of the Pacific plate moving over a hot-spot in the mantle beneath the plate. This hot-spot is periodically active, and when it is, a volcano is formed above it. As the plate moves, each volcano in turn becomes inactive or extinct.

17.17 Judging from the map of the Hawaiian Islands in Fig. 3.16, in which direction is the Pacific plate moving?

To the northwest.

17.18 What is the origin of fold-belt mountains (see Problem 16.27) according to plate tectonic theory?

They form during convergence of two plates; the light continental portions of the plates collide and are deformed and metamorphosed.

17.19 What kind of plate movement resulted in the formation of the Himalayas and the Appalachians?

These are fold-belt mountain ranges, the result of the convergence of two plates and the eventual collision of two continental blocks. The Himalayas formed when the Indian subcontinent collided with Asia, and the Appalachians formed when North America and Europe collided. Realize that subduction of intervening oceanic crust must have occurred before the continental blocks collided.

17.20 What is the rate of plate movement?

The rates are relatively slow, 1 to 10 cm/year.

17.21 Using a velocity of 10 cm/year, how far would a plate move in 1000 years? In 1,000,000 years?

100 m in 1000 years and 100,000 m (i.e., 100 km or about 62 mi) in 1 million years. Note that although the rate is slow, the total movement with time is truly significant.

17.22 Refer back to Fig. 3.16. The rocks on the island of Hawaii are the youngest on the islands, for they are still forming via active volcanism. The rocks on the island of Oahu are 3 million years old. Is the rate of movement calculated in the previous question compatible with these data?

Yes. Oahu is about 300 km from Hawaii. (A movement of 100 km per million years means that the 3-million-year-old volcanic island of Oahu should be about 300 km from the present active volcanism on Hawaii.)

17.23 Where do most of the world's earthquakes occur?

Along plate margins (Fig. 3.2). Note that they occur along all three main types of plate boundaries.

17.24 Where are most of the world's volcanoes, relative to the plate boundaries? (Refer back to Fig. 3.2.)

They are along two types of plate margins, divergent and convergent.

17.25 Problem 3.22 referred to the "ring of fire" around the Pacific, a belt of active volcanism. How does plate tectonic theory explain this "ring"?

Subduction zones are now present around much of the Pacific. See Fig. 17.3.

17.26 Where are most of the world's major young mountain ranges?

Along plate margins. They are either fold-belt mountain ranges (the result of the collision of continental blocks) or volcanic mountains (the result of subduction, melting, and volcanism). (Many fault-block mountain ranges, as in the Basin and Range Province of southwestern United States, are young, too, but they are small ranges.)

17.27 What is paleomagnetism?

Paleomagnetism is "ancient magnetism." When magmas are extruded, or when sediment accumulates, those crystals or grains that contain iron come to rest so that their electron fields are oriented in the earth's magnetic field. The north-seeking poles of the grains point toward the magnetic north pole and are inclined at the proper dip (depending upon the latitude) so as to actually point toward the position of the north pole along the shortest path through the earth. Therefore, basalts (flows and dikes) and red sedimentary rocks (i.e., with hematite) will indicate where the magnetic north pole was during the time the rock was formed. This magnetism recorded in the rocks is called *thermoremanent magnetism,* or *TRM.*

17.28 Why are paleomagnetic studies of value to plate tectonic theory?

By combining these paleomagnetic data with radiometric age data, the location of the magnetic north pole at different times in earth's history can be determined.

17.29 What is a polar wander path?

A *polar wander path* or curve is a line connecting the positions of the north pole for rocks of different ages. A polar wander path has been constructed for each continent.

17.30 How did polar wander paths indicate that the continents have moved relative to each other?

If the paths for each continent had been identical, it would have indicated that the continents had been stationary throughout time. That is not the case. Instead, each continent has a different path, indicating that the continents have moved relative to each other.

17.31 What is meant by a "reversed magnetic polarity"?

Paleomagnetic studies indicate that the earth's magnetic polarity reverses itself from time to time. That is, the magnetism preserved in rocks (Problem 17.27) indicates that the north-seeking magnetic poles of the minerals have from time to time pointed toward the south rather than toward the present north. Why the earth's magnetic field reverses itself is not well understood. Recall that the earth's magnetic field is thought to be generated by convection in the earth's liquid outer core. One "explanation" is that the core is behaving as a bistable flip-flop circuit.

17.32 Figure 17.6 shows stripes of ocean floor basalts with alternating magnetic polarities. Study this figure, and suggest why this finding was of basic importance to plate tectonic theory.

You should have noted that the stripes on opposite sides of the central ridge are mirror images. Not only are the polarities the same in corresponding stripes, the widths of the corresponding stripes are similar. This has been interpreted to mean that as basaltic lava pours out along a spread center, it assumes the magnetism of the earth's magnetic field at that time, including either normal or reversed polarity. As spreading continues, the stripe of basalt is split and new lava forms in between, again assuming the imprint of the earth's magnetic field. This provides important evidence that spread centers exist and that new crust is continually being formed along the spread centers (the divergent boundaries).

17.33 (*a*) How old is the oceanic basaltic crust?

The oldest crust found is less than 200 million years old.

(*b*) Why is there no older crust, since the earth is probably 4600 million years old?

It has been subducted and recycled into the mantle.

17.34 What are the relative ages of the stripes of rock shown in Fig. 17.6?

Rock stripes C and C′ are the youngest, and A and A′ are the oldest. This relationship is found on opposite sides of all oceanic ridges, with the youngest stripes nearest the ridges and the older stripes correspondingly farther away from the ridges.

17.35 By observing Figs. 1.4 and 3.2, describe the Mid-Atlantic Ridge.

It is a great mountain range, longer than any on land (65,000 km, or 40,000 mi, long and 2400 km, or 1440 mi, wide), it rises more than 3 km (1.8 mi) above the adjacent abyssal plains, and it has a down-dropped (i.e., faulted) rift-valley as much as 2.5 km (1.5 mi) deep and 32 to 48 km (19 to 29 mi) wide along its crest. It is a zone of active volcanism, active earthquakes, and high heat flow.

17.36 What is an exotic terrane, and what is continental accretion?

Exotic terranes are small fragments of crust (either oceanic or continental) that have been welded to or joined to the continental margins during convergent plate motions. (Another name for these fragments is suspect terranes. Some are called minicontinents.) *Continental accretion* is the process whereby continents grow larger by having volcanic arcs or other terranes welded to their margins during subduction.

17.37 How important are such exotic terranes?

It is now suggested that much of westernmost North America, including all of Alaska, consists of a large number (perhaps 200) of such fragments.

17.38 Why is the asthenosphere (see Figs. 17.5 and 17.6) so important in plate tectonic theory?

Because it is the "weak" zone along which the overlying lithospheric plates move. (See Problems 15.14 and 15.15 for a review of the asthenosphere.)

17.39 What is a Wilson cycle?

A *Wilson cycle* is the opening of a new ocean by rifting, its growth as plate divergence continues, and its closure as the plates change motion and converge and culminate in the collision of the continental blocks. Thus, it can be thought of as the "birth and death" of an ocean.

17.40 Why is plate tectonics called a "grand unifying theory"?

Because it explains the earth's earthquake belts, volcanic belts, young mountain belts, trenches, island arcs, and many other features.

Supplementary Problems

17.41 Where are the low and high heat flow values in the ocean basins?

Heat flow values are higher over the oceanic ridges and lower over the trenches, relative to the rest of the ocean floor.

17.42 Where in the ocean basins do negative gravity anomalies occur, and why?

Over the trenches. This is because lighter-weight oceanic crustal rocks from the earth's surface are being moved downward into the mantle. This results in less pull on a gravity meter, relative to adjacent areas where there is more effect and a higher gravity reading from the denser mantle rocks.

17.43 What is the difference in the compositions of lavas extruded along divergent (spreading) boundaries (e.g., the Mid-Atlantic Ridge) compared to those in volcanic arcs along convergent plate boundaries, and why this difference?

Along divergent boundaries, the lavas are basaltic, the product of the partial melting of mantle rock (peridotite) as pressure is released along the boundary by faulting. In the case of convergence, the subducted plate of basalt and sediments is partially melted at high temperatures and pressures as it descends, and the resultant magma is andesitic in composition (intermediate between basalt and rhyolite). The line separating these andesites from other volcanics (e.g., it "encircles" the Pacific Ocean basin) is known as the *andesite line*.

17.44 Some earthquakes along plate boundaries are shallow, some are at intermediate depths, and some are deep. Why? (See Problem 14.23.)

Recall that most earthquakes are caused by movement along faults. Along a divergent plate boundary, the crust is thin and near the surface, and all earthquakes are shallow. Along a subduction zone, however, the oceanic crust is being moved down to greater depths and as the slab moves along faults, earthquakes occur. Therefore, there are shallow, intermediate, and deep earthquakes along a subduction zone. The earthquakes are deepest under the volcanic arc and shallowest at the trench. On Fig. 17.5, the small X's indicate the foci of earthquakes.

17.45 How deep are the deepest known earthquakes?

About 700 km (427 mi) beneath island arcs.

17.46 In geosynclinal theory (Chapter 16), the eugeosyncline is situated seaward of the miogeosyncline and contains volcanic rocks (especially pillowed basalts), some bedded chert, and other sedimentary rocks. The sediments may be easily derived from the adjacent landmass, or from a magmatic arc, but can the volcanic rocks be explained without a zone of volcanoes in the eugeosyncline?

Yes. In plate tectonic theory, the trench deposits can be compared to the eugeosyncline deposits. The pillowed basalt and the chert that form by volcanism and hydrothermal activity along spread centers may be scraped off during subduction, and mixed with sediments.

17.47 Where is the sediment that is eroded from a magmatic (volcanic arc) deposited? Refer to Fig. 17.5.

It may be deposited in the fore-arc basin between the arc and the trench, as well as in the back-arc basin between the arc and the adjacent landmass. Some may reach the trench to be interbedded with volcanics and chert scraped off of the descending oceanic plate, as in the previous problem.

17.48 Geosynclinal theory was around for about a century before plate tectonic theory was developed. In the last chapter, you learned that major mountain ranges developed on the sites of major geosynclines. The mechanism for this was always the weakest part of the geosynclinal theory of mountain building. See Fig. 17.5 and try to relate the illustrated plate boundary to geosynclinal theory by renaming certain parts of the figure as the miogeosyncline and the eugeosyncline.

The trench could be the eugeosyncline, and the fore-arc basin could be the miogeosyncline. (This is a simplified approach. While such comparisons have some validity, it is a complicated picture.) Nevertheless, plate tectonic theory can in many instances be considered a recent addition to geosynclinal theory, for it provides the mechanism of collision of continental masses as the cause of the uplifted mountains.

17.49 Figure 17.7 shows the area of the Arabian Peninsula and Africa. Of what significance is this area to plate tectonic theory?

It is an area of geologically young crustal rifting that is still going on. It may be the location of a hotspot, with the three cracks forming at about 120° to each other. Two (the Red Sea and the Gulf of Aden) are spreading more than the third (the system of East African Rift Valleys). The former two may be the first steps in the development of a new ocean basin, and the Rift Valleys may be a "failed arm" (an *aulacogen*).

17.50 What is a transform fault?

Ridge systems (i.e., spreading centers) are cut by numerous fracture zones that are oriented perpendicular to the ridges, as in Fig. 17.8. The fracture zones are dormant fault zones. As a spreading center forms, *some parts* of these dormant fracture zones are activated by the spreading motion, and strike-slip (i.e., horizontal) movement occurs. These reactivated segments of the faults are called *transform faults*.

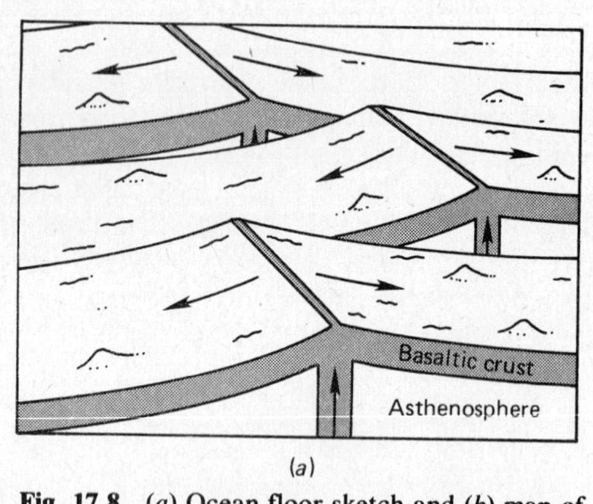

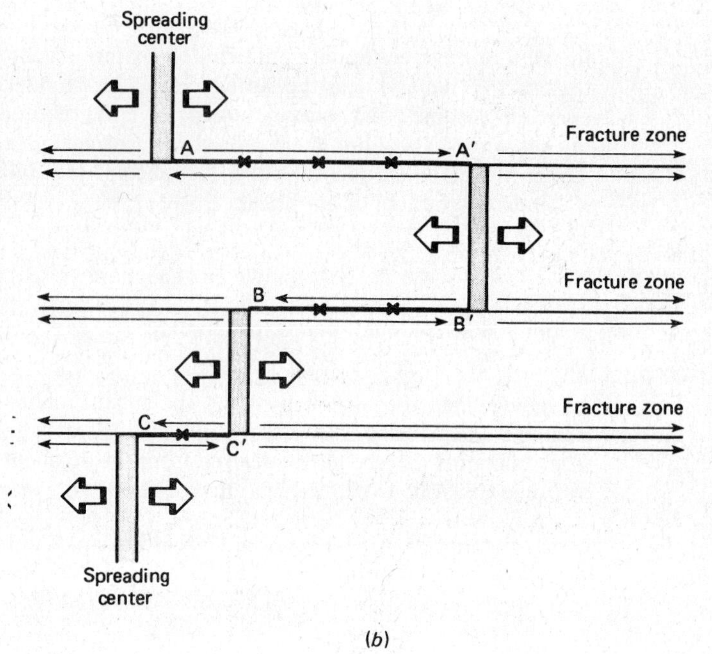

Fig. 17.8 (*a*) Ocean floor sketch and (*b*) map of a segmented ridge or spread center (divergent plate boundary) with perpendicular fracture zones. The broad arrows represent spreading from segmented spread center. The thin arrows show the movement of each block in more detail. X's represent earthquakes. (See Problem 17.51 for interpretation.)

17.51 Study the arrows on Fig. 17.8 that show the directions of movement on both sides of each segment of the spread center. Note that arrows on the opposite sides of a fracture zone are either indicating similar or opposite directions of movement, depending upon where that part of the fracture zone is located relative to the axis of the spread center. Thus, there is relative motion along only one part of a given fracture zone, and that is the part that can be called a transform fault. Locate three transform faults on Fig. 17.8.

The three segments marked by the darker lines are the transform faults (A–A′, B–B′, C–C′).

17.52 How were these transform faults on the ocean floor recognized? (*Clue:* Think earthquakes!)

Earthquakes are caused by movement along faults, and only the transform faults have earthquakes along them—the other parts of the fracture zones are dormant. The small X's indicate earthquakes.

17.53 See Fig. 17.9, which shows the San Andreas fault of California. You have learned (Problem 16.22) that it is a strike-slip fault. Now, with your knowledge of plate tectonic theory, state *why* there is strike-slip movement along the San Andreas.

The San Andreas is a transform fault exposed on land. Note from Fig. 17.3 that the East Pacific Rise comes into North America in the vicinity of California. The movement on the San Andreas is like that shown in Fig. 17.8. Note the two ends of spreading segments shown on Fig. 17.9.

17.54 What relationship, if any, is there between the Appalachian Mountains of eastern North America and the Caledonide Mountains of Scotland and Norway?

They are the same elongate mountain range formed by the collision of North America and Europe; this was one of the collisions that led to the formation of Pangea. When Pangea broke up, this once continuous range was fragmented. This also shows that plate tectonic motions were taking place *prior* to the breakup of Pangea. Later chapters will place such earlier motions in a historical perspective.

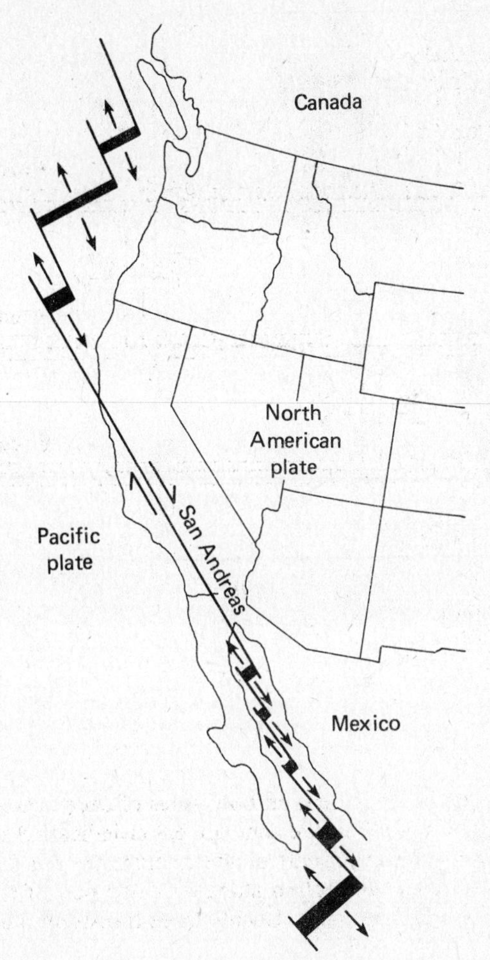

Fig. 17.9 The San Andreas fault and the East Pacific Rise (thick dark lines) with other transform faults also shown. (See Problem 17.53 for interpretation.)

17.55 What is the theoretical explanation of *how* a continent splits apart and an ocean basin develops between the two continental portions?

A series of aligned hot-spots, each with spreading along two of the three cracks or arms that develop above each hot-spot, results in an ocean basin.

17.56 The Sierra Nevada of California today consists largely of a great composite batholith. What was the origin of this range?

It is a magmatic arc on the continent, originating in the same way as the Andes of South America, the result of subduction of an oceanic plate. A further complication in the case of the Sierra Nevada is that there is a large normal fault forming the eastern side of the range.

17.57 Why is Iceland (Fig. 17.10) of great interest to volcanologists?

It is a site of active volcanism that is located on the Mid-Atlantic Ridge, and it is accessible for study. The lava flows there are samples of the type of lava poured out all along spreading centers.

17.58 Note that Fig. 17.10 is a geologic map of Iceland, with the relative ages of the volcanic rocks given in the legend in the standard manner, with the youngest rocks in the top box and the oldest in the lowest box. Describe the geographic relationships of rocks of different ages on Iceland, and relate this age relationship to plate tectonic theory.

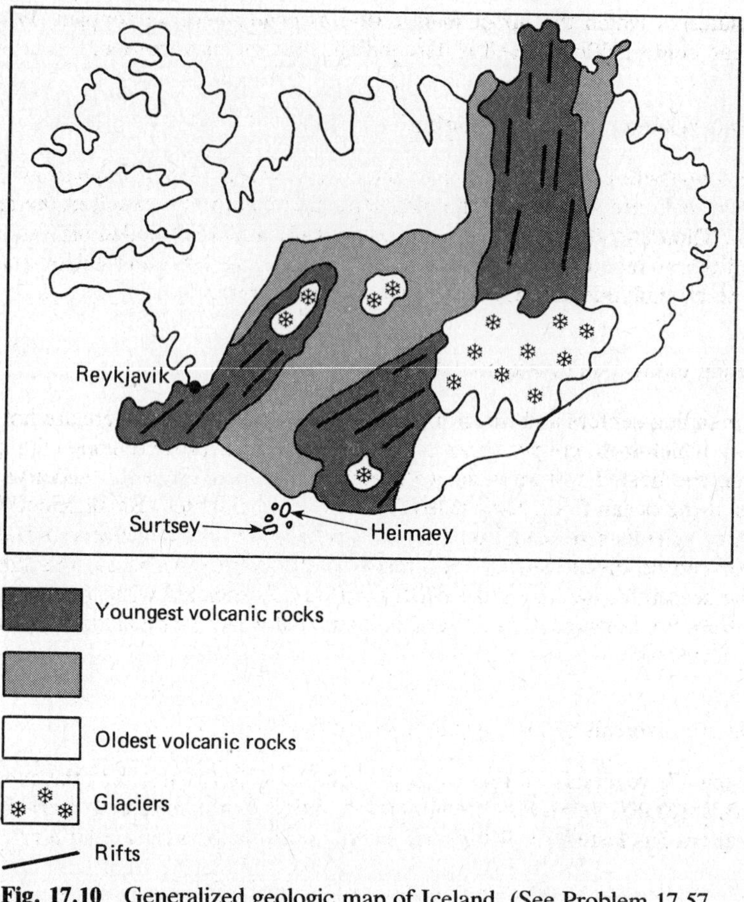

Fig. 17.10 Generalized geologic map of Iceland. (See Problem 17.57 for interpretation.)

The youngest rocks are in the center of the island, and the oldest are on the edges. This verifies the spreading nature of the Mid-Atlantic Ridge.

17.59 What is MORB, and why is it of great interest to igneous petrologists?

MORB stands for *mid-oceanic ridge basalt,* and it is the product of the partial melting of the mantle. Thus it helps to define the process of magma evolution.

17.60 Assume you are a structural geologist interested in major zones of thrust faulting. Where on earth would you go to study such zones?

To convergent plate boundaries.

17.61 What is the driving mechanism for plate movement?

This is the hard question. Basically it is assumed to be convection cells in the upper mantle. Such cells would theoretically be the result of heat transfer, much as boiling water is moved by convection in a pot; the hot water rises and the cooler water is carried downward. Convection is a process of heat transfer via mass movement. In the case of the solid mantle, which behaves plastically over time, the less-dense hotter mantle rock from deeper in the mantle rises, and cooler mantle rock from the top of the mantle sinks. Beneath the lithospheric plates, the mantle is moving laterally from the point at which the heat is rising (i.e., the plume). Because the mantle rock is now near the surface, it becomes cooler and denser and eventually sinks. As it is moving laterally, it provides the mechanism for moving the overlying lithospheric plates to which the upper mantle (the asthenosphere) is coupled. Where a cell is rising, there is a spread center. Where a cell is descending, there is convergence.

17.62 Are ore deposits related to plate boundaries?

Yes. The big low-grade copper deposits of the world, the porphyry coppers which are mineralized plutons of felsic rock, are all found in the magmatic arcs that are the result of the melting of subducted oceanic plates. Theoretically, the low copper content in basalts is "boiled off" as the subducting plate is melted, and this is redeposited from hot fluids in cracks and intercrystalline spaces in the plutonic rock bodies. Many molybdenum deposits have a similar origin.

17.63 Are ores forming today along plate boundaries?

Yes! Along spreading centers and transform faults on the ocean floor, there are hot springs and "black smokers" from which iron, copper, lead, and zinc are being deposited along with silica, as oxides and sulfides. When the heated waters laden with metals dissolved from the underlying basalts meet the colder waters at the ocean floor, their dissolved load is precipitated. Small scientific submarines (e.g., the Alvin) carry scientists to such spots where this activity can be observed and photographed and where samples can be taken. Since "the present is the key to the past," this kind of information is valuable in the search for metals on the earth's surface. Associated with some of these hot water sites on the ocean floor are colonies of plants and animals that were totally unknown at such depths, that is, at thousands of meters.

17.64 How common are reversals of the earth's magnetic field?

There have been 171 reversals of polarity in the last 76 million years, with the average duration of the normal intervals 420,000 years and the average duration of the reversed interval 480,000 years. The present normal era has lasted 700,000 years, so we appear to be overdue for a reversal.

17.65 Is there any evidence that the earth's magnetic field will reverse again?

Yes. The strength of the earth's magnetic field has decreased 6 percent in the last 100 years, and perhaps 50 percent in the last 2500 years. Presumably the earth's magnetic field diminishes to a very low strength, and when it builds up again, it is reversed.

17.66 How are the magnetic polarity reversals dated, since it is difficult to obtain rock samples from the ocean floor that are suitable for radiometric age dating?

The reversal patterns have also been worked out for rock sequences (especially basaltic lava flows) on land. These rocks can be dated, and the paleomagnetic data from land are matched with paleomagnetic oceanic data.

17.67 Coral reefs have been found on some isolated mountains in Alaska. How did they get there?

Prior to plate tectonics, they would have been explained as the result of a warmer climate at that northern latitude. Now it seems more likely that this area, which is small, is an exotic terrane, a coral atoll from the ocean floor, joined to North America during plate convergence.

17.68 What is a Benioff zone?

A Benioff zone is the zone of earthquakes that exists along a subducting plate, getting deeper under the volcanic arcs (Fig. 17.5). It is named after H. Benioff, a geophysicist who first noted the existence of the zone.

17.69 What is blueschist, a rock found along convergent plate boundaries?

Blueschist is a special type of metamorphic rock containing a blue amphibole called glaucophane. This mineral forms at high pressures and low temperatures. The high pressure is the result of compression, and the low temperature is the result of the cool descending basaltic slab. Some of this material is then brought up into the fore-arc basin by thrust faulting.

17.70 What is obduction?

Obduction is a process that may be thought of as the opposite of subduction, in which one plate is forced down under another plate. In obduction, one plate stays at the same level and another plate *overrides* it. This is the situation where a continental block collides with another continental block.

17.71 Why are the Tibetan Plateau and the Himalayas so high?

Because this is a collision zone between two continental blocks, India and the Asian continents. Obduction has occurred.

17.72 Ophiolite sequences—what are they, and why do geologists look for them?

Ophiolite sequences consist of pillowed basalts, gabbros, dike swarms with dikes intruding into earlier dikes, chert, and deep oceanic sediments. Such sequences of rock are commonly scraped off (or obducted) from a descending slab of rock along subduction zones, and if found, they are a good indicator of the previous existence of a subduction zone.

17.73 On Fig. 17.4, all three diagrams show the crust plus the upper mantle as 70–100 km thick. What are the two layers, together, called?

This is the lithospheric plate that overlies the asthenosphere and theoretically moves upon and over this ‘‘soft’’ zone.

17.74 Why is the island of Surtsey (Fig. 17.10) of special interest?

Because it is a recent example of volcanism along the mid-Atlantic Ridge. In 1963, the new volcanic island of Surtsey was born in the ocean just south of Iceland. The volcanic activity lasted four years and was studied by volcanologists throughout its active life.

17.75 Why is the island of Heimacy (Fig. 17.10) of special interest?

It is the most recent example of volcanism in Iceland. The volcano Kirkju erupted in 1973. The islands volcanoes had been quiet for 6,000 years. The pyroclastics buried part of the fishing village of Vestmannaejar. It was feared that a lava flow would seal off the harbor from the ocean, but instead a more protected harbor was the result. Inhabitants sprayed cold water on the front of the basalt flow in hopes of cooling and thereby stopping it. It did stop!

Chapter 18

Time, Fossils, and the Rock Record

INTRODUCTION

As late as the early 1800s, most people thought the earth was created in the year 4004 B.C., based on the Biblical genealogy from Christ back to Adam, as worked out by Archbishop Ussher of Ireland in 1654.

The age-dating of rock units can be either relative or numerical (radiometric). Relative dating was developed first. It was not until the advent of radiometric age dating in the mid-1900s that rather firm numbers on the age of the earth and on events in earth's history were obtainable.

Most natural scientists of the 1700s and early 1800s were also theologians, and Ussher's date for the creation of the earth was generally accepted. However, many early scientists were reading a long history from the rocks. James Hutton of Scotland in the late 1700s, based upon decades of field work including the observations of geologic processes, of unconformities, and of dikes that had obviously intruded as hot magmas, stated, "There is no vestige of a beginning and no prospect of an end." Although it was not yet called uniformitarianism, Hutton championed an ever-changing and dynamic earth, saying that processes at work today were at work in the past. This implied slow but steady processes and a long time for the processes to be working. The Church accused him of heresy.

Abraham Gottlob Werner of Germany (and his many students), also in the late 1700s, had a different outlook than did Hutton. Werner thought that many of the rock types on earth, including basalt and granite, had been precipitated out of a universal ocean. Because he did not emphasize a great age for the earth, his ideas were more in keeping with a divine creation, the Noachian Deluge, and "catastrophism."

Various approaches for determining the age of the earth were tried, including determination of the rates of sedimentation, salt content of the oceans, and cooling rates of hot steel spheres used as models of the earth. Some estimated the age of the earth at 100 m.y.

RELATIVE AGE DATING

Two key principles—superposition and cross-cutting relationships—give the relative ages of rock units. The *principle of superposition* can be stated as follows: In a sequence of strata (i.e., sedimentary rocks or lava flows) that have not been overturned by crustal deformation, the older layers are on the bottom and the youngest are on the top. If an igneous rock unit in the form of a dike, stock, or batholith cuts across another rock, the igneous rock unit is said to have a *cross-cutting relationship* to the other rock and is younger. Also, pieces of one rock included in another give the age relationships, for the former must have existed first in order to be contained in the latter. The study of fossils, described below, also provides a means of determining the relative ages of the strata in which the fossils occur. The geologic time chart was developed on the basis of fossiliferous sedimentary strata.

NUMERICAL AGE DATING

Numerical age dating is based on isotopic (radiometric) methods. Statistical average rates of spontaneous decay (i.e., radiometric decay) of the unstable nuclei of certain isotopes (see Problem 2.11) can be observed in the laboratory. It has been determined that this decay goes on at a constant rate for a given isotope, regardless of changes in chemical or physical environment.

To determine the age of a given mineral or rock, the ratio of parent to daughter atoms is multiplied by the decay rate for that isotope. For example, the parent isotope ^{238}U decays to the daughter product ^{206}Pb. The amount of each is accurately measured with a mass spectrometer. (A closed system, in which no U or Pb is added or removed, is assumed.) The decay rate can be simply expressed by the *half-life* for that isotope, the time it takes for half of that isotope to decay to the daughter product. The half-life of ^{238}U is 4.5 b.y. (billion years). Every 4.5 b.y., one-half of the ^{238}U will decay. Thus, after a period of one half-life, half of the parent is gone; after a time equal to another half-life, half of the remaining parent has decayed, leaving ¼ of the original amount.

Igneous rocks and minerals, when dated, give the age of crystallization and hence the actual age of the rocks and minerals. Metamorphic rocks will usually give the age of metamorphism, for the increased temperature and pressure may have driven off some daughter product, thereby "resetting the atomic clock." Dating of most sedimentary rocks gives the age of the source material and not the date at which the sediment accumulated.

Perhaps the best-known radiometric isotope is ^{14}C, used to date organic remains. It decays to ^{14}N, a gas, and this readily escapes. Therefore, in the ^{14}C method, only the remaining parent is measured and compared to the amounts of common ^{12}C. This assumes an original uniform ratio of the amounts of the C isotopes. The half-life of ^{14}C is 5730 years, so every 5730 years one-half of the remaining ^{14}C is changed to nitrogen. Since ^{14}C constitutes less than 1 percent of the carbon content of the atmosphere and of any living organic material, after a relatively few half-lives it is present in such small quantities that it cannot be accurately measured. Thus the system can be used only for material that is no older than about 50,000 to 60,000 years.

GEOLOGIC TIME SCALE

The geologic time scale that is in use today evolved in the early to mid-1800s on the basis of the superposition of layers and their contained fossils. Strata were correlated with each other based on the fossils they contained; that is, they were assumed to be of similar age if they contained similar assemblages of fossils. Nearly all the early work involving strata and fossils was done in western Europe, and therefore the time scale (Fig. 18.1) contains many names based on geographic areas of western Europe (e.g., the Devonian from Devonshire, England) as well as names based on Latin words (e.g., Paleozoic meaning "ancient life"). Sedgwick, Murchison, and Lyell were major contributors to the time scale.

Now radiometric dates are available for all the boundaries within the time scale. The age of the earth is estimated at 4.6 b.y., even though the oldest rocks yet dated on earth are 3.8 b.y. The 4.6-b.y. date is based on meteorites and moon rocks. (One reported date from rocks in Australia is 4.2 b.y.).

FOSSILS

Fossils can be defined as traces or remains of prehistoric life. About 200 years ago, scientists noted that different sets of sedimentary strata contained different fossil assemblages. They determined that in many cases, the assemblages were unique. Many life forms became extinct, never to reappear in overlying strata, being replaced by new and different life forms. Thus the concept of a faunal succession was developed, based on countless observations.

Generally, for an organism to be fossilized, it must possess hard parts (bones, teeth, shell) and be buried "immediately," before scavengers, weathering, and agents of abrasion can destroy the remains. Therefore, the most abundant fossils are marine shelled organisms that were buried by sedimentational events in a sedimentary basin that was subsiding rather than being uplifted and eroded. Fossils are useful in dating strata; this will be discussed in the next chapter.

Relative time			Radiometric time, millions of years before present
Era	Period	Epoch	
Cenozoic	Quaternary	Holocene	0.01 (10,000 yr.)
		Pleistocene	1.5
	Tertiary	Pliocene	12
		Miocene	20
		Oligocene	35
		Eocene	55
		Paleocene	65
Mesozoic	Cretaceous	Numerous epochs recognized	130
	Jurassic		185
	Triassic		230
Paleozoic	Permian	Numerous epochs recognized	265
	Pennsylvanian*		310
	Mississippian*		355
	Devonian		413
	Silurian		425
	Ordovician		475
	Cambrian		570
Proterozoic	Late		900
	Middle		1600
	Early		2500
Archean	Late		3000
	Middle		3500
	Early		4600?

*The Mississippian and Pennsylvanian together make up the Carboniferous of Europe.

Fig. 18.1 The geologic time chart.

Solved Problems

18.1 What is the difference between relative age dating and numerical age dating?

Relative age dating just gives the relative ages of the rock units, whereas numerical age dating gives an age in years.

18.2 What is the underlying basis of radiometric age dating?

Some isotopes are radioactive. That is, their nuclei continually decay, emitting either alpha particles (two protons and two neutrons), beta particles (electrons), or gamma rays (like x-rays). The rate of decay is uniform for each isotope, regardless of physical and chemical conditions.

18.3 How are parent and daughter isotopes utilized in radiometric age dating?

The parent is the radioactive isotope that is decaying. The decay scheme eventually stops, with a daughter product as the final, stable isotope. For example, ^{238}U is the parent and ^{206}Pb is the daughter; the decay actually involves about a dozen steps, with many other isotopes formed as intermediate steps to ^{206}Pb.

18.4 What is the half-life of a radioactive isotope?

The half-life is the time it takes one-half of the isotopes present in a given mineral or rock to decay or change to a daughter product. The half-life of ^{238}U is 4.5 b.y.

18.5 If a geochronologist accurately measures the amount of ^{238}U and ^{206}Pb in a mineral and determines that there is only one-half as much U as there was originally, how old is the mineral?

If half of the U has been converted to Pb, it has been decaying for a time equal to its half-life, or 4.5 b.y.

18.6 A scientist is determining the age of a recently discovered piece of bone and ascertains that the amount of ^{14}C present is only ⅛ the original amount that should have been there, based on the fixed ratio of ^{12}C, ^{13}C, and ^{14}C in all living material. How old is the bone?

The half-life of ^{14}C is 5730 years. Therefore, in 5730 years, ½ would decay, leaving ½. In the next 5730 years, half of the remaining ½ would decay, leaving ¼ of the original. In the next 5730 years, half of the remaining ¼ would decay, leaving ⅛ of the original. Thus the time involved in decay is 3 half-lives (3×5730), or about 17,190 years. The piece of bone is 17,190 years old.

18.7 What would be the age of a piece of charcoal that has gone through 10 half-lives, and how much of the original ^{14}C would still be present?

57,300 years would be the age of the charcoal (10 times 5730 years). Only a tiny fraction of the original amount of ^{14}C would be present. As 1/2 decays during every half-life, the remaining fractions through 10 half-lives would be 1/2, 1/4, 1/8, 1/16, 1/32, 1/64, 1/128, 1/256, 1/512, and 1/1024. There will always be a bit remaining.

18.8 Referring to the previous problem, why is the ^{14}C dating method only good as far back as 50,000 to 60,000 years, or about 10 half-lives?

^{14}C constitutes less than 1 percent of the carbon in an organism. This is a small amount, and as the amount decreases, it becomes harder and harder to measure accurately.

18.9 What rock types can be dated by the ^{14}C method?

The method is used on materials that contain C. Therefore, a young limestone could be dated. Other common rock types cannot. The method is most used for organic remains—wood, charcoal, bone, teeth, and shells.

18.10 Radiometric age dates are not "absolute" or certain. What are some sources of uncertainty that result in a discordance of age dates?

Isotopic age dates, while usually reliable, yield numbers that have some uncertainty to them. If the system is not a closed system, parent or daughter may be added or removed, thereby obviously changing the ratios and the isotopic dates. There may also be some uncertainty in laboratory measurements. Most isotopic age dates are given with an uncertainty figure, such as 320 + or − 6.5 m.y.

18.11 Name the commonly used radioactive isotopes and their daughter products.

PARENT	DAUGHTER
^{238}U	^{206}Pb
^{40}K	^{40}Ar
^{87}Rb	^{87}Sr
^{14}C	^{14}N*

**Note:* N is not retained and measured.

18.12 Which type of rock—igneous, sedimentary, or metamorphic—can be dated by radiometric means to obtain the actual age of the rock? What would dates of the other two types indicate?

Radiometric dates of igneous rocks yield the date of crystallization of the magma, and hence the age of the rock. A date on a metamorphic rock may yield the age of metamorphism rather than the age of the original rock. Dating a sedimentary rock will give the original age of the source rocks from which the fragments in the sedimentary rock were derived.

18.13 In Fig. 18.2, what are the relative ages of formations A through G, based on superposition?

A is the oldest, and each successive unit is younger, with G the youngest.

18.14 In Fig. 18.2, what can be said about the ages of the sedimentary formations based upon the two igneous dikes, each of which has been eroded prior to deposition of the overlying sediment?

Formations A, B, and C are older than 220 m.y., formations D and E are younger than 220 m.y. and older than 60 m.y., and F and G are younger than 60 m.y.

18.15 Which formations in Fig. 18.2 are Mesozoic in age?

D and E.

18.16 Which formations in Fig. 18.2 are Cenozoic in age?

F and G.

18.17 Which formations in Fig. 18.2 are Paleozoic in age?

Only A, B, and C could be Paleozoic in age, but they could also be of Precambrian age, based on the given data.

18.18 Which formation in Fig. 18.2 shows different sedimentary facies?

Formation D has a conglomerate facies, a sandstone facies, a shale facies, and a limestone facies.

18.19 When did folding occur in the area of Fig. 18.2?

Before 220 m.y. ago.

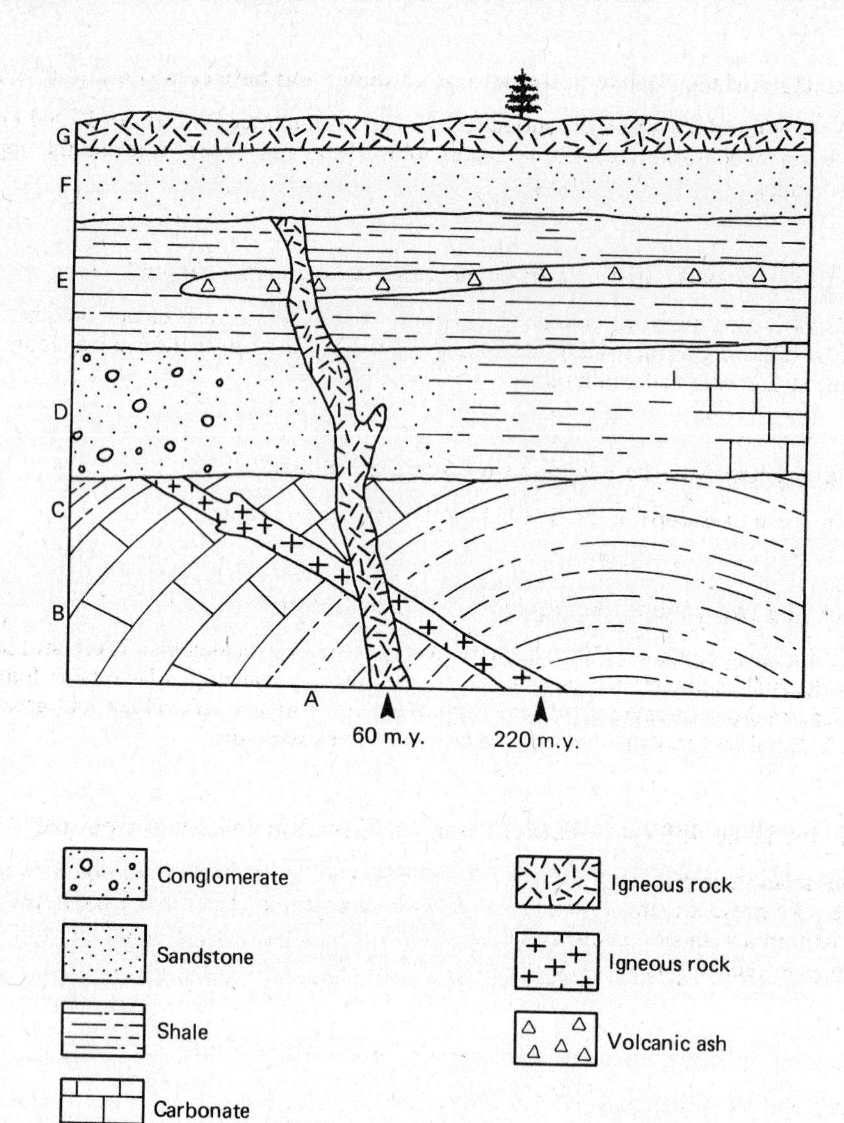

Fig. 18.2 A diagrammatic cross-section showing relationships of various sedimentary and igneous rock units. Note the legend and the given ages.

18.20 Which sedimentary lithology in Fig. 18.2 could be dated with radiometric methods to obtain the date of deposition, and why can it?

The volcanic ash bed in formation E could be dated. Crystals of zircon or feldspar in the ash would have crystallized just before the volcanic explosion that spread the ash over this area. A volcanic ash, while a sediment, is in one sense an igneous rock, with the daughter product starting to accumulate in the crystal structures after the rock formed.

18.21 If the volcanic ash of the previous question were dated at 150 m.y., in what geologic period was it deposited?

Jurassic.

18.22 What are the surfaces between formations C and D and between formations E and F called?

They are unconformities which are buried surfaces of erosion. (See Problem 16.38 and Fig. 16.13.) Note that the lower unconformity clearly truncates folded beds and a dike, whereas the upper one clearly truncates a dike.

18.23 On what basis was the geologic time chart or time scale developed?

On the fossils which the strata contained. A faunal succession was worked out, and then it was possible to say, e.g., "this organism lived in the Silurian Period." Wherever that fossil was found thereafter, the containing strata were called Silurian.

18.24 Where and when was the geologic time scale developed?

Mostly in Western Europe in the first half of the nineteenth century.

18.25 What are the two general prerequisites for fossilization?

That the organism contain hard parts and that it is buried rather quickly. There are, however, exceptions to the first requirement. For example, extinct woolly mammoths have been found preserved in areas of permafrost (permanently frozen ground), with hair and skin intact and green grass in their mouths. Animal footprints and worm burrows are other exceptions.

18.26 What is the basic difference between uniformitarianism and catastrophism?

Uniformitarianism implies that geologic processes at work today, sometimes slowly, were at work in the past (i.e., the present is the key to the past). Catastrophism implies rapid, catastrophic processes. Actually, uniformitarianism encompasses catastrophes such as volcanic eruptions and earthquakes, so these two concepts are not really diametrically opposed to each other.

18.27 Who is known as the Father of Geology?

Sir James Hutton of Scotland (1726 to 1797).

18.28 Why is Hutton the Father of Geology?

He championed the concept of uniformity of earth processes through time by observing processes in action and comparing the resulting sediments to sedimentary rocks. He realized that the earth must be very old. He correctly interpreted dikes as having been the result of the intrusion of hot magma and postulated a hot interior for the earth.

18.29 What is the estimated age of the earth, and what are the oldest radiometric age dates on rocks?

4.5 to 4.6 b.y., although the oldest rocks dated so far are 3.8 b.y. The age is based on meteorites and moon rocks, for scientists are in agreement that all the bodies in our solar system originated at about the same time. (However, a few zircon crystals from a quartzite in western Australia have been dated by the uranium-lead technique at 4.2 b.y. This is intriguing, for it indicates that the zircons were eroded from a source rock 4.2 b.y. old and deposited in a sandstone that was later metamorphosed to a quartzite.)

18.30 List the five eras of geologic time, oldest at the bottom and youngest at the top, and place numerical dates at each boundary.

Present	- - - - - - - - - - - - - - -
	Cenozoic
65 m.y.	- - - - - - - - - - - - - - -
	Mesozoic
230 m.y.	- - - - - - - - - - - - - - -
	Paleozoic
570 m.y.	- - - - - - - - - - - - - - -
	Proterozoic
2500 m.y.	- - - - - - - - - - - - - - -
	Archeozoic (Archean)
4600 m.y.	- - - - - - - - - - - - - - -

18.31 Geologic periods are subdivided into smaller portions of geologic time. What are these called?

Epochs.

18.32 *How* is a bone, tooth, or a piece of wood fossilized?

These materials, on a microscopic scale, are quite porous. After they are buried by sediment, groundwater can carry solutions through these pores, and compounds such as silica, calcium carbonate, or pyrite may be precipitated out of solution. Thus the fossilized material is more solid and heavier than the original. This process is called *permineralization*.

18.33 How is a sea shell fossilized?

The shell may be preserved exactly as it was. However, any pores are commonly filled with calcite, and if the shell was originally made of the metastable mineral aragonite (which has the same chemical composition as calcite but a different crystal structure), it will recrystallize to calcite.

18.34 Most of the petrified wood of the geologic record is associated with volcanic ash. Why?

Much volcanic ash consists of fine particles of volcanic glass which lacks a crystal structure and is metastable. With time, the glass "devitrifies" or deglasses, and forms small crystals. Explosive volcanic material is commonly felsic in composition (see Chapter 3), and during this devitrification process, excess silica is released into the percolating groundwaters. If trees were buried by the ash, they become permineralized. (The original cellulose material may still be present, but the pores are all filled with silica.)

18.35 Is a mummy from an Egyptian pyramid a fossil?

No. It is not "prehistoric," for the body was mummified during historical times.

18.36 Is a sea shell on the beach a fossil?

No. It is modern and not prehistoric.

18.37 Are worm tracks or dinosaur footprints preserved in rocks fossils?

Yes. A *fossil* is defined as traces or remains of prehistoric life. These would be traces of the original animal. The study of trace fossils is a subfield of *paleontology*, the study of fossils.

18.38 What is the age of the oldest sediment found on the floors of the ocean basins (i.e., deep ocean sediments)?

150 to 200 m.y.

18.39 Refer to the previous problem. Why are the deep-sea sediments so young compared to the ages of rocks on land and to the age of the earth?

Because sediments on the ocean floor are subducted where oceanic plates sink beneath other plates.

18.40 What do inclusions (xenoliths) of, say, volcanic rocks in a granite indicate about the relative ages of the two rock types?

The volcanic rocks are older and were intruded by the granitic magma which engulfed pieces of the volcanic rock. Some of the volcanic rock may have been totally melted and assimilated by the magma.

Supplementary Problems

18.41 Geochronologist Jones dated orthoclase in an arkose sandstone by using the K-Ar method, and arrived at an age of 325 m.y. Geology student Snodgrass happily announced, "This sandstone in my thesis area was deposited 325 m.y. ago and is late Paleozoic in age." What should you tell her?

You should inform her that the 325-m.y. date is the age of the granite from which the orthoclase grain was eroded, and not the time of deposition of the sandstone. All Snodgrass really knows from that date is that the sandstone is younger than 325 m.y.

18.42 Which sedimentary lithology, other than volcanic ash, can be dated radiometrically to give the age of the sediment?

A glauconitic sediment would yield the age of the sediment if the glauconite were dated. Glauconite is a potassium-bearing micaceous mineral that *forms* on the sea floor at the sediment-water interface, usually by the interaction of ions in the seawater with preexisting clay minerals. Glauconite is dated by the K-Ar method, but because Ar (argon) is a gas and glauconite has a layered, micaceous structure from which the argon can escape, dates on glauconite give a minimum age for the sediment. (If all of the argon were still there, the date would be older.) One other sedimentary rock can be dated—limestone, by the ^{14}C method, but the method is only good back to about 50,000 to 60,000 years.

18.43 If an igneous sill (see Problems 3.18 and 3. 19) were present in a sedimentary rock sequence, what information would it reveal about the age of the sedimentary sequence?

The sill has to be younger than the sedimentary rocks. Therefore, it would give a minimum age for the sedimentary rocks; the sedimentary rocks could be slightly older or much older.

18.44 Who is generally given credit for first stating the principle of superposition, and when did he do it?

Nicholas Steno of Denmark, in 1669.

18.45 Radioactive decay of isotopes produces energy, which is manifested as heat, according to the equation $E = mc^2$. (See Problem 2.15.) In relative terms, how much heat is produced by the radioactive decay of ^{238}U today, as compared with the early days of earth's history 4.5 b.y. ago?

About half as much. The half-life of ^{238}U is 4.5 b.y. In the 4.5 b.y. since the earth was formed, one-half of the ^{238}U has decayed. Therefore, today there is only half as much ^{238}U present to decay and give off heat. Because there are other major radioactive isotopes, such as ^{40}K with a half-life of 1.3 b.y., the amount of heat generated during earth's early days was substantially higher than at present. Thus the crust must have been hotter, and the geothermal gradient (the temperature increase per unit of depth) must have been higher than today's geothermal gradient.

18.46 How do geochronologists attempt to verify their radiometric dates?

By using three different decay schemes and comparing the results. When the radiometric results from isotopes with different decay rates (half-lives) agree, it not only validates the date for the rock but also validates the entire radiometric dating concept and procedure. In some instances, fission-track dating or thermoluminescence, two other techniques, can be used as checks.

18.47 Are all geologic periods subdivided into epochs, or are epochs used only in the Tertiary and Quaternary periods?

All periods are subdivided, but only those of the Cenozoic are commonly used in introductory courses. Most periods have a three-part subdivision into early, middle, and late, but there are also names for each.

18.48 Explain the unusual fossilization process of insects in amber.

Insects were trapped in the sticky sap that exuded out of a tree wound such as a broken branch. More sap eventually covered the insects. Decay allowed the volatiles in the organism's body to escape, leaving a thin black carbon film lining the cavity. (Therefore, the insects cannot be removed intact.) This sap hardened and became the commonly clear to yellowish material called amber. (The word *amber* is even used to describe a honey-yellow color.)

18.49 What were some early interpretations about the significance of fossils?

A common early interpretation was that fossils were animals that had died and were buried during the great flood described in the Bible in the book of Genesis. Another was that fossils were "works of the devil." A third was that fossils were "cast off creations" of the Creator. Some thought that fossils had grown in the rock, much like crystals can grow in a rock, after the rock layer was deposited.

18.50 In 1669, Nicholas Steno, a Dane living in Italy, stated three fundamental principles regarding the interpretation of sedimentary strata. What were they?

The principle of superposition, the principle of original horizontality, and the principle of original continuity to the edge of the depositional basin.

18.51 What is *fission-track dating?*

It is a relatively new technique of isotopic age dating. During the radioactive decay of ^{238}U, most particles given off by the nucleus are alpha particles (helium nuclei consisting of two protons and two neutrons). Very rarely, one of these helium nuclei undergoes further splitting or fission, yielding two high-energy particles which move extremely fast through the containing mineral, capturing electrons from atoms of other elements. Their paths show up as short lines after the mineral is etched; the more lines within a given area, the longer the ^{238}U present in the minerals has been decaying and the older the sample is. Measurement of the amount of ^{238}U is necessary, and then an age can be calculated from a decay constant.

18.52 How can glacial varves be used for dating?

Glacial varves are annual layers of sediment, each composed of a darker clayey layer and a coarser (silty to sandy) lighter-colored layer. The former layer is the winter layer deposited out of suspension when the lake is frozen; it is dark because of organic material. The lighter layer is the summer layer, deposited when the lake is unfrozen. By counting the number of varves in a sequence, the duration of the varve accumulation (i.e., the number of years the lake existed) can be determined.

18.53 How are tree rings used in dating?

Each year a tree grows a new layer of wood just under the bark. By taking a small core from the tree (or by cutting it down), the rings can be counted and the age of the tree determined. Some redwoods and bristlecone pines are older than 4000 years!

18.54 What is a depositional sequence?

A *depositional sequence* is an unconformity-bounded sequence of related sedimentary rock units. Major unconformities can be found at the same position in the rock column on most or all of the continents. They are therefore worldwide and must be related to worldwide changes of sea level. A major drop in sea level will cause a marine regression (withdrawal), and a major rise in sea level will cause a marine transgression (advance).

18.55 What are possible explanations for the worldwide sea level changes described in the previous problem?

Either changes in the volume of water in the oceans or changes in the volume of the ocean basins (i.e., the "containers"). The former can be caused by glaciation; when there are large glaciers on land, sea level drops. However, many sea level rises and falls are too great to be explained by glaciation. The best explanation is that the volume of the ocean basins has changed.

18.56 How can the volume of the ocean basins, as mentioned in the previous problem, change?

Plate tectonics provides an explanation. When there is a rapid spreading at spread centers such as the Mid-Atlantic Ridge, there is much hot basaltic crust on the ocean floor, and it has a greater volume than cooled basaltic rock. Therefore, rapid spreading causes sea level rises, and slow spreading (cooler oceanic crust) causes sea level drops.

18.57 What are worldwide sea level curves?

Interpretations of the unconformities between depositional sequences have resulted in sea level curves. Because these are worldwide, they constitute a kind of relative dating technique.

18.58 What is a formation, and how does it relate to time boundaries?

A *formation* is a mappable rock unit, the basic unit of geologic mapping. A formation may have formed at different times in different places, and therefore may not be the same age throughout its extent. For example, if a sandstone formed in a slowly transgressing sea, it is younger in that area where the sea reached last. It is said to be a time-transgressive formation.

18.59 What are systems, series, and stages?

By definition, the rocks deposited during a period, an epoch, and a smaller subdivision called an *age*, respectively. These are time-stratigraphic units or time-rock units, and their boundaries perfectly correspond with the time boundaries on the geologic time chart. (They should! The time chart was erected on the basis of the fossils in the rock record.)

18.60 What is the polarity reversal scale?

The reversals of the earth's magnetic poles through time (see Problems 17.31, 17.64, 17.65, and 17.66) allow for correlation of sequences of normally and reversely polarized rocks, and this constitutes a special type of relative time scale called the *polarity reversal scale*. The reversals can be assigned numerical dates by radiometric dating of the rocks or correlative rocks.

18.61 Who was the first scientist to assign the earth a numerical age greater than the Biblically derived age of about 6000 years?

Buffon, a French scientist, in 1749 experimented with the cooling rates of steel spheres and, based on a hot origin for the earth, arrived at an age of 75,000 years.

18.62 What is Walther's principle, commonly referred to as Walther's law (Fig. 18.3)?

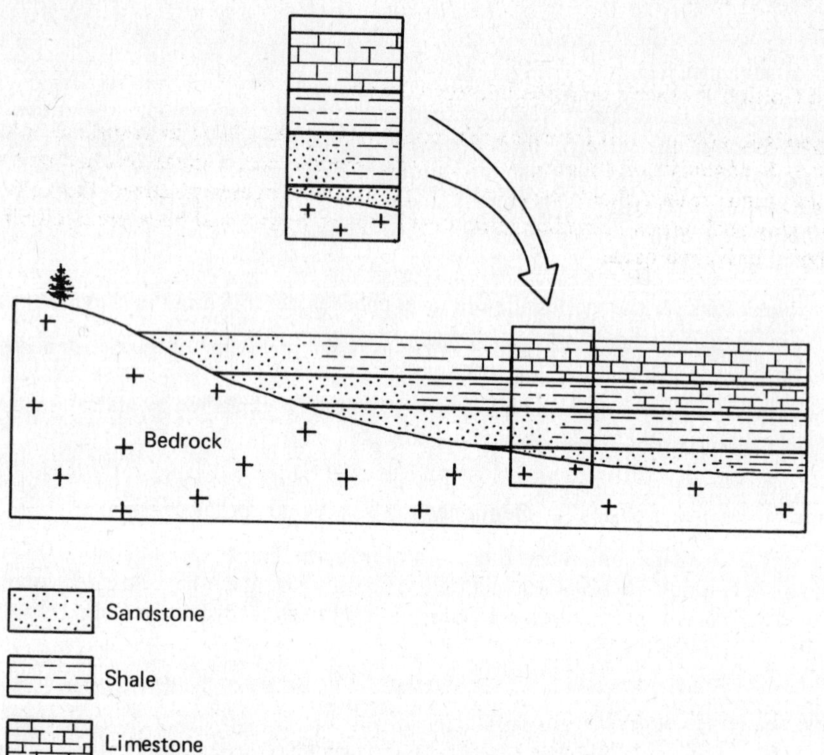

Fig. 18.3 Diagrammatic representation of Walther's law. The horizontal lines represent time lines.

Walther's law can be stated as follows: Sedimentary facies found in a vertical sequence (i.e., one on top of another) are related laterally in the same order. This applies only if the sequence is undisturbed and continuous, with no faults or unconformities in the sequence.

18.63 What does Fig. 18.3 imply about the location of sedimentary facies relative to the shoreline?

That sands were deposited near shore and muds further out, and calcium carbonate was being precipitated still further from shore where the mud and sand were not being deposited.

18.64 What does Fig. 18.3 imply about the ocean in that area—was the sea transgressing or regressing?

Transgressing, for what are probably successively deeper water sediments were deposited upon what are probably shallow water sediments. Therefore, the shoreline was advancing over the land, and each successive environment was eventually located at that same place, one after the other and one on top of the other.

18.65 In which situation—transgression or regression—might there be the best chance that the sediments will be buried and preserved?

Transgression, for as a sea retreats (regresses), the former areas of deposition will be above sea level and subject to erosion.

18.66 What were the main contributions of Sir Charles Lyell to the field of geology?

He wrote the first good geology text, *Principles of Geology*, in 1830; it was revised 12 times between 1830 and 1875. He championed uniformitarianism, but to an extreme that included uniformity in the intensity (i.e., rate) of geologic processes through time, something that Hutton, who originated the idea, had never intended.

18.67 What was Abraham Gottlob Werner's contribution to geology?

He organized the geologic column into four geologic sequences. From oldest to youngest, they were Primitive, Transition, Secondary (or Floetz), and Alluvial. Werner was a great teacher at Freiberg Academy in Germany and trained many early geologists. He also overemphasized the role of the ocean—he thought many rock types, including granite, schist, and basalt, had been precipitated out of the water of an original universal ocean.

Chapter 19

Development of Life and the Fossil Record

INTRODUCTION

Fossils have long been recognized as evidence of ancient life. Leonardo da Vinci about 1500 A.D. correctly identified fossils found in rocks of northern Italy, far from the sea, as marine shells. Furthermore, he found them in many different rock layers, indicating many depositional and fossilization events. In the late 1700s and early 1800s, fossil collecting became a favorite pastime of many people in western Europe (Fig. 19.1).

Some made serious studies of fossils; for example, William "Strata" Smith in England in the late 1700s and early 1800s could look at a fossil and tell which sedimentary rock unit it had been collected from. Smith is now known as the "Father of Stratigraphy," which is the study of strata and their fossils. He could relate fossils to sedimentary layers, even though he hadn't collected the fossils—he was correlating strata. Thus it was realized that there is a succession of fauna in the stratigraphic record, and the *law of faunal succession* was formulated. At the same time as this work was going on in England, Cuvier and Brongniart in France were coming to similar conclusions.

The *principle of fossil correlation* was a product of these studies; like assemblages of fossils are of like age, and the strata in which they are found can be correlated as being of about the same age. However, not all fossils are useful for correlation. The best *index* or *guide fossils* have four main characteristics: they are limited vertically in the rock record (i.e., in time), widespread, distinctive, and abundant. Many of the best index fossils were floaters (pelagic) or swimmers, rather than forms that were dependent upon a certain type of bottom sediment for their habitat (i.e., facies-dependent).

ORGANIC EVOLUTION

The abundant fossil data collected by many workers eventually culminated in the theory of evolution as put forth by Charles Darwin in his *Origin of Species* in 1859, based on the premise of natural selection (i.e., "survival of the fittest"). We now know that random chemical changes (mutations) in genes (segments of the nucleic acid, DNA) result in new characteristics. If such a new characteristic helps the organism to better adapt to the environment, the characteristic may be perpetuated by natural selection. Darwin was not the first to espouse evolution; he brought together many ideas put forth by others and added a quarter century of his own observations of both fossil and living forms that supported organic evolution. In opposition to evolution was the literal interpretation of the Bible, with one creation and one demise of many creatures in Noah's flood.

Figure 19.2 depicts a generalized evolutionary tree of animal life. Organic evolution provides the best explanation of the fossil data which include thousands upon thousands of independent observations. Thus it is an example of the scientific method, whereby scientists have observed, gathered data, and come to the most objective conclusions based upon the data.

As new data come in, details of evolutionary theory will change. Evolution can be considered a doctrine, and that is not likely to change. However, models explaining *how* evolution proceeded will undoubtedly change, as they have in the past, in order to explain the new data. For example, according to the relatively new concept of *punctuated equilibrium,* species appear rather suddenly in the rock record and do not change much before becoming extinct. This idea helps to explain why the fossil record commonly does not contain (or geologists have not yet found them) series of fossils with gradual changes between species (i.e., "missing links"). This does not mean that there are no

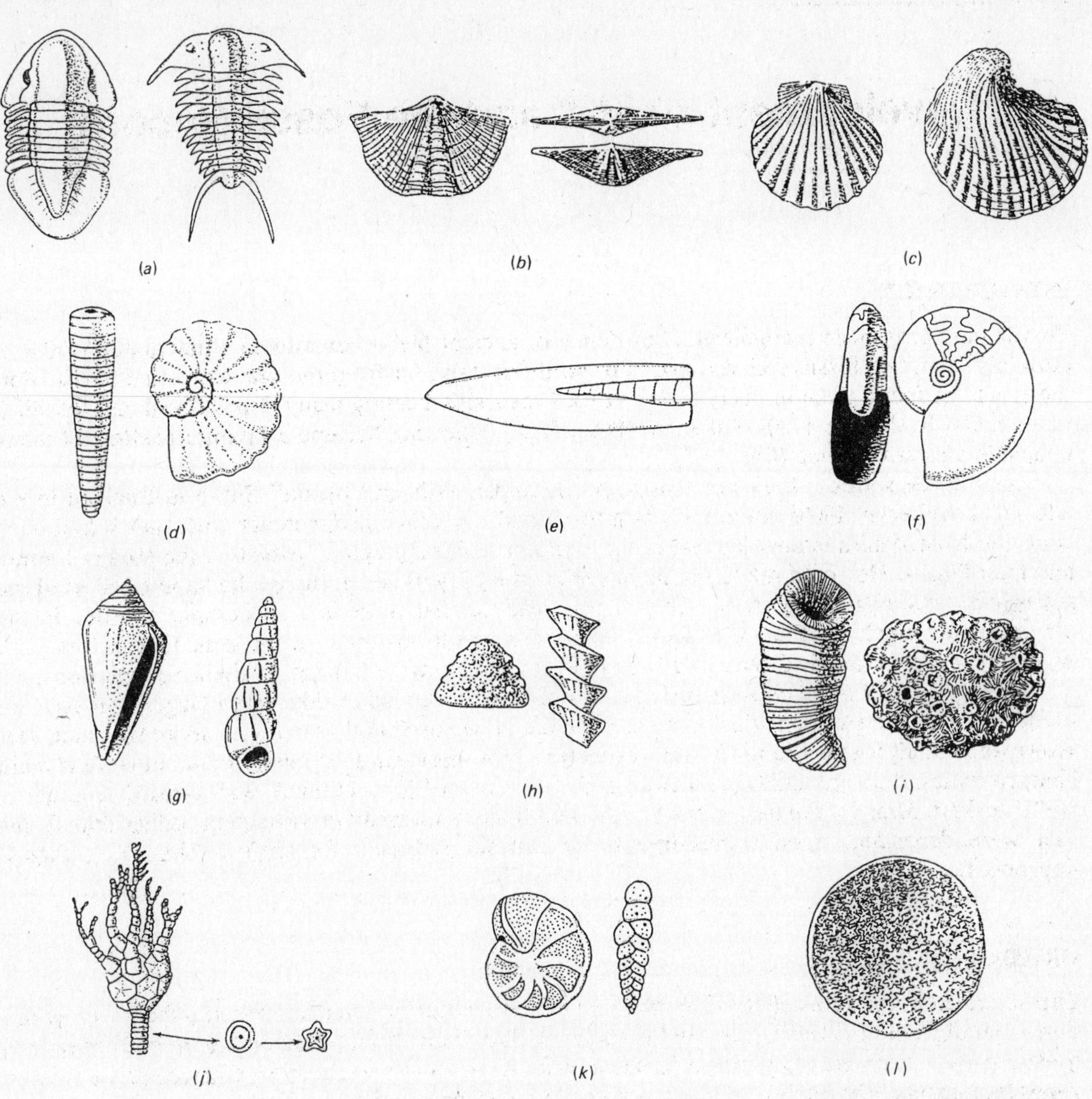

Fig. 19.1 Sketches of representatives of the main groups of invertebrates. A = trilobite; B = brachiopod; C = pelecypod (clam); D = nautiloid cephalopod; E = belemnoid cephalopod; F = ammonoid cephalopod (ammonite); G = gastropod (snail); H = bryozoan; I = coelenterata (coral); J = crinoid; K = protozoa (microscopic in size); L = sponges. (*From Ojakangas and Darby, 1976.*)

missing links; *Archeopteryx,* the first bird with feathers but retaining some reptilian characteristics, is a fine example (the best?) of a missing link. Nor does punctuated evolution mean that there is not evidence of the gradual evolution of some species; scientists have documented several examples, including the ammonite cephalopods, some pelecypods, and the horse.

THE FOSSIL RECORD: THE SPOTTY PRECAMBRIAN

The fossil record in Precambrian rocks is sparse. Only a few dozen fossil localities are known, in rocks ranging in age from 3500 m.y. (western Australia and Zimbabwe, Africa) to the 600- to 700-m.y. soft-bodied *Ediacara* fauna of southern Australia.

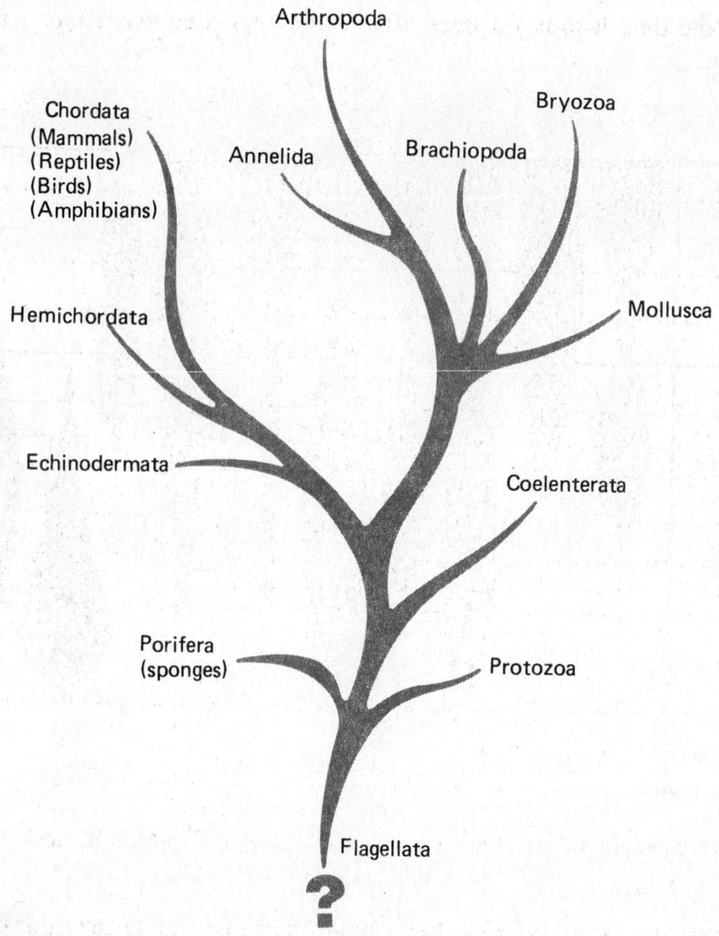

Fig. 19.2 Generalized evolutionary tree of animal life.

Most of these fossil localities contain bacteria and blue-green algae. The latter commonly build *stromatolites,* reeflike mounds of laminated sediment. The bacteria, many microscopic blue-green algae, and some unidentified forms are preserved in black cherts that have escaped recrystallization.

THE FOSSIL RECORD: THE PROLIFIC PHANEROZOIC

There are many types of fossil invertebrates (Fig. 19.1). These, along with microscopic single-celled animal fossils such as protozoa (there are other single-celled fossils, too) are the most useful to geologists for correlations and as environmental indicators because they are so abundant throughout the Phanerozoic (Cambrian to Recent) record.

There are documented evolutionary trends in various invertebrate groups. For example, it is known that nautiloid cephalopods evolved into ammonoid cephalopods. Different fossil groups are useful for correlations in specific parts of the rock record, as shown in Fig. 19.3. Trilobites are found only in the Paleozoic. Therefore they are useful in a broad way in correlation—one could immediately say the age of the rocks in which they are found is Paleozoic, between about 230 to 570 m.y. old. There are about 600 genera of trilobites in the Cambrian alone, and thousands of species, so they are valuable for more detailed correlations as well. Many individual species of invertebrates have short ranges; many are restricted to geologic periods or even portions of periods, and are useful for detailed correlations. For example, the Jurassic Period is about 55 m.y. long (see Fig. 18.1). The Jurassic rock record (i.e., the Jurassic System) has been subdivided into 3 series, 12 stages, and 45 zones on the basis of detailed studies of ammonites, which showed rapid evolution and extinction.

Therefore, the average time it took for each zone to be deposited was about 1 m.y., a short time, geologically speaking.

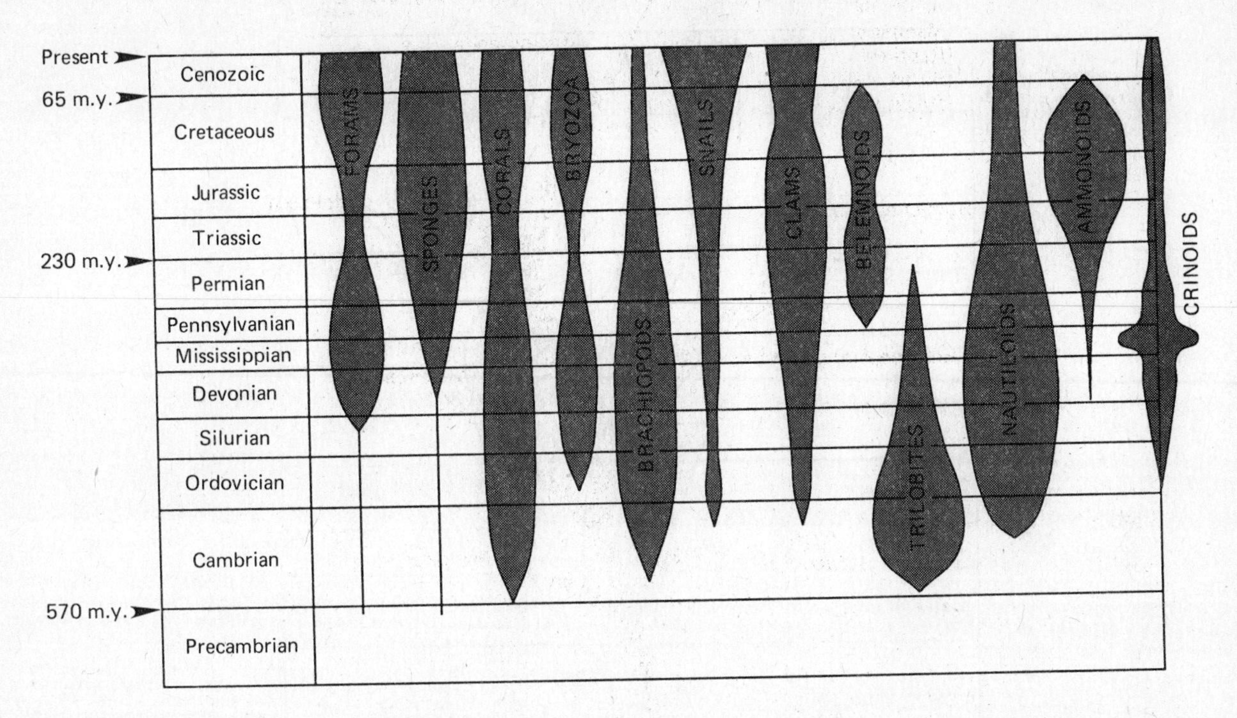

Fig. 19.3 Generalized geologic ranges of main invertebrate groups. Width of the lines indicates abundance.

The vertebrate record (see Fig. 19.4) is probably more intriguing than the invertebrate record, for it includes the position of mammals in the tree of life. The ranges of the vertebrates are given in Fig. 19.4.

The fossils record of hominids is a spotty one, consisting of fragments of skulls and other bones, but one that is being added to each year. Hominids are primates. The various types of primates descended from a common primate ancestor in the early Cenozoic, about 60 m.y. ago. Thus we are related to many other primates.

Solved Problems

19.1 Was Charles Darwin the first to write about evolution?

No. His grandfather, Erasmus Darwin, wrote a book entitled *The Descent of Mammals,* several years before Charles Darwin's *Origin of Species* in 1859. About the same time Darwin was developing his theory, A. R. Wallace came to much the same conclusions. In fact, they published a paper together to present their ideas in 1858.

19.2 What is the study of stratigraphy?

Stratigraphy is simply the study of strata and their contained fossils.

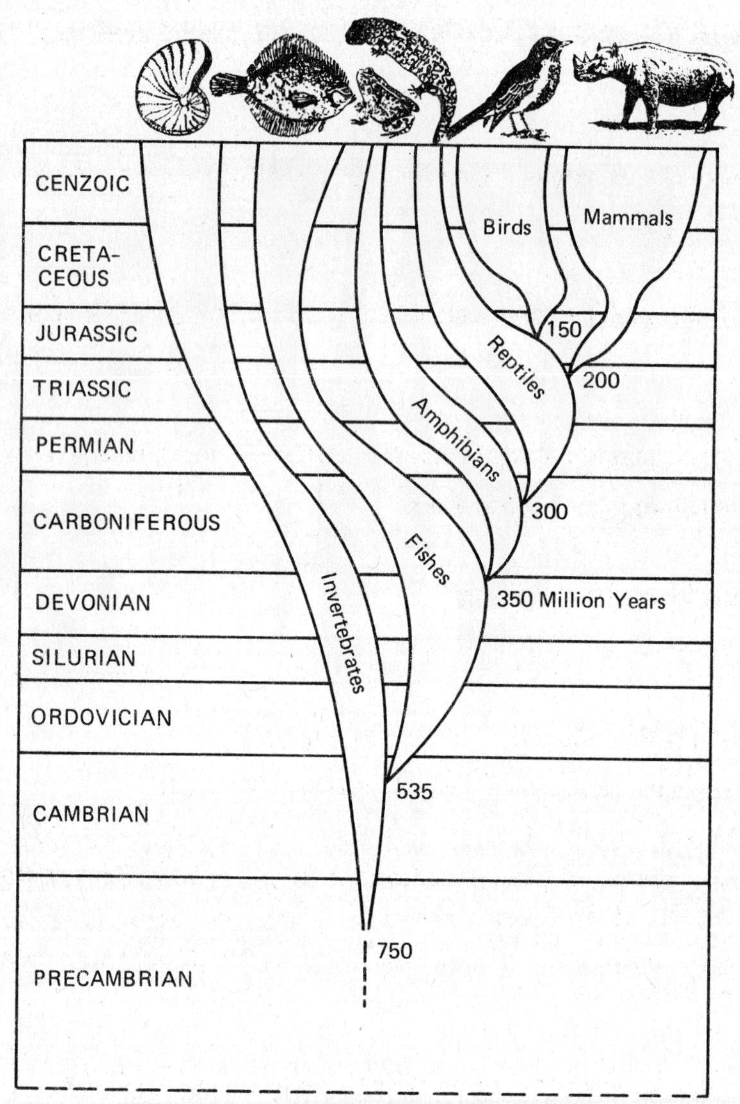

Fig. 19.4 Generalized geologic ranges of vertebrate groups. (*From Newell, N. D., Why Scientists Believe in Evolution, American Geological Institute, 1984.*)

19.3 Who is the Father of Stratigraphy and why?

William Smith of England. He was one of the first to use fossils for correlating rock units in his work as an engineer and surveyor.

19.4 How does natural selection, as proposed by Darwin, enter into the evolution of organisms?

As mutations occur, the new characteristics that are advantageous to the organism result in that organism's surviving better. This is natural selection. The surviving organisms reproduce and thus pass on their genetic characteristics.

19.5 Which *era* might be called the "age of invertebrates"? Refer to Fig. 19.4.

Paleozoic. However, by the end of the Paleozoic, fishes, amphibians, and reptiles had also evolved.

19.6 Which era might be called the "age of reptiles" or the "age of dinosaurs"? Refer to Fig. 19.4.

Mesozoic.

19.7 Which era might be called the "age of mammals"? Refer to Fig. 19.4.

Cenozoic.

19.8 Which part of the geologic time scale might be called the "age of humans"?

Quaternary.

19.9 In which geologic period did many varieties of fishes, including large armored types, live?

Devonian. It is known as the "age of fishes."

19.10 When did land plants first appear?

Silurian.

19.11 When did grasses first appear?

Miocene, based on the presence of grass pollen in the rock record.

19.12 Fred Flintstone, the caveman comic character, had a friendly Brontosaurus as a companion. Possible?

No. Dinosaurs became extinct at the end of the Mesozoic (65 m.y. ago), and hominids have been present only for the last few million years.

19.13 Which two animal groups evolved from the reptiles? (See Fig. 19.2.)

Birds and mammals.

19.14 What is the evolutionary sequence of vertebrates from fish to mammals?

Fish to amphibians to reptiles to mammals.

19.15 In which period did fishes with backbones first appear?

Ordovician, or possibly Late Cambrian.

19.16 In rocks of which period do we find the earliest amphibians?

Devonian.

19.17 In rocks of which period were the earliest fossil forests found?

Devonian.

19.18 In rocks of which period were the earliest reptiles?

Pennsylvanian.

19.19 How old are the earliest mammals, small rodentlike creatures?

Late Triassic. (The first placental mammals were Cretaceous in age.)

19.20 We have a good fossil record of one common mammal. Which one?

The horse, from Eohippus in the Eocene to the modern horse in the Pleistocene and Present.

19.21 Assume a fossil assemblage from a shale bed contains pelecypods, gastropods, brachiopods, and trilobites. What is the age of the bed? (See Fig. 19.3.)

Based on the given evidence, the age of the rock would be Paleozoic. The trilobites are restricted to the Paleozoic—the other fossils as groups lived throughout the Phanerozoic. (If the genus and species of each of the named classes of fossils were known, the age could be known with more precision, perhaps even down to a geologic period. If the brachiopods were a spirifer type, for example, the age of the bed would be late Paleozoic.)

19.22 Assume a fossil assemblage from limestone contains crinoids, brachiopods, and bryozoans. What is the age of the limestone?

The data are not specific enough to say anything more than that the age of the bed is Phanerozoic (excluding Cambrian, for crinoids and bryozoans, originated in Ordovician time) (Fig. 19.3).

19.23 Assume a fossil assemblage from a shale unit contains remains of clams, snails, flying reptiles, and swimming reptiles. What is its age?

Based on the presence of flying and swimming reptiles, the rock is Mesozoic in age.

19.24 How common is fossilization? That is, what portion of the fauna and flora are preserved in rocks as fossils?

Probably only a very small proportion is preserved. Recall that, in general, fossilization requires that the organisms possess hard parts and be buried "immediately" (see Problem 18.25). There are many soft-bodied organisms living today, including worms, slugs, jellyfish, many insects, most plants, and many other groups. Each of these groups lived in the geologic past but they are rarely preserved because of the lack of hard parts. Immediate burial, before decay or destruction by scavengers or by abrasion, is also unusual. (This is an important factor in the lack of "missing links." If they existed, their chances of preservation were slim even if they possessed hard parts and were immediately buried. Besides, they must then be exhumed and discovered.)

19.25 How many species of organisms have been described?

1.5 million. Who knows how many living species are undescribed? Obviously, many fossil species are also undiscovered and hence undescribed.

19.26 How do geologists know that grasses did not appear until the Miocene?

By the study of microscopic pollen preserved in the rock record. None is pre-Miocene in age.

19.27 What are the subdivisions of the animal kingdom, from phylum down to species?

Phylum
    Class
        Order
            Genus
                Species

19.28 Which two of the invertebrate groups of Fig. 19.1 are the most abundant in the oceans today?

Two classes of the phylum mollusca, the pelecypoda (clams) and the gastropoda (snails).

19.29 Is the following a true statement? "Once extinct, a species never reappears."

Yes. Organic evolution is irreversible. The fossil record solidly substantiates this statement.

19.30 How do mutations occur?

Mutations are chemical changes in genes, which are segments of the nucleic acid commonly known as DNA. DNA can duplicate itself, following an internal genetic code that controls the proteins that that organism can assemble. A mutation occurs if the new DNA does not perfectly duplicate itself as the cell divides, or if the various segments of DNA are out of order or missing, or if cosmic radiation alters the ordering within the DNA.

19.31 Define species.

A *species* is defined as the major subdivision of a genus, regarded as the basic category of biologic classification, composed of related individuals that resemble one another and are able to breed among themselves, but are not able to breed with members of another species.

19.32 Formations A, B, and C of Fig. 18.2 (yes, Chapter 18) contain fossil trilobites. What age are these formations?

Paleozoic, because trilobites lived only in the Paleozoic.

19.33 In which units on Fig. 18.2 should you search for primitive fish?

A, B, or C. Fish evolved in the early Paleozoic, and only these units could be Paleozoic in age.

19.34 In which formations on Fig. 18.2 should one look for the remains of dinosaurs?

Formations D and E, those of Mesozoic age.

19.35 Formation G of Fig. 18.2 is a lava flow that has just been dated radiometrically at 10 m.y. An archeologist has just found primitive tools associated with large mammal remains in the top part of formation F and is elated to no end! Why is he jumping up and down with excitement?

These would be the oldest human remains ever found! To date, the oldest are perhaps 5 m.y., so these would be at least twice as old.

19.36 See Fig. 19.1. How can a brachiopod be distinguished from a pelecypod?

The shell of a brachiopod is bilaterally symmetrical. That is, if the shell is cut into two pieces with the cut made down the middle of both valves, the two sides will be mirror images of each other. On the other hand, if such a cut is made on a pelecypod, the two halves will not be similar. (The pelecypod used as the symbol of a major oil company is really not quite symmetrical, either.) In many pelecypods, the two valves (shells) are mirror images of each other.

19.37 Refer to Fig. 19.1 without looking at the figure captions. Which of the drawings is a: (1) Trilobite? (2) Brachiopod? (3) Nautiloid cephalopod? (4) Ammonoid cephalopod? (5) Bryozoan? (6) Pelecypod? (7) Gastropod? (8) Crinoid? (9) Belemnoid cephalopod? (10) Coelenterata (coral)? (11) Protozoa? (12) Sponge?

1. a	7. g
2. b	8. j
3. d	9. e
4. f	10. i
5. h	11. k
6. c	12. l

19.38 What is the concept of *punctuated equilibrium?*

This concept can be stated as follows: Species appear rather suddenly in the rock record and do not change much before becoming extinct. This helps to explain situations where there is a lack of a series of fossils showing gradual changes between species (i.e., the so-called missing links).

19.39 In which formation in Fig. 18.2 should you search to find ammonites?

In D and E, for ammonites are found in strata of Mesozoic age.

19.40 An absent-minded professor found fossil remains of *Archeopteryx,* the earliest bird, but cannot recall in which formation of Fig. 18.2 he found them. Can you help?

He must have found them in formations D or E, the only Mesozoic units present. *Archeopteryx* is Jurassic in age. (However, he could have theoretically, at least, found them in another formation, but if so, the range of this fossil would be changed.)

19.41 Which formation of Fig. 18.2 might contain mammal remains?

Formation E, for it is the only sedimentary unit that is Cenozoic in age, other than the lava flow. (See Problem 19.35.)

19.42 When and why did fossils first become abundant in the rock record?

Animals started secreting phosphatic and calcium carbonate shells in early Cambrian time, and by the end of the Ordovician, all major invertebrate groups were represented by abundant shelled forms.

19.43 What are the characteristics of primates?

They are mammals. They have five-digit feet with opposed thumb and finger to grasp with, and both eyes are on the front part of the skull, which results in stereoscopic vision.

19.44 What are the identifying characteristics of humans (*Homo*)?

Large brain size and bipedalism (walking on hind legs). To these, some would add the ability to make tools, but of course this cannot be ascertained from fossil remains unless tools are intimately associated with the fossils.

19.45 What identifies a skull as "human"?

Several characteristics, the most important of which is the size of the brain cavity. Modern humans (*Homo sapiens*) have a brain size of about 1400 cm^3, as compared to Neanderthal Man (also *Homo sapiens,* 1300 cm^3), 1000 cm^3 for *Homo erectus,* 800 cm^3 for the earliest humans (*Homo habilus*), and only 500 to 600 cm^3 for *Australopithecus*. Gorillas have brains that are only about one-third the size of a human brain.

19.46 How old are the oldest human remains?

The answer to this question has been changing repeatedly in recent years as new discoveries of hominids are made in eastern Africa. The oldest finds, by Mary Leakey, are 3.6 to 3.8 m.y. old. However, some other remains may possibly be as old as 5 m.y. Hominid primates (human ancestors) are certainly as old as 8 m.y. and perhaps as old as 14 m.y.

19.47 Who are the Leakeys?

Louis Leakey, his wife Mary Leakey, and their son Richard Leakey were all scientists who spent their careers searching for hominid remains in east Africa. Richard is still active.

19.48 Which fossils are the earliest animal fossils found thus far?

The *Ediacara* fauna from southern Australia. These fossils are imprints of soft-bodied organisms, an unusual fossil occurrence.

Supplementary Problems

19.49 What is the history of the horse, with regard to the North American fossil record?

The horse apparently evolved in North America, with the small *Eohippus* bones found in western North America. However, by Late Pleistocene, about 8000 years ago, the horse had become extinct in North America. The Spanish conquistadors in the 1500s reintroduced the horse to North America when they left their horses here rather than bring them back to Europe when they had completed their exploration for cities of gold.

19.50 What caused extinctions of animal groups at various times in earth's history?

This is a big question, too big to answer well here. However, there are several possibilities. Many groups may have become extinct due to climatic changes which affected the food supply, or to competition from other animals that were better adapted to the environment (natural selection). Other factors appear to have been the movement of continents into different latitudes, or the contraction of marine environments (i.e., broad seas retreating from the continents) that theoretically would have increased competition for a more areally limited habitat. It has been suggested that diseases may have killed off animal groups, but this idea is especially difficult to verify. Some time-specific extinctions are the subjects of the next questions.

19.51 Why did the dinosaurs, and many marine organisms as well, become extinct at the end of the Mesozoic?

The factors listed in the previous question may have been important. A very interesting theory involves catastrophe, as proposed by Alvarez and Alvarez. A very large stony asteroid (i.e., a large meteorite or comet core) may have hit the earth, sending up a great dust cloud that prevented sunlight from reaching the earth for months or years and thus affecting plant growth and the entire food chain. (And the source of vitamin D from the sun would have been cut off, too!) Perhaps it caused raging forest fires—there is evidence of much charcoal at this horizon in the rock record. But the big evidence is the "iridium anomaly" in a thin layer (less than 1 in thick) at many localities in the world right at the Mesozoic-Cenozoic (Cretaceous-Tertiary) boundary. The iridium content at this stratigraphic horizon is 10 to 100 times above the normal very low quantities. Iridium is a rare element on earth, and an extraterrestrial source is the explanation. In 1989, two amino acids that are absent on earth but are present in the carbonaceous chondrite type of meteorite, were discovered at the boundary in Denmark; this adds further support to the hypothesis.

19.52 Name another catastrophic event that could have affected the amount of sunlight reaching the earth.

A large volcanic eruption. Even rather small eruptions in recent years (such as the 1982 eruption of a volcano in Mexico) affected radiation and plant growth in many places on earth.

19.53 Many Pleistocene mammals became extinct in North America between 6000 and 12,000 years ago, including the horse, the camel, a pronghorn, a bison, the ground sloth, the wooly mammoth, and the saber-toothed tiger. Why?

Climatic changes after the last glacial advance was once thought to be the cause of these extinctions, but why didn't such extinctions occur after the first three Pleistocene ice advances? A more recent theory is called "prehistoric overkill." According to this idea, people in North America (paleo-Indians) caused their extinction, not by just killing the animals for food (which was difficult with primitive weapons), but by disturbing the food cycle. If these people competed with carnivores for the weak and young herbivores during hard winters, and succeeded, it could have caused the demise of some carnivores. This in turn could have upset the food chain, causing the rapid expansion of certain herbivores at the expense of other less well adapted herbivores.

19.54 What is the biggest cause of extinctions of plants and animals today and in the last few hundred years?

Human activity has either directly (by killing) or indirectly (by destroying natural habitats) caused myriad extinctions. The obvious victims are the large animals, but small animals (including insects) and plants have suffered the most.

19.55 What was distinctive about the animal life of South America until the Pliocene, and what caused extinctions there?

There were no carnivores in South America until the Panama landbridge was established by volcanism and plate movements in the Pliocene. Then, carnivores moved south, and many animals (giant anteaters, giant ground sloths, and others) became extinct. The armadillo successfully migrated northward into North America.

19.56 Figure 19.3 shows the time ranges of various animal groups. How can they be useful for dating the sedimentary rocks in which they are found when the range of each group is so long?

This is a generalized diagram for the total time range of each phylum or group of organisms. Smaller subdivisions, especially genus and species, have short time ranges and provide much more precise information. A given species, for example, may have lived only in Late Cambrian time, and another in Middle Jurassic time.

19.57 The principle of fossil correlations allows strata containing like assemblages of fossils to be correlated (i.e., assigned like ages). Are there exceptions to this principle?

If fossils are eroded from an older stratum and redeposited in a younger stratum, there could be an error in correlation. However, geologists should be aware of this problem and be sure the fossils they use in correlation do not show signs of abrasion and redeposition.

19.58 What is the significance of the Burgess Shale?

It is a Middle Cambrian unit in the Canadian Rockies of British Columbia that contains a rich fauna of soft-bodied organisms that usually are not preserved because of the lack of hard parts. The fauna is a deep-water one, consisting mostly of worms and arthropods, plus sponges and jellyfish. Some plants are also present. Usually, only burrows or tracks (trace fossils) of such organisms are preserved.

19.59 How do the following three classes of chordata (animals with a protected spinal nerve chord)—amphibians, reptiles, and mammals—differ from each other?

Amphibians need water for eggs and larva. Reptiles are independent of water, as their eggs can be laid on land. Mammals are warm-blooded, give birth to live young, suckle their young, and are hairy. (The above characteristics, of course, cannot be used to distinguish fossil remains.) Details in limb and hip structure, tooth structure, and other features are diagnostic to a specialist.

19.60 In earth history, a few events might be classified as the most important events to happen in organic evolution. What are they?

The development of photosynthesis in plants. Early plants were anaerobic heterotrophs that ate amino acids from the "primeval soup." Later, autotrophs utilized the process of photosynthesis to manufacture their own food (hence the food cycle that is so basic to all organisms) and to produce oxygen as a waste by-product that enabled animals to develop.

The development of the DNA molecule, which enabled genetic variation to be perpetuated.

The development of eucaryotic cells (i.e., those with a nucleus and some organelles, which are structures within cells that process food, allow for the elimination of wastes, initiate reproduction, and other bodily functions).

19.61 Give a general equation for photosynthesis, the process whereby inorganic materials are converted into food for plants.

$$CO_2 + H_2O + \text{light} \xrightarrow{\text{chlorophyll}} CH_2O + O_2$$

19.62 What are the oldest fossilized remains, and where are they?

Microscopic blue-green algae and bacteria. The algae also resulted in the construction of *stromatolites,* mound- and reeflike laminated structures composed of sediment trapped by the mucilageous (sticky) algal colonies. The oldest finds are in the 3.5-b.y.-old Pilbara block of northeastern Australia. Other stromatolites probably of about the same age have been found in Zimbabwe, Africa.

19.63 What does ontogeny recapitulates phylogeny mean?

Ontogeny is the development of the individual, whereas *phylogeny* is the history or evolution of an organism's ancestors. The above expression means that the ontogeny of an individual reflects a history of the evolution of the organism. That is, the various life stages of an organism, from conception onward, indicate the life history of the organism.

19.64 What is adaptive radiation?

Adaptive radiation is a rapid expansion of an animal or plant species into an unoccupied ecological niche. More technically, it is diversification by means of the development of new species into a new environment and the subsequent specialization by new species in the new environment.

19.65 What is "creation science," and why is it not really science?

"Creation science," or creationism, states that the book of Genesis in the Bible explains the creation of the earth and its inhabitants. That is, God created it all, rather quickly. Thus, the scientific method is ignored. Recall that the scientific method involves the identification of a problem, gathering data, the formulation of an objective hypothesis, and finally testing the hypothesis by experiment or observation to see if it is indeed a valid hypothesis. There is no conflict between religion and science.

Chapter 20

Origins and the Long Precambrian

ORIGIN OF THE CRUST, MANTLE, AND CORE

Our solar system, of which earth is one very small part, probably originated as a disk-like cloud of gases and dust (a solar nebula), of which the sun remains as the "core." Part of the cloud condensed to form planetesimals which collided and grew into the four dense terrestrial interior planets (Mercury, Venus, Earth, and Mars). The gaseous outer planets formed from the more diffuse and gaseous portions of the nebula. All this may have occurred 5 to 6 b.y. ago. By 4.5 to 5.0 b.y., the Earth had essentially formed. An abundance of planetesimals bombarded the Earth until about 3.9 to 4.0 b.y. ago.

Heating up of the earth during this accretion, perhaps about 4.7 to 4.6 b.y., led to a large-scale melting and internal differentiation of the earth into an iron-rich core and a silicate-rich mantle and the more familiar crust. All these internal "layers" of the earth are elaborated upon in Chapter 15. The crust has been in a state of change ever since it first formed, and much of the contents of this book deals with changes on or in the crust.

ORIGIN OF THE HYDROSPHERE AND ATMOSPHERE

Outgassing of the earth's interior via volcanism probably produced the earth's hydrosphere and much of its atmosphere. This early atmosphere may have been composed of methane, ammonia, and water vapor. The oxygen of the present atmosphere, however, was probably produced both by the photoelectric dissociation of water and most importantly, as a by-product of photosynthesis of early plants (algae).

OLD ROCKS, MOUNTAIN BUILDING, AND STABLE PLATFORMS

The oldest radiometric dates on earth's rocks are 3.8 b.y., on some rocks from southern Greenland and Labrador, but individual zircon crystals from a quartzite in western Australia have been dated at 4.2 b.y. Numerous other localities, including the Minnesota River Valley in southwestern Minnesota, contain rocks 3.5 b.y. old. Probably very few rocks will give older dates because the earth's crust has been reworked so intensively by weathering, erosion, metamorphism, and melting.

All continents have nuclei of Precambrian shields composed of Precambrian rocks (older than about 570 m.y.). The best known of these shields is the Canadian Shield (Figs. 20.1 and 20.3). The older parts of these shields contain an abundance of plutonic and volcanic rocks dated at 2.7 b.y., with many dates between 2.7 and 3.2 b.y. In North America, 2.7-b.y.-old granitic rocks formed during the Kenoran (Algoman) orogeny, or mountain-building event, that included deformation and metamorphism of the slightly older volcanic and sedimentary rock units. The emplacement of all these granites gradually transformed earlier thinner-crusted protocontinents into thick (35 to 60 km) continents. Erosion of the mountains by about 2.4 b.y. ago produced stable platforms upon which seas could transgress and upon which shallow water sediments such as quartz sandstones and carbonates could accumulate for the first time in quantity. Minor unique and important rock units include glacial deposits from a 2.3 to 2.2 b.y. glaciation.

After the 2.7 b.y. activity, the next major concentration of age dates on the world's shields is about 1.8 b.y., the age of the Hudsonian (Penokean) orogeny of North America. This synchroneity of dates on the various continents may be related to early manifestations of worldwide plate move-

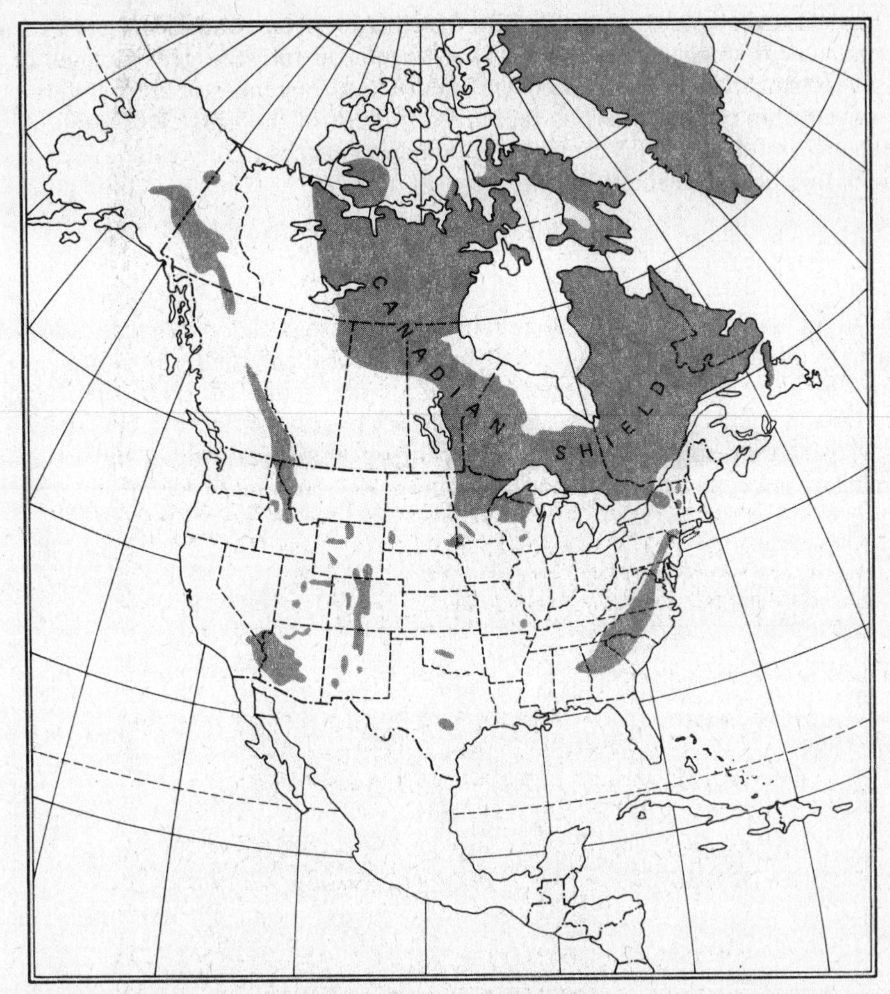

Fig. 20.1 North America showing areas of Precambrian rocks. The large area of Precambrian rocks is the Canadian Shield. (*From Ojakangas and Darby, 1976.*)

ments. There is excellent evidence on several continents of subduction having occurred at 1.8 b.y., but evidence is not as convincing for older rocks.

Following each of the above orogenies, or periods of mountain building, there was a long period of erosion (as long as hundreds of millions of years) which produced beveled platforms, in places covered with fluvial (river) deposits of quartz-rich sands that were the end product of weathering coupled with wind abrasion. Some of the quartz-rich sands were deposited in the seas that transgressed unto the eroded and low-lying continents. The main product of weathering of rocks, clay, was probably largely blown away by winds and deposited in the oceans.

THE LATER PRECAMBRIAN

The North American continent started to split apart along a line extending from the Lake Superior region to Kansas about 1.1 b.y. ago. This was evidently the same type of spreading process that broke up Pangea at the beginning of the Mesozoic (see Chapter 17). However, the spreading stopped after less than 100 km (60 mi) of separation. Flood basalts in the rift zone, which is called the Midcontinent Rift, may be locally as thick as 15 km in the Lake Superior region. The volcanism was followed by sedimentation, mostly as redbeds on alluvial fans and in braided river systems that flowed into the depression.

On the eastern and western margins of the North American continent, Late Precambrian sediments accumulated in elongate depressions that developed into the geosynclines of the east and west, the Appalachian and Cordilleran geosynclines. Thick sequences of unfossiliferous sedimentary rocks pass upward into the fossiliferous sedimentary rocks of the Early Paleozoic. Also in the sedimentary sequences of the Late Precambrian at numerous places in the world are deposits of at least two glaciations that occurred about 800 m.y. to 600 m.y. ago.

SUMMARY

Not mentioned in the above historical review are new suggestions that various continents or microcontinents collided with the Canadian Shield, thereby enlarging the continent. Whether the continent grew by accretion or not, there is at least a crude zoning of ages, with the oldest near the center of the shield (Fig. 20.2). The Precambrian history of any continent is obviously long and complicated. However, the antiquity of the rocks, coupled with overprinting by younger events and burial by younger rocks, makes the history difficult to decipher in detail.

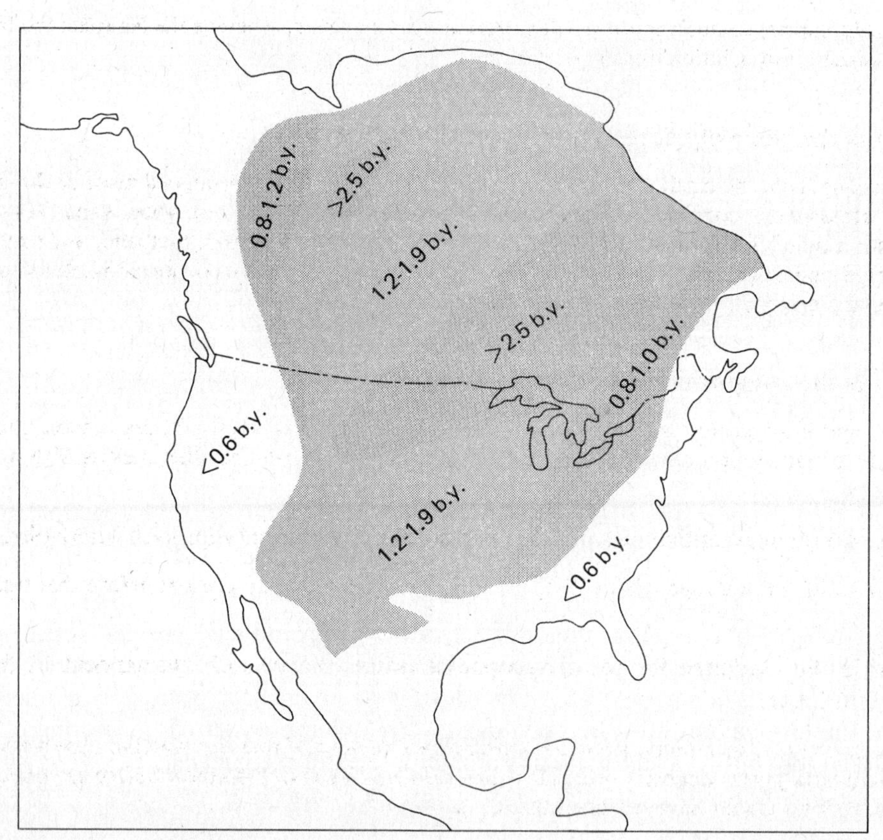

Fig. 20.2 Generalized zones of different-aged rocks of the North American continent. (*From Ojakangas and Darby, 1976.*)

Solved Problems

20.1 What is the generally accepted origin of the earth—hot, cold, or both?

Both origins are involved in the *nebular hypothesis* in which a spinning gas cloud differentiated into denser terrestrial planets near the sun and less dense gaseous planets further out. The denser planets formed by the aggregation of planetesimals (previously condensed solid bodies), perhaps with the lighter gaseous fraction "blown away" by the "solar wind," whereas the gaseous planets formed from condensation of the less dense and more gaseous outer parts of the nebula.

20.2 Where did earth's water probably come from?

Probably from the outgassing and condensation of water from the earth's interior via volcanism. (Recall from Chapter 2 that many minerals contain water.)

20.3 Why is seawater salty?

The Cl (chlorine) came from the earth's interior via volcanism, whereas the Na (plus the Mg, Ca, and K) were placed into solution during weathering.

20.4 What is the generally accepted origin of the atmosphere?

Outgassing from the earth's interior during volcanism could have produced most of the gases if earth's early atmosphere consisted of methane (CH_4), ammonia (NH_3), and water vapor (H_2O). Oxygen is unique; it may have formed in part by the dissociation of water vapor, methane, and ammonia. A more accepted and important possibility is that the oxygen was produced during photosynthesis by early plants and given off as a by-product.

20.5 What is the composition of the earth's atmosphere?

78 percent nitrogen (by volume), 21 percent oxygen, 0.9 percent argon, and traces of other gases as well. Depending upon other factors, water vapor may be present in quantities as high as 4 percent.

20.6 When do most scientists assume the earth developed an oxygen-rich atmosphere?

About 2.2 to 2.0 b.y. ago. (However, some oxygen was certainly present before that time.)

20.7 What is the evidence for the development of an oxygen-rich atmosphere at that time (see Problem 20.6)?

The worldwide distribution of Precambrian iron-formations of that age, and the observation that detrital uranium-gold-pyrite deposits are older than that and probably formed before an oxygen-rich atmosphere, which would have oxidized them.

20.8 How long was the Precambrian?

It includes earth's history from its origin at about 4.5 or 4.6 b.y. ago until the beginning of the Paleozoic, 570 m.y. ago. This encompasses about ⅞ of earth's history.

20.9 What are the two main subdivisions of the Precambrian?

The Archean and the Proterozoic eras. The boundary between them is generally accepted as 2.5 b.y.

20.10 How old are the oldest Precambrian fossils?

About 3.5 b.y. They are stromatolites from Zimbabwe and western Australia.

20.11 How old are the oldest dated rocks on earth, and where do they occur?

3.8 b.y., from southern Greenland and Labrador. However, a date from western Australia is 4.2 b.y.

20.12 Why was the study of Precambrian rocks generally avoided by early natural scientists, in favor of the Phanerozoic rocks?

Because of deformation, metamorphism, and the general lack of fossils.

20.13 Are there mineral deposits in Precambrian rocks?

Yes! Many! Most of the world's iron is in Proterozoic rocks, and much of the world's uranium, copper, zinc, nickel, chromium, and other elements are also in Precambrian rocks.

20.14 Are Precambrian rocks found on all continents?

Yes. (See Fig. 20.3.) Note that much of eastern Antarctica, based on the rocks exposed along the coast, is also Precambrian in age.

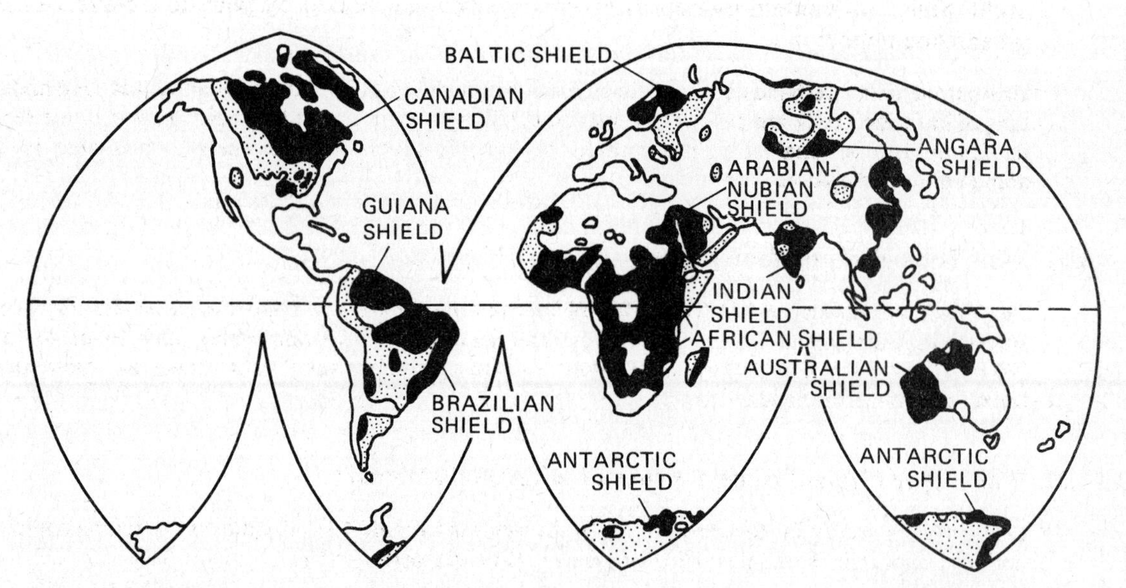

Fig. 20.3 Generalized map of Precambrian shields of the world. Dark pattern represents exposed rock; dotted pattern represents shield rocks beneath a thin cover of younger sedimentary rocks. (*After Heller et al., Earth Science, McGraw-Hill, Webster Division, New York, 1979.*)

20.15 What is a general term for the large areas of Precambrian rocks that occur on all continents?

Shields. They constitute continental nuclei.

20.16 Two major Precambrian orogenies seem to be worldwide. When did they occur?

About 2.7 and 1.8 b.y. ago.

20.17 What is the likely origin of the great volumes of granite in the shields of the world?

Partial melting of rocks in the crust or upper mantle. These could be sedimentary, metamorphic, or igneous rocks.

20.18 Quartz sandstones are not common in the Archean portions of continental shields. Why not?

Because stable platforms are necessary for the prolonged and deep weathering that eliminates most minerals other than quartz, and large platforms did not form until the late Archean. In addition, wind and shore processes would be effective on a platform.

20.19 Quartz sands are difficult to "make" today. What factor would have made them easier to make in the Precambrian?

The lack of vegetation, and hence unchecked wind abrasion. Recall from Chapter 5 that wind is the best agent for rounding sand grains, better than water processes. Wind abrasion would cause feldspars to fracture along cleavage, and fine-grained rock fragments would break along mineral boundaries; silt would be the product.

20.20 The Canadian Shield of North America, the Guianan and Brazilian Shields of South America, the Fennoscandian and Angara Shields of northern Europe, the Ukrainian Shield, the Indian Shield, the African Shield, the Arabian-Nubian Shield, the Australian Shield, and the Antarctic Shield all contain numerous greenstone belts separated by granitic terrane. What are greenstone belts?

Greenstone belts are bands of volcanic-sedimentary rocks, generally a few hundred km long and a few tens of km wide. They are called "greenstone belts" because many of the rocks are green, the result of metamorphism of original volcanic rocks to green minerals, mainly chlorite, actinolite (a green amphibole), and epidote.

20.21 What is banded iron-formation, sometimes abbreviated as BIF?

The major iron-formations of the world, all Precambrian in age, are composed of alternating bands of silica-rich (cherty) and iron-rich minerals (mainly hematite and magnetite), and hence are called "banded iron-formations." They are sedimentary rocks; some consist of thin bands or laminations and others contain thick bands.

20.22 Which type of fossil is most abundant in the Precambrian?

Stromatolites are by far the most common Precambrian fossil. The only others are microscopic forms and rare impressions of soft-bodied organisms. (See Problem 19.62.)

20.23 What is the evidence for Precambrian glaciation?

Diamictites (matrix-rich conglomerates) associated with laminated beds that contain oversized dropstones that obviously were dropped into a low-energy environment provide the best evidence. Other evidence includes striated surfaces beneath diamictites and striated stones in the diamictites. If the diamictite was likely deposited directly by glacial ice, then the term *tillite* (lithified till) might be applied by some workers (see Problem 10.46).

20.24 What is the Midcontinent Rift?

A down-faulted long and narrow portion of the North American continent, extending from the Lake Superior region to Kansas. It originated when North America started to split apart along a spread center about 1.1 b.y. ago. It contains basaltic lava flows and a thick sequence of red sedimentary rocks deposited in braided stream and alluvial fan environments. (See Fig. 20.4).

20.25 What is the theory of *continental accretion?*

The theory that continents have grown through time by the addition of material around the edges. This accreted material may be island arcs formed during subduction of an oceanic plate, or other types of suspect or exotic terranes that slammed against the continental margins during plate movements (see Problems 17.36 and 17.37).

20.26 Where are the oldest rocks in North America (3.8 to 2.7 b.y.)?

Generally speaking, in the center of the continent, in south-central Canada and the Wyoming-Montana area. Small areas of 3.5- to 3.8-b.y.-old rocks occur in Minnesota, Michigan, and Labrador. (See Fig. 20.2.)

20.27 Where are the youngest rocks in North America, in general?

Along the continental margins. (See Fig. 20.2.)

Supplementary Problems

20.28 Much of the Midcontinent Rift is buried beneath Phanerozoic rocks and has no surface expression. How is its presence determined in those areas?

By geophysical methods. The basalts of the rift have high gravity and strong magnetic signatures.

20.29 Where are rocks of the Midcontinent Rift best exposed?

Lake Superior region. (See Fig. 20.4.)

20.30 Geologists who study the lava flows of the Midcontinent Rift have commonly journeyed to Iceland. Why?

To eat bread baked in the steam from the active volcanic area. Also, Iceland sits astride a modern, active spread center, and is about the only part of the spread center that is easily accessible without submarines. Therefore, the rocks of the ancient spread center can be compared to a modern analog. There are, in fact, striking similarities.

20.31 Are all the Precambrian rocks of the various continents found in the shields?

No. In North America, for example, Precambrian rocks are exposed in the cores of many uplifted areas (mountains and domes) and even in the bottom of the Grand Canyon. A shield is generally defined as the large exposed area of Precambrian rocks of a continent. In reality, the Precambrian rocks of outlying areas are connected at depth to the shields; younger rocks have simply covered the older rocks. (See Fig. 20.1.)

20.32 In northwestern Australia, zircons from a quartzite have been dated by the U-Pb radiometric method as 4.2 b.y. old. What does this imply?

It means that the zircon crystals crystallized that long ago, probably in a granitic rock, and that the rock was weathered and eroded to produce quartz sand and zircon grains. It means, if the age is correct, that granitic rocks of that age were uplifted and exposed by erosion.

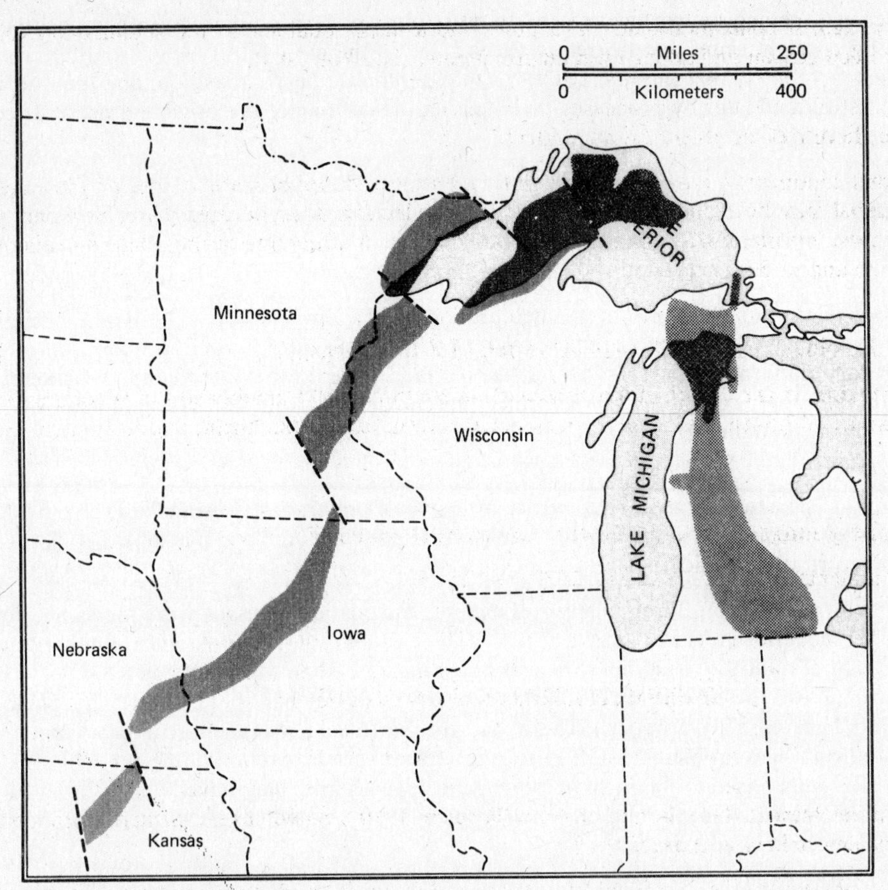

Fig. 20.4 The Midcontinent Rift of Late Precambrian age. (*From Ojakangas and Darby, 1976.*)

20.33 What age are most gneiss belts of the world?

Archean.

20.34 How do Proterozoic sedimentary rocks differ from Archean sedimentary rocks, in general?

While some are very similar, the Proterozoic contain quartzites and carbonates, deposited in platformal stable environments as quartzose sands and carbonates. Such rocks are not very common in the Archean portions of the rock column.

20.35 Have geologists found the "original crust"?

No. Recall that the earth is about 4.5 b.y. old, based on the dates of moon rocks and meteorites, whereas the oldest dated rocks on earth are about 3.8 b.y. old. Therefore, the original crust has not been found.

20.36 Why is it highly unlikely that the "original crust" will ever be found?

Because of extensive cratering on earth by meteorites about 4 b.y. ago and because of uplift, weathering, erosion, magmatism, and metamorphism.

20.37 Would the "original crust" likely be granitic or basaltic?

Many decades ago, geologists assumed the early crust must have been granitic in composition, composed of the light-weight scum that floated to the surface as the early earth melted and differentiated. Today it seems more likely that the first crust was mafic (basaltic) in composition, and that the granitic continental crust formed later by processes including partial melting of the original crust and recycling of early formed sediments.

20.38 Holmes of England in the 1950s compared geologic time to a six-story building, in which only the top story was nearing completion. The lower five stories have only the basic girders but no walls or other details. Each floor, he said, represents 600 m.y. of time. Why aren't the lower floors finished?

The upper floor represents about 600 m.y., the duration of Phanerozoic time. The Phanerozoic is relatively well known, for the Phanerozoic rock record contains fossils, and its rocks are less deformed than the older Precambrian rocks and are nearer the surface. The lower five floors, representing the Precambrian rock record, commonly are deformed and metamorphosed, and do not contain the fossils that allow for detailed studies of ages and environments.

20.39 Geologists use a basic axiom in their studies, the principle of uniformitarianism (see Problem 18.26). Is this principle—that the present is the key to the past—valid for the Archean part of the rock record? If not, in which regard is it invalid?

It is not valid in every respect. Surely many surficial and magmatic processes were the same, but there were some differences with regard to other aspects. The earth was hotter in Archean time because of a greater abundance of radioactive isotopes then. (For example, the half-life of U^{238} is 4.5 b.y., so back in earliest Archean time at the birth of the earth, there was twice as much of that isotope available to decay and to give off heat as a by-product.) Therefore, the earth had a higher heat flow and a higher geothermal gradient. (See Problem 18.45.) Also, the atmosphere in Archean time probably had much less oxygen than today's (see Problem 20.4), so chemical weathering may have been different. Finally, there was no land vegetation until Silurian time (Problem 19.10), so wind action must have been much more important in surficial processes.

Chapter 21

Earth History: The Paleozoic

INTRODUCTION

The Paleozoic, the earliest era of the Phanerozoic, began with a profusion of life forms in the seas. The resulting fossil record in the sedimentary rock column provides a record of transgressions and regressions of generally shallow seas (i.e., epeiric or epicontinental) upon the platforms (cratons) of Precambrian rocks. The warm water fossils indicate that the equator extended across the middle of the North American continent for most of the Paleozoic era. The sedimentary rocks are also the documentation for uplifts (mountains, domes, and arches), for the subsequent erosion of these once topographically prominent features produced the sediments.

In Late Precambrian time, the Appalachian and Cordilleran Geosynclines were present along the eastern and western margins of the North American craton (Fig. 16.12), and they continued to exist throughout the Paleozoic. The continental interior was largely a craton that was relatively "stable" compared to the rapidly subsiding geosynclines. However, on the craton were broad topographically positive features called *arches* or *domes* that were rising during sedimentation, with the result being thinner sedimentary sequences compared to the more stable parts of the craton. Also on the craton were *basins* that subsided during sedimentation and that consequently received greater thicknesses of sediments than did the arches and domes. Even the "stable" parts of the craton subsided slightly during sedimentation, thus leading to burial and preservation of the sediments. The relative amounts of sediments are shown diagrammatically in Fig. 21.1.

CAMBRIAN PERIOD

During the Cambrian Period, the seas slowly advanced out of the two geosynclines onto the craton, and by latest Cambrian time, may have covered much of the United States, Canada, and Mexico. Trilobites and brachiopods dominated the fauna, although most major classes of invertebrates were already represented in the upper Cambrian sediments.

The low-lying Transcontinental Arch (Fig. 16.12) shed sand into the Cambrian sea. The sand was commonly mature quartz sand from the peneplaned, weathered, and wind-abraded surface of Precambrian rocks that included quartzose sandstones. Also, minor amounts of silt and clay were deposited. Further from the landmass, carbonate precipitation dominated.

The Cambrian was a tectonically quiet time. Gondwanaland may have already formed, but the northern continents of Pangea (i.e., Laurasia, composed of North America, Europe, and Asia) had not yet assembled.

ORDOVICIAN PERIOD

Late Ordovician time saw one of the greatest marine transgressions onto the North American continent, with seas probably covering the entire craton. The dominant Ordovician sedimentary rock is limestone, indicating warm seas and a general lack of terrestrially derived clastic sediments. Quartz sandstone units, especially the widespread middle Ordovician St. Peter Sandstone that was deposited in a transgressing sea and lies unconformably upon the lower Ordovician carbonates, are probably products of the recycling (i.e., erosion and redeposition) of Cambrian sandstones.

Ordovician marine limestones are generally quite fossiliferous; there were many more species than in the Cambrian Period, and all major animal groups were present. Trilobites, brachiopods,

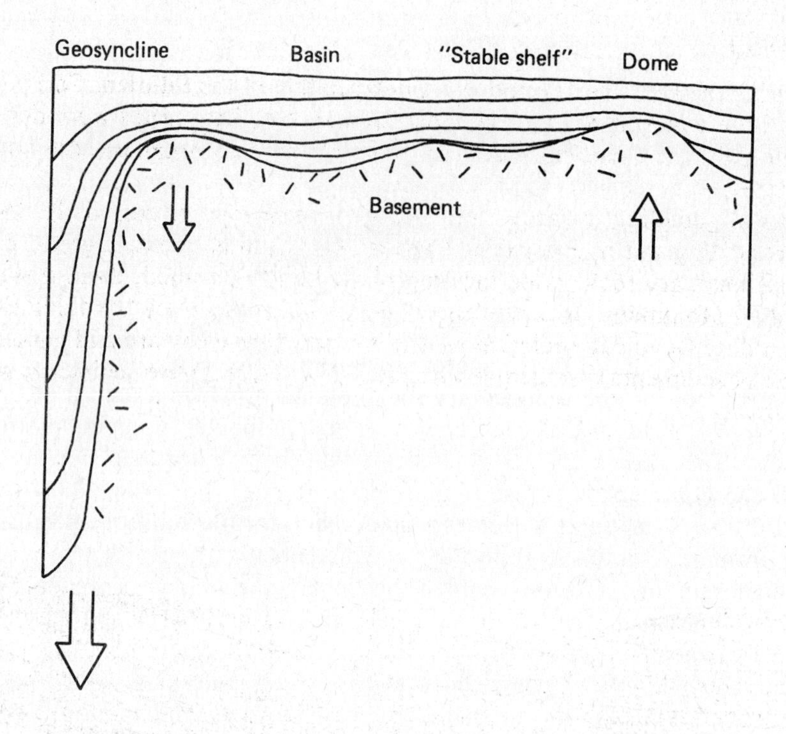

Fig. 21.1 Diagrammatic representation of the relative amounts of sediment that accumulate in geosynclines, in basins, on domes, and on the "stable shelf." The sediment thickness is directly related to the amount of subsidence.

bryozoans, and graptolites were common. Nautiloid cephalopods as long as several meters were the largest life forms, and early fish were also present.

The Taconic Mountains rose out of the Appalachian Geosyncline in Late Ordovician time, the first recorded mountain building within the North American geosynclines. The mountains consisted of metamorphosed sedimentary rocks and igneous rocks, probably in a series of thrust faults caused by the collision of North America with a volcanic arc or some other terrane (see Problem 17.36). Subsequent erosion of this highland resulted in great quantities of clastic sediments along a 1600-km-(2000-mi) long zone. This is an example of a clastic sedimentary sequence used to interpret the former existence of a topographically high source terrane. Some volcanic detritus was eroded from volcanoes that formed as a result of subduction. Fine volcanic ash from volcanic eruptions was blown westward and now constitutes thin bentonite (altered volcanic ash, now clay) in the carbonate sequences of midwestern United States.

SILURIAN PERIOD

Carbonate rocks dominate the Silurian rock column, for shallow warm seas probably covered much of North America. However, clastic sediments were still being deposited in the Appalachian region as erosion of the Taconic Mountains continued. Iron-bearing sediments of Silurian age are mined in Alabama and extend all the way to New York. Evaporite deposits as thick as 1500 m, mainly halite and gypsum, formed in the Michigan Basin and in other smaller basins (Fig. 16.12).

Fossils, including the earliest coral reefs, are abundant in the limestones. Large eurypterids, scorpion-like swimming animals, dominated the seas. The first land plants appeared.

DEVONIAN PERIOD

Sedimentation during the Devonian Period was similar to that of the Silurian. Carbonates were dominant, coral reefs were widespread (e.g., in Alberta), and more evaporites were deposited. A widespread black shale unit, the Chattanooga Shale, accumulated in a broad epicontinental sea in east-central United States.

A new mountain-building episode, the Acadian orogeny, occurred in the Appalachian Geosyncline of New England and adjacent Canada. During this event, granitic plutons were intruded, and the sedimentary rocks were metamorphosed and deformed; these document the existence of the Acadian Mountains. Independent evidence is provided by the thick Catskill Delta of New York state, a Late Devonian sediment wedge that coarsens eastward and passes westward into a marine sequence of sediments that include turbidites (Fig. 21.2). These sediments were shed by the Acadian Mountains.

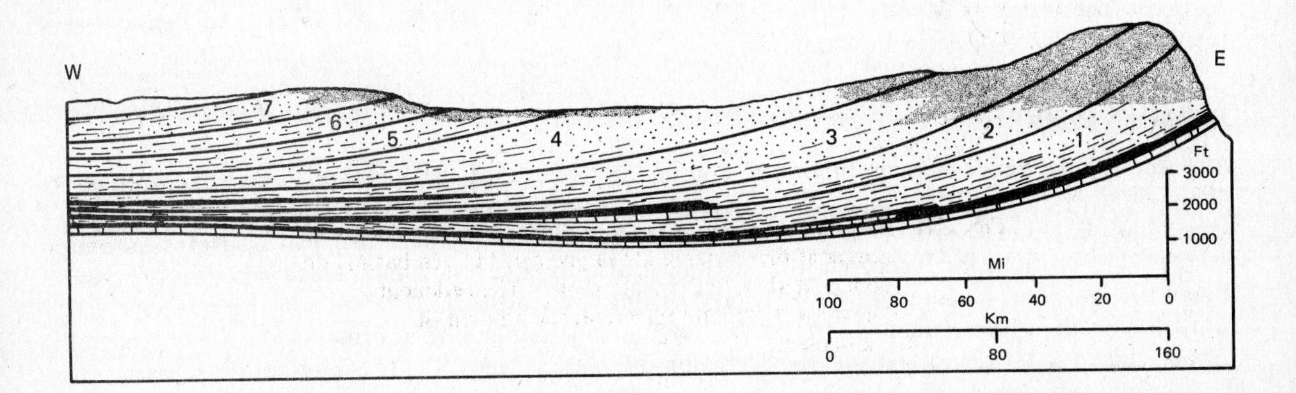

Fig. 21.2 The "Catskill Delta" in the Devonian rock column of New York. The subhorizontal lines represent time lines. The numbers indicate successive sequences of sediments, now sedimentary rocks. (*After Dunbar, C. O., and Rodgers, John, Principles of Stratigraphy, Wiley, New York, 1956.*)

The Caledonian orogeny of northern Europe (Great Britain and Norway) is interpreted as essentially the same orogenic event as the Acadian orogeny. This widespread orogeny seems to be the result of the collision of Europe and North America.

Fossils in the Devonian rocks include the oldest fossil forest (in the fluvial redbeds of the Catskill Delta), an abundance of armored fishes, and the first amphibians. Invertebrates continued to be abundant.

MISSISSIPPIAN PERIOD

The Mississippian Period is the last period in which carbonate deposition was widespread in the shallow seas that covered much of North America. The dominant fossils in the Mississippian limestones are crinoids and certain brachiopods. Minor feldspathic sands were derived from the vicinity of the Appalachian Geosyncline.

A small landmass destined to become much larger throughout the remainder of the Paleozoic and the Mesozoic eras rose out of the Cordilleran Geosyncline. This highland may have been the product of eastward-directed thrust faulting that resulted in stacked thrust fault slices of rocks that then provided a source for clastic sediments.

PENNSYLVANIAN PERIOD

The most distinctive sediments of the Pennsylvanian Period are the coal-bearing *cyclothems* of central United States. In western Europe, the Mississippian and Pennsylvanian Periods (as used in North America) together make up the Carboniferous, named for the carbon-rich (i.e., coal) layers. Each of the 50 or so cyclothems is a thin (commonly a few tens of meters thick) sequence of nonmarine and marine lithologies; this alternation was the result of numerous transgressions and regressions related to sea level changes. These were probably caused by glacial advances and retreats on Gondwanaland. The coals formed in widespread swamps in the mixed environment between the rivers and deltas and the shallow marine environments. The forests and associated plants of the swamps were inhabited by amphibians. The first reptiles appeared and insects were abundant. Most invertebrate groups continued to thrive in the seas, and one group of single-celled foraminifera (the fusulinids) were prominent.

In the four-state area of Colorado, New Mexico, Arizona, and Utah, another landmass rose out of the Cordilleran Geosyncline. It has been called the Ancestral Rockies or the Colorado Mountains. Erosion of these highlands produced an abundance of alluvial fan and river deposits, mostly redbeds. The Ouachita Mountains formed in the vicinity of Oklahoma, Arkansas, and Texas, the result of northward-directed thrusting.

PERMIAN PERIOD

Starting in later Pennsylvanian time, and continuing throughout the Permian Period, the Appalachian orogeny formed the Appalachian Mountains; the southwestward continuation formed the Ouachitas. In effect, North America was tilted westward, and the seas that had transgressed and regressed throughout the Paleozoic were no longer able to cover the higher eastern half of the continent that was now "high and dry." Major thrusting toward the west in the Appalachians and toward the north in the Ouachitas formed the mountainous regions; this thrusting and folding of the Appalachians was the result of the collision of Africa and South America (i.e., parts of Gondwanaland) with North America.

Seas were therefore restricted to western United States. In southwest Texas and adjacent New Mexico, a very large reef complex developed. Phosphatic sediments were deposited along the western side of the North American craton where nutrient-rich cold bottom waters were upwelling to meet warmer waters of the continental shelf. Redbeds and evaporites accumulated in western United States. Pangea was finally assembled, the result of numerous collisions of continents and microcontinents.

The Permian saw the further development of reptiles, including some mammal-like forms. At the end of the Permian Period, half of all invertebrate marine families (including all trilobites), and maybe 90 percent of all species, became extinct. Why this happened is a big question.

Solved Problems

21.1 What is the meaning of the word *Paleozoic?*

Ancient life.

21.2 Where were the two main geosynclines of North America during the Paleozoic?

The Appalachian Geosyncline was located on the present site of the Appalachian Mountains and joins the smaller Ouachita Geosyncline on its southern end. The Cordilleran Geosyncline was located along the present site of the Rocky Mountains. (See Fig. 16.12.)

21.3 Was the "stable craton" of the continental interior completely stable (i.e., neither rising nor subsiding)?

No. Some parts (the basins) were subsiding during the times they were covered by seas and receiving sediment. Other parts (the domes and arches) were rising during marine sedimentation. Other parts of the shield were "stable," but nevertheless were subsiding slowly. (See Figs. 16.12 and 21.1.)

21.4 What are the thicknesses of sediments deposited on the craton, as compared to the geosynclines? (See Figs. 16.12 and 21.1.)

On the craton, from zero to a few thousand meters in general. In the geosynclines, the thicknesses are 10,000 m or more. On the domes and arches, thicknesses are less than on the shelf proper, and in the basins, the sediments are thicker than on the shelf.

21.5 Where was the equator relative to North America during the Paleozoic?

The equator crossed North America about at its center. However, North America was oriented differently, relative to today's equator. The equator crossed in a southwest-northeast direction, relative to the present orientation of the continent.

21.6 What is the evidence for the equatorial position of North America during the Paleozoic?

The evidence is of three types. One is the presence of a diverse warm water fauna (including corals) preserved as fossils, a second is the abundance of evaporite deposits (see Problem 5.53), and a third is paleomagnetic evidence (see Problems 17.27 through 17.32).

21.7 Where were the oceans relative to the North American continent during the Paleozoic?

They repeatedly advanced onto the continent and repeatedly withdrew.

21.8 Why did the seas advance onto the North American craton?

Minor advances may have been related to the melting of glaciers in the polar latitudes. Bigger transgressions (advances) were probably related to times of rapid spreading of the ocean floors from spread centers. (See Problem 18.56.)

21.9 When did mountain building occur in the Appalachian Geosyncline?

In the Ordovician (Taconic Mountains), Devonian (Acadian Mountains), and Permian (Appalachians and Ouachita Mountains).

21.10 What caused the sedimentary sequence that accumulated in the geosyncline to be uplifted to become the mountains of the previous problem?

Collisions with Europe and perhaps with minicontinents as well.

21.11 What is the evidence for mountain building at the times mentioned in Problem 21.9?

Metamorphic and plutonic rocks, fold belts, and thrust faults in the present Appalachians, when integrated with the fossil ages and radiometric dates, indicate that mountain-building processes were at work.

21.12 In the Appalachian region, there are major conglomerate and sandstone units of Ordovician and Devonian age, essentially sitting amid carbonate sections. What do they indicate?

They indicate the presence of high source areas and constitute additional evidence for the existence of mountains in the Appalachian Geosyncline in Ordovician and Devonian times.

21.13 What rock type dominates the Ordovician rock column, especially in the central part of North America, and what does it indicate about paleogeography?

Limestones with abundant fossils. They are indicative of shallow, warm seas and the general absence of highlands in most of North America, as no coarser clastics (e.g., sand and pebbles) were being supplied to the seas.

21.14 When did mountains rise out of the Cordilleran Geosyncline?

The present Rocky Mountains did not appear until early Cenozoic time and will be discussed in Chapter 23. However, the "Ancestral Rocky Mountains" rose out of the "four-corner area" of Colorado, New Mexico, Utah, and Arizona in Late Pennsylvanian time and were eroded throughout the rest of the Paleozoic. Also, a landmass was rising farther west, beginning in Mississippian time. (This landmass will be mentioned again in the next chapter.)

21.15 What are cyclothems, and when were most of them deposited?

Cyclothems are thin cycles of marine and nonmarine sediments deposited in the vicinity of the ocean shoreline in the midcontinent during Pennsylvanian time. There are about 50 such cycles in central United States, and nearly all include a bed of coal. The cycles are generally only a few meters thick.

21.16 How were the cyclothems of Problem 21.15 formed?

In each cycle, river sediments and organic-rich beds that accumulated in swamps (and after burial became coal beds) were covered by marine sediments as the sea transgressed. Withdrawal of the sea resulted in some erosion (i.e., a minor unconformity), and then the sea advanced again. This repetition is thought to be due to the waxing and waning of glaciers on Pangea, which at that time was positioned near the South Pole.

21.17 Were coral reefs important during Paleozoic time?

Yes. Small "patch" reefs are abundant in Silurian rocks, and large reefs exist in the Devonian rocks of Alberta, Canada, and in Permian rocks of West Texas and New Mexico.

21.18 Redbeds are common in the Permian of western United States. What do they indicate about environment?

Redbeds are red due to the presence of hematite—even small amounts will make a sediment red. They are the result of the oxidation of iron-bearing minerals such as magnetite, biotite, or hornblende. Most oxidation occurs on land during exposure of the sediment to the air, and so most redbeds are nonmarine deposits. However, redbeds are a problem in that it is difficult to determine when a sediment became red—were the original sediments red, were they oxidized while being deposited, were they oxidized after deposition (during diagenesis), or were they reddened by weathering during their present exposure? Each or all are possible.

21.19 When did the Appalachians and Ouachitas finally attain their greatest stature, and why?

In latest Pennsylvanian and Permian time, when South America and Africa collided with North America.

21.20 When did Pangea come into existence?

This is a difficult question that is still being researched by geologists. The southern part of Pangea (Gondwanaland) may have formed in Early Paleozoic time, for it was certainly intact by Late Paleozoic time. However, the northern part (Laurasia, consisting of North America, Europe, and Asia) probably was not added to Pangea until latest Paleozoic or later.

21.21 When was the first widespread advance of the seas out of the geosynclines onto the craton?

Late Cambrian time.

21.22 When was the greatest sea advance onto the North American continent?

In Ordovician time.

21.23 During which period were the great sequences of carbonates deposited in the middle part of the continent?

Mississippian.

21.24 What were the first two groups of animals to dominate the Paleozoic seas?

Trilobites and brachiopods in the Cambrian.

21.25 What types of sediment were deposited in the mid-United States during Late Cambrian time, and why?

Sands dominated in the north, derived from the Transcontinental Arch (see Fig. 16.12). Farther away from the landmasses, to the south, carbonate deposition dominated.

21.26 In the Mid-Ordovician rock column of the mid-United States, dominated by carbonates, is a very pure mature quartz sandstone called the St. Peter Sandstone. How might it have formed?

By erosion and recycling of Cambrian sandstones located to the north in northern United States (e.g., southern Minnesota and southern Wisconsin). The sea withdrew in early Ordovician time and then readvanced into the interior in middle Ordovician time. The exposed sands deposited in the sea during late Cambrian time were exposed to erosion by both water and wind during the time the sea had withdrawn. When the sea readvanced, it further reworked the widespread quartz sands.

21.27 What is the Chattanooga Shale?

A widespread Devonian-Mississippian black shale unit in the mid-United States. It contains only a few fossils (graptolites and conodonts) that evidently dropped onto the black mud bottom.

21.28 Where were the seas on the North American continent during the Permian Period?

In the West, for in the East the Appalachian Mountains were rising out of the Appalachian Geosyncline. This, in effect, raised eastern North America above sea level.

21.29 Summarize Paleozoic life.

Invertebrates dominated. The first primitive vertebrates were fish in the Late Cambrian or Early Ordovician. Amphibians appeared in the Devonian, and reptiles in Early Pennsylvanian time.

21.30 When were most invertebrate groups finally present in the seas?

By the end of the Ordovician Period.

Supplementary Problems

21.31 What is the Catskill Delta (Fig. 21.2)?

It is a thick sequence of Devonian sedimentary rocks deposited westward of the Acadian Mountains.

21.32 Describe the sedimentary facies of Fig. 21.2.

Conglomerates were deposited near the uplifted mountains, as alluvial fan deposits. Sands were deposited farther west, as both river sands and shallow marine sands. Marine shales and carbonates were deposited still farther to the west.

21.33 Note the subhorizontal lines on Fig. 21.2, through the various rock types. What do they represent?

Time lines, based on fossils; they separate rock sequences of different ages.

21.34 Describe the relationship of sedimentary rocks during time interval 2 of Fig. 21.2.

Sedimentary facies (see Problem 5.27) are particularly well shown by the rocks deposited during time interval 2, as well as by the rocks of other intervals. Note that conglomerate, sandstone and shale of the same age are found in time interval 2.

21.35 How are the Catskill Delta of New York and the Old Red Sandstone of Great Britain (first studied by James Hutton in the late 1700s) related?

They were deposited on opposite sides of the Acadian Mountains. Both include a redbed continental facies.

21.36 Of what interest, other than for the study of fossils and paleogeography, are the Devonian reefs of Alberta?

Much of Canada's petroleum comes from the porous reef complex.

21.37 Arches and domes commonly have less complete sedimentary sequences than do basins or geosynclines. Why?

Because of uplift and erosion during sedimentation.

21.38 As Early Paleozoic seas transgressed repeatedly onto the North American continent, the sand deposition that was so important in Early Cambrian time was replaced by carbonate deposition. Why?

Carbonate can form beds only where it is not masked by abundant terrigenous sediment (mud, silt, sand). The widespread abundance of limestones (and dolomites) suggests that the original low-lying sources of sand were covered by the transgressing sea, were already worn down to sea level by erosion, or both.

21.39 What happened to many forms of invertebrate life at the end of the Permian?

They became extinct.

21.40 Suggest a cause for the great extinction of life at the end of the Paleozoic.

Climatic change has been proposed. Perhaps a better explanation is that mountain building due to the collisions of continents reduced the total area of shallow seas as the seas withdrew from the lands. This regression, which probably lasted a few tens of millions of years, would have drastically reduced the amount of shallow water habitat, and in the ensuing competition for living space, the fittest groups of organisms survived.

21.41 Which invertebrate groups became extinct at the end of the Paleozoic?

One notable group is the trilobites; none survived into the Mesozoic, as can be seen on Fig. 19.3. It has been estimated that as many as 90 percent of all living species of marine invertebrates living at this time perished.

21.42 What is the Caledonian orogeny of Europe?

It is a mountain-building event in northwestern Europe that culminated in the Mid-Paleozoic. Rocks that are the product of this orogeny are best exposed in Norway and Scotland. It corresponds to the Acadian orogeny of North America.

21.43 What is the probable origin of the Ural Mountains of the U.S.S.R?

The Urals lie at the boundary of the European craton and the Siberian (i.e., Asian) craton and are the result of the collision of these two landmasses in Late Paleozoic time. It is quite likely that ocean floor between these two masses was consumed during plate movements.

21.44 When and where was the Hercynian mountain-building event (i.e., orogeny) of Europe?

In southern Europe in Late Paleozoic time. The Hercynian Mountains stretched from Spain and England to the Black Sea. It was partly due to the collision of Europe and North America. However, the younger Alpine orogeny has greatly obscured the details of the Hercynian event.

Chapter 22

Mesozoic: The Age of Reptiles

INTRODUCTION

The Mesozoic era is probably best known to most people as the "age of dinosaurs." Actually, the "age of reptiles" is a better designation, for not all reptiles were dinosaurs ("terrible lizards"). Geologically speaking, a more important Mesozoic event occurred that affected the entire world. This was the breakup of Pangea which had finally been assembled in Late Paleozoic time.

In North America, the eastern half of the continent was "high and dry" after the Appalachian orogeny formed the Appalachian Mountains. Consequently, eastern North America experienced erosion rather than sedimentation during most of the Mesozoic (Fig. 22.1).

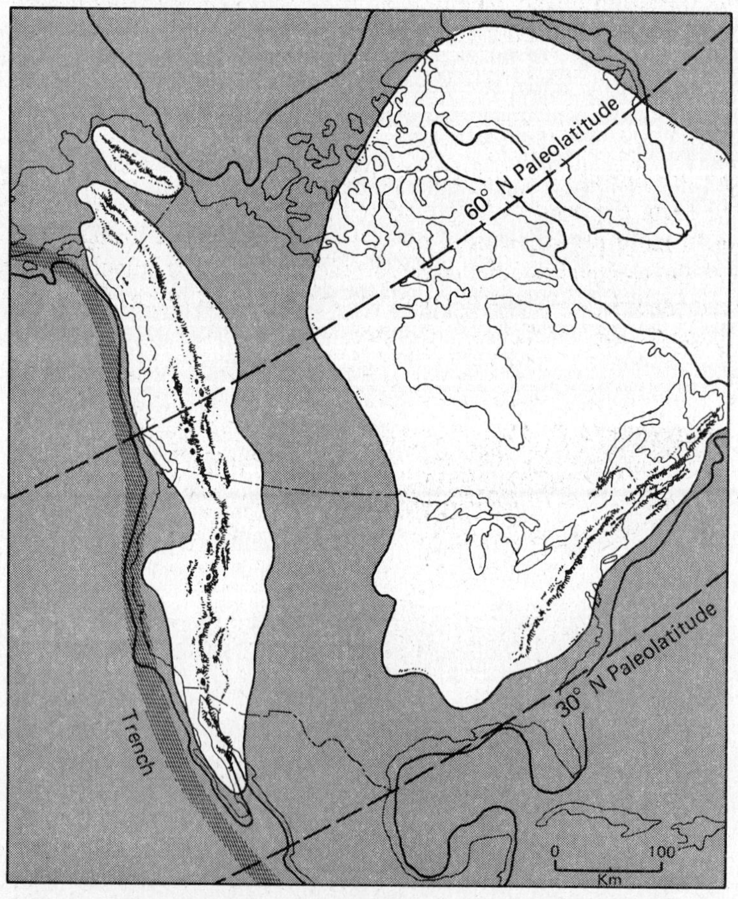

Fig. 22.1 Map showing North America during Cretaceous time, about 100 m.y. ago. Light areas are landmasses, shaded areas are under the seas. Note volcanoes and trench in the west. The dark line around the coast is the edge of the shelf. (*After Dott, R. H., Jr., and Batten, R. L., Evolution of the Earth, McGraw-Hill, New York, 1988.*)

However, western North America was where the action was, as the Cordilleran orogeny resulted in various mountain ranges. Thrust faulting, volcanism, and plutonism were widespread. All these events were related to plate tectonics and, more specifically, to subduction along the western edge of the continent. Western North America also grew by accretion, as numerous suspect terranes "docked" (joined) North America along its Pacific Coast before the end of Mesozoic time.

THE TRIASSIC

The Triassic Period saw the first signs of the breakup of Pangea. In eastern United States, big rift-type valleys formed as North America began separating from Africa. These were mostly half-grabens that subsided and filled with sediment that consisted mostly of fluvial (river) redbeds (sands and coarser clastics) and finer-grained gray lake sediments. Interbedded with the sediments in these Triassic and Jurassic troughs are basaltic lava flows; some were subaerial and others extruded under water. Mafic dikes cut the sequences. The mafic magmas were the result of the extensional tectonics that allowed mafic magmas generated in the mantle to reach the surface.

In western North America, the dominant sediments were continental redbeds (as in the Permian), but marine limestones also were deposited. The landmass that was formed in western North America in Mississippian time (see Problem 21.14), as a product of the Antler orogeny, continued to grow as the thrusting of the Sonoman orogeny of Permian-Triassic time raised large areas and sediment was shed both to the east and to the west off of this landmass.

JURASSIC

Western North America during the Jurassic Period was the site of some unique sedimentation. Quartz sand, eroded from exposed older sandstone units to the east, was redistributed by winds over broad areas and finally deposited and preserved as large sand dunes (Fig. 22.2). Finally, the seas readvanced over the western interior in the first large-scale readvance since early Permian time, and fossiliferous limey sediments were deposited over the wind-deposited sands, thus forming the widespread Jurassic limestones of the western interior.

Fig. 22.2 Sketch of large-scale cross-bedding.

The large landmass that had been rising out of the geosyncline since Mississippian time continued to rise as intrusions related to the subduction along western North America were emplaced in what is now the Sierra Nevada during the Nevadan phase of the Cordilleran orogeny. This landmass was being drained by numerous eastward-flowing rivers, and a broad, swampy, alluvial plain was formed in latest Jurassic time. Evidently the site of abundant vegetation, dinosaurs flourished and their remains are well-preserved in numerous places, including Dinosaur National Monument in northeasternmost Utah.

THE CRETACEOUS PERIOD

In Middle Cretaceous time, about 100 m.y. ago, the seas made their last great advance across the low-lying continental interior of Canada and the United States, reaching as far eastward as eastern Minnesota. Various types of sediment were deposited in the sea, including sandstone, shale, and abundant chalk. To the west, nearer the large landmass that had now attained a size of over 500,000 km^2 (about 200,000 mi^2) and that was the major source of sediment, rivers carried pebbles and sand seaward (eastward), depositing much of the sediment prior to reaching the shoreline but also forming large deltas in the sea. This sea transgressed and regressed several times, as evidenced by the distribution of coals (formed in swamps on deltas) and marine fossils. Westerly winds deposited fine-grained volcanic ash layers in the black muds of the sea. The thickness of Cretaceous sediments in what is now the Rocky Mountains is great; nearly 9,000 m (30,000 ft) were deposited during Cretaceous time alone, resulting in thicknesses characteristic of a geosyncline.

Meanwhile, back to the west of the great landmass that was the source for most of the clastic sediments of the west, a new coastal basin, the Pacific Geosyncline, had formed. In it, 9,000 to 18,000 m (30,000 to 60,000 ft) of sediment accumulated, mostly during the Cretaceous. The source area was the Sierra Nevada magmatic arc, including the original geosynclinal sediments, the volcanic rocks, and the granitic rocks that were emplaced into the volcanic arc. Figure 17.5 can be used to depict the West Coast at this time.

This magmatic arc was the result of subduction along a converging plate boundary. The same plate motions resulted in numerous microcontinents (volcanic arcs, volcanic plateaus, and other terranes) being added to or accreted to western North America. Because the collisions were not right-angle collisions, as is implied by cross-sections such as Fig. 17.5, there was an important lateral motion as well, resulting in many elongated terranes separated by strike-slip faults. Most of Alaska is thought to consist of such terranes, and there may be more than 100 such accreted terranes along western North America. (Recent work is indicating that similar accretion may have occurred in the Appalachian region during the Late Paleozoic.) Also, such accretion occurred along the western margin of the Pacific Ocean; parts of Siberia and most of China may be accreted terrane.

Solved Problems

22.1 When did the breakup of Pangea begin?

In Triassic time, continuing into Jurassic.

22.2 What evidence in North America is indicative of the start of this breakup?

The rift-valleys (half-grabens) of eastern North America.

22.3 Are these rifts shown on Fig. 22.1?

No, they are not shown; some of them were under the Cretaceous seas at this time.

22.4 What do rifts indicate about the overall tectonic framework—was it extensional or compressional?

Rifts indicate that extension, rather than compression, was the dominant regime. That is, forces were "pulling apart" Pangea, rather than "pushing" parts of Pangea together.

22.5 Is there a modern example of such rift-valleys?

Yes. The East African Rift Valleys. (See Fig. 17.7.)

22.6 What types of rocks were formed in these valleys?

Alluvial fan, river, and lake sediments. Basaltic rocks include cross-cutting dikes, subaerial massive lava flows, and some pillowed (subaqueous) lava flows.

22.7 Most of the sediments in the rifts are river deposits. What color might you expect them to be?

A good guess would be red, due to the continental (and therefore oxidizing) environment of deposition. Actually, many are brownish rather than reddish, because of the presence of hydrated iron oxides (goethite) rather than hematite. Many older buildings of the eastern cities are built of this "brownstone."

22.8 What fossils are locally abundant in the brownstone?

Dinosaur footprints.

22.9 The Triassic Petrified Forest of Arizona is world famous. What is the present chemical composition of the forest, and how did it form?

The logs are now composed largely of fine-grained SiO_2. The porous cells of the wood were filled with silica after the logs were buried in sediment. Such petrifaction is commonly an indication of the presence of abundant silica-rich volcanic ash, which releases the silica as it is weathered and as it devitrifies (i.e., "deglasses," or becomes crystalline on a microscopic scale).

22.10 What is probably the most abundant sedimentary structure in the Jurassic quartzose sandstones of the West?

Large-scale cross-beds, as in Fig. 22.2.

22.11 See Fig. 22.2. What does the scale suggest (but not prove) about the environment of deposition?

The large size of the cross-beds is suggestive of deposition in large Eolian (wind) sand dunes.

22.12 The sands of Fig. 22.2 are commonly quartzose. Why?

The sources were probably older quartzose sandstone units. Thus, the sand was being recycled.

22.13 If you looked at the sands of Fig. 22.2 with a microscope, what features might be indicative of a history of wind abrasion?

Well-rounded grains, excellent sorting, frosted grain surfaces (due to repeated impacts), and a grain size capable of being moved by the wind (i.e., not coarse sands).

22.14 Could running water produce the features listed in the previous problem?

Probably not. Wind is the best agent for rounding sand grains, as the grains hit each other without the cushioning effect of water. River sands are commonly less well sorted.

22.15 What other paleogeographic data can be derived from sandstones such as those in Fig. 22.2?

Study of the cross-beds can give the paleowind patterns.

22.16 When was the first major Mesozoic sea advance (marine transgression) onto the interior of the North American continent?

In the Jurassic Period.

22.17 What was the source of most of the Mesozoic sediment?

A large landmass in western North America that began rising out of the Cordilleran Geosyncline in Mississippian time and continued to grow throughout the Mesozoic.

22.18 The Upper Jurassic Morrison Formation of western United States (especially in Utah, Colorado, and Utah) is world famous. Why?

Because of the abundance of dinosaur remains. It has probably yielded more remains than any other rock formation in the world.

22.19 What kind of rocks make up the Morrison Formation?

The sediments are mostly river deposits, both channel sands and finer-grained floodplain deposits. They were deposited on a vast plain of river deposits formed by eastward-flowing streams draining the highland to the west.

22.20 What is the name of the rock formation found at Dinosaur National Monument?

The Morrison!

22.21 Rocks of which Mesozoic period contain an abundance of fine-grained limestone called chalk?

The Cretaceous. Chalk beds, for example, are widespread in Kansas. These contain numerous marine fossils, including large fish and large flying reptiles.

22.22 What is the evidence that the dominant winds in western United States were westerly (from west to east) during the Cretaceous?

There is volcanic ash interbedded in the black shales of Cretaceous age. The only place volcanism was going on was farther to the west, so that had to be the source of the volcanic ash. (See Fig. 22.1.)

22.23 What is the origin of the Sierra Nevada?

Subduction of an oceanic plate beneath the continental plate (See Fig. 17.5).

22.24 Granitic rocks are abundant in the Sierra Nevada. Why?

As the sediments and sea floor basalts carried downward by subduction were partially melted by the high temperatures at depth, granitic magmas were formed. Some reached the surface, forming explosive volcanoes and resulting in volcanic sediments, but much of the magma cooled at depth, forming large plutons that together make up the huge Sierra Nevada composite batholith.

22.25 How much sediment was eroded off of the large western landmass?

The landmass covered at least 500,000 km^2 (about 200,000 mi^2) at its zenith. Some geologists have estimated that as much as 16 km (10 mi) of rock may have been eroded off of that area.

22.26 What would the total volume of sediment be, based on the previous problem?

Two million cubic miles, or about 8 million cubic kilometers! Some geologists have suggested that about half was shed to the east and half to the west.

22.27 How thick were the Cretaceous sediments in the Cordilleran Geosyncline?

The Cretaceous sediments are locally 9000 m (30,000 ft), or about 5 mi, thick. If we add in the Triassic, Jurassic, and the Late Paleozoic sediments, the huge volume of Problem 22.26 does not seem impossible.

22.28 How is the Sierra Nevada related to this large western landmass?

The Sierra Nevada is one part of this landmass.

22.29 How important was subduction in western North America during the Mesozoic?

It was occurring along the western margin of North America throughout the Mesozoic era and is the underlying cause of much of the history of western North America.

22.30 When was the last great advance of the seas onto the North American continent?

During the Late Cretaceous Period. (See Fig. 22.1.)

Supplementary Problems

22.31 How does an "Andean-type" continental margin relate to Mesozoic history of western North America?

The present tectonic picture along the west coast of South America (i.e., the "Andean-type" margin) is a present-day analog of what western North America was like during the Mesozoic.

22.32 Do the margins of South America and North America have similar ore deposits?

Yes. The batholiths generated during subduction are the host rocks for large porphyry-copper deposits. These are low-grade but of great size. The copper of Chile and Arizona are examples.

22.33 What caused the Cretaceous seas to advance onto North America, and onto all other continents as well?

Such a worldwide transgression is best explained by events on the ocean floors (see Problems 18.55 and 18.56). Rapid sea floor spreading which is accompanied by increased volumes of oceanic ridges in effect "displaces" so much water that it encroaches upon the continents.

22.34 What technique can be used to determine the temperature of the ancient oceans, such as those of the Mesozoic?

Oxygen isotope studies. Two isotopes of oxygen, ^{18}O and ^{16}O, are found in the carbonate of fossil shells. Modern shells show that ^{18}O is most abundant in cold waters and is less abundant in warmer waters. Thus a measure of the ratio compared to common ^{16}O gives the water temperature. For example, Cretaceous fossils from North America indicate water temperatures such as are found today in the Caribbean.

22.35 Could the information of the above problem be interpreted to mean that the equator was closer to North America than it is today?

Yes. However, paleomagnetic information indicates that North America was not closer to the equator. Therefore, the interpretation that the world climate may have been warmer in the Cretaceous seems likely.

22.36 Cretaceous batholiths, similar to the Sierra Nevada, are found all around the Pacific Ocean. Why?

Subduction, of course.

22.37 What is the significance of all the mafic (basaltic or diabasic) dikes that cut the Triassic-Jurassic rocks of the basins of eastern North America?

They were probably feeder dikes to the volcanic rocks of the basins.

22.38 How many suspect terranes were accreted to western North America during the Mesozoic?

Many. Estimates range from 50 to 200.

22.39 What is Wrangellia?

Wrangellia is one of the best-studied accreted terranes of western North America. It consists of Triassic rocks and oceanic basalts. Paleomagnetic and fossil evidence suggest the rocks were formed near the equator; today Wrangellia includes four long slivers of rock located between northern United States and Alaska.

22.40 When were the suspect terranes accreted to western North America?

During the Mesozoic (and perhaps Paleocene) when subduction was active.

22.41 In general, what is the evidence for such suspect terranes being present today in locations other than the locations in which they formed?

Paleomagnetic data give the latitudes at which such terranes were formed. Fossils can indicate something about latitude. Rock types may also do this. Adjacent terranes may have totally unrelated rock types, with different structural histories.

22.42 What might have elevated the large western landmass that was the source of so much sediment during the Mesozoic?

A likely possibility is eastward-directed thrust-faulting, which elevated a number of thrust sheets, perhaps one on top of the other. The Sonoman orogeny may have been such an event, in large part.

22.43 What is a foreland basin?

A *foreland basin* is a rather recently defined type of basin that is formed in front of a thrust-faulted terrane in which several thrust sheets are stacked up one upon the other. This elevated mass weighs down the crust beneath, and the underlying mantle rock ''flows'' outward. The result is an asymmetrical basin in front of the stacked thrust blocks, and sediments will accumulate in this basin. Logically, most of the sediment will be derived from the elevated and thrusted terrane, although some may come into the basin from the opposite (continent) side of the basin.

22.44 Could the thick sequence of sediment which accumulated in the Cordilleran Geosyncline have accumulated in a foreland basin?

Yes. See the previous two problems.

22.45 What is the evidence that some of the mafic lavas in the Triassic-Jurassic half-grabens (rift basins) of eastern North America were extruded under water?

Some of them are pillowed.

22.46 What is the evidence that the sediments in the rift basins of eastern North America were *not* deposited in a marine environment?

The lack of marine fossils and the abundance of dinosaur tracks.

22.47 When did the Alps of Europe form?

During the Alpine orogeny of latest Cretaceous and Early Cenozoic time.

22.48 What was the history of Gondwanaland during the Mesozoic?

Essentially, it was a large supercontinent during earliest Mesozoic (it had assembled in the Paleozoic), and similar rocks and fossils were deposited on all five of the southern continents. The breakup of Pangea started earlier between North America and Europe, with the rift basins of eastern North America a main result, than it did in Gondwanaland. The breakup of Gondwanaland continued into Mid-Cenozoic time. (See Fig. 17.4.)

22.49 Do the Alps contain deformed rocks as in the Appalachians of North America?

Yes! Some of the most complexly deformed rocks on earth are found here. There are recumbent folds, thrust faults, and nappes (see Chapter 16). Many of these structures are exposed in three dimensions in this high-relief mountain system.

22.50 What are the White Cliffs of Dover (England)?

Fine-grained limestone (chalk) was deposited in Late Cretaceous time over much of Europe. These are the most famous and best-exposed deposits.

Chapter 23

The Cenozoic Era

INTRODUCTION

The last 65 m.y. of earth history are very important, as the present shape of the landscape is largely a product of Cenozoic events, including weathering and erosion. North America was essentially "high and dry" during the Cenozoic (relative to sea level), except for the coastal plains on the East and Gulf coasts, where the oceans continued to encroach upon the low-lying margins of the continent. The last sea in the continental interior was Paleocene (a relic of the Cretaceous seas), and it retreated from western United States to the Gulf of Mexico.

Major events in North America during the Cenozoic were the final uplift of the Rocky Mountains, the normal faulting that formed the Basin and Range Province of southwestern United States, the outpouring of lavas to make the Columbia Plateau, and the uplift of the Pacific Coast Ranges. The glaciations of the Great Ice Age of the Pleistocene (the last 1 or 2 m.y.), certainly a great Cenozoic event, will be covered in the next chapter.

CORDILLERAN OROGENY

Mountain building in western North America culminated with the final uplift of the Rocky Mountains in Eocene time. Because mountain building was long and complicated and is actually a series of sometimes overlapping events, it is summarized here so that all events from late Paleozoic to Cenozoic are included.

The *Antler orogeny* of late Devonian to Mississippian time covered a long and narrow belt from Nevada to Canada. The *Sonoman orogeny* followed in Late Permian to Early Triassic time. During the time from Late Jurassic to Eocene, mountain building of the Cordilleran orogeny moved eastward "like a wave." The *Nevadan event* occurred in the vicinity of the Sierra Nevada in Late Jurassic time. The *Sevier event* occurred in Late Cretaceous time in the Rocky Mountain region. The final phase, the *Laramie event,* included numerous uplifts as well as major eastward thrusting of geosynclinal rocks in Latest Cretaceous and Early Cenozoic (Eocene) time in the Rocky Mountain region.

After periods of erosion and the formation of peneplains, there was intermittent broad upwarping (epeirogenic uplift) in Early, Middle, and Late Cenozoic time. The last uplift, in Pliocene-Pleistocene time, gave the Rockies their present imposing elevations. Several stream valleys are cut deeply into mountain ranges and plateaus, indicating that these ranges and plateaus were uplifted in areas where a peneplain had formed. As uplift proceeded, the rejuvenated streams continued to erode their valleys more deeply, forming many of the prominent canyons of western United States. These streams are said to be *superimposed streams,* as they were imposed from above.

The broad mountainous region of western North America, along with volcanism, seems to be related to subduction along the western margin of the continent. This manifestation far inland may be related to a very gently dipping subduction zone that resulted in the great width of the orogenic area.

INTERMONTANE BASINS AND SEDIMENTATION

The Laramie event formed numerous uplifts (arches and mountain ranges) and numerous downwarped basins between them. Many of these intermontane ("intermountain") basins had no rivers flowing out of them, so they rapidly filled with sediment carried in by rivers on all sides. In some basins, the Cenozoic sedimentary pile is as thick as 6000 m. Lakes formed in some, perhaps the most famous being Lake Gosiute and Lake Uinta in the Green River and Uinta Basins, respectively; these are noted for their Eocene fossil fish but even more for the great reserves of petroleum tied up in oil shale (see Problem 7.16 and Fig. 7.6).

Off to the east of the Rocky Mountains, the "gangplank" of fluvial sediments formed as the mountains were eroded (especially during Oligocene time) by eastward-flowing river systems. This large, broad zone of merged alluvial plains and lake deposits is currently being eroded away; the famous Badlands of South Dakota, with abundant fossils of Oligocene mammals, are part of this terrane. Similarly, the sediment that was deposited in the intermontane basins is also being eroded away.

BASIN AND RANGE PROVINCE

In a large portion of southwestern United States and adjacent Mexico, normal faulting (block faulting) occurred in Late Cenozoic time (Oligocene through the present) in an extensional tectonic regime that probably was and is related to plate tectonics along the western margin of the United States. About a hundred small fault-block mountain ranges were formed; all trend north-south. Volcanism was widespread, and some of North America's most recent volcanism has occurred in this region. Between the uplifted small mountain ranges are low, down-faulted basins, some with only internal drainage (i.e., no rivers flowing out of them) and saline lakes.

COLUMBIA AND SNAKE RIVER PLATEAUS

In the region of Oregon-Washington-Idaho, a thick sequence of basalt, commonly over 1.6 km (1 mi) thick and locally twice as thick, was extruded as numerous subaerial lava flows (see Fig. 3.14). The major volcanism occurred during Middle Cenozoic time but has continued to the Recent; some flows of the Snake River Plain are only about 2000 years old.

THE PACIFIC COAST RANGES

The Coast Ranges, so named because there are many individual ranges along the western coast of North America, rose out of the Mesozoic Pacific Geosyncline in Mid-Cenozoic time as the geosyncline became the site of various small uplifts and down-warps. Nonmarine sediments were deposited on the flanks of uplifts, and marine sediments were deposited in lower basins. The sediments in places were very thick; for example, the Ventura Basin of the Los Angeles area received 15,000 m (49,000 ft) of sediment.

Solved Problems

23.1 When were the Rocky Mountains uplifted?

Late Cretaceous and Early Cenozoic, but they were eroded during the Cenozoic to a low-relief surface. A broad epeirogenic uplift and subsequent erosion of the Rocky Mountain region in Late Cenozoic time produced the present topography.

23.2 What is the origin of the Grand Canyon?

Late Cenozoic epeirogenic (broad) uplift resulted in the Colorado Plateau. The ancestral Colorado River was flowing across the area prior to uplift, and downward cutting by the river kept pace with the uplift. Hence, the Colorado River is a superposed (or "rejuvenated") stream. The canyon may be as old as 10 m.y.

23.3 What is the origin of the Sierra Nevada?

The Sierra Nevada is a magmatic arc formed over a subduction zone in Late Jurassic to Late Cretaceous time (see Problems 22.23 and 22.24). However, it was eroded to a low topography in Cenozoic time, with streams even flowing westward across the region. In Late Cenozoic time, block faulting and uplift in the eastern Sierra Nevada formed the present range with its steep eastern face.

23.4 How much movement has occurred along the eastern side of the Sierra Nevada?

The Sierra Nevada rises more than 3000 m (10,000 ft) above Owens Valley on the east. However, total movement along this normal fault is estimated at 7700 m (25,000 ft). The top of Mt. Whitney is 14,495 ft above sea level.

23.5 What is the Basin and Range Province, and how did it form?

The *Basin and Range Province* is a broad region of extension in southwestern United States and Mexico, containing more than 100 small uplifted fault-block mountain ranges with down-faulted basins between them. Its western limit is the east front of the Sierra Nevada, and its eastern limit is the western side of the Wasatch Range of Utah.

23.6 What is the origin of Death Valley?

Death Valley is a down-faulted block between the uplifted Panamint Range to the west and the Amargosa Range to the east. Its lowest point is 282 ft below sea level.

23.7 What is the origin of the Cascade Range of the Pacific Northwest?

These volcanoes (see Fig. 3.13) are the product of on-going subduction of ocean floor rocks moving eastward and downward beneath western North America. This is the only part of the western coast of North America, except Alaska, where subduction is still continuing.

23.8 How are the oil shale deposits of western United States (see Fig. 7.5) related to the Cenozoic history of North America?

The oil shales of the tristate area of southwestern Wyoming, northeastern Utah, and northwestern Colorado are Eocene lake beds formed in intermontane basins after the Rocky Mountains were uplifted.

23.9 How did the oil in the oil shales of the previous problem form?

The incomplete decay of organic material derived from living organisms in the Eocene lakes. The "oil" is really not oil, but kerogen, solid organic compounds that can be volatilized at high temperatures.

23.10 What is the origin of the Coast Ranges along the Pacific Coast of North America?

They are a series of small mountain ranges which rose out of the Pacific Geosyncline that was present west of the Sierra Nevada throughout the Mesozoic. They consist of both marine and nonmarine sedimentary rocks. Most, but not all, are elongated parallel to the coast.

23.11 Why are the Rocky Mountains so high and rugged compared to the Appalachians?

They are considerably younger, with their last broad upwarping (epeirogenic uplift) in Late Cenozoic time, and hence they have not been as extensively eroded as the Appalachians, which have essentially been eroding since the end of the Paleozoic, 225 m.y. ago. However, the Appalachians have also been epeirogenically uplifted. Therefore, the present elevations may be thought of as the net result of original mountain building, epeirogenic uplift, and erosion.

23.12 What is the origin of the sedimentary rocks exposed in the Badlands National Monument?

They are river and lake beds deposited as streams flowed eastward off of the then-young Rocky Mountains in Mid-Cenozoic (Oligocene) time.

23.13 Why are there hot springs and geysers in Yellowstone National Park?

Yellowstone is a site of Late Cenozoic volcanism. It is situated over a hot-spot in the earth's crust, where heat is escaping from the mantle. The origin of such hot-spots is in question, but what is probably a similar younger hot-spot is beneath Hawaii. The Yellowstone hot-spot may once have been situated in the Pacific Ocean basin, before plate movements resulted in the hot-spot being overridden by North America.

23.14 What are the Columbia and Snake River Plateaus?

These are broad areas of plateau basalts which constitute sequences of lava flows that have a total thickness of 3000 m or more. The flows extruded from long fissures. (See Problem 3.38 and Fig. 3.14.)

23.15 Are the mountain ranges of the Basin and Range Province the result of extension or compression, and what is the evidence?

Extension, as the faults are normal faults which are the result of extension.

23.16 In general, why is the Basin and Range Province the site of abundant volcanism?

The extensional stresses opened up faults which allowed magmas to rise from the lower crust and the upper mantle. Although the reasons for the extensional regime are debatable, it seems to be related to plate movements.

23.17 The Cordilleran orogeny formed the mountains of western North America. Were the forces compressive or extensional?

The Cordilleran orogeny was a compressional event, actually a series of events. Only the small mountain ranges of the Basin and Range Province, plus its bordering Sierra Nevada and Wasatch Ranges, are the result of extension.

23.18 What became of the thick piles of sediment deposited in the Cordilleran Geosyncline during the Paleozoic and the Mesozoic?

They were lithified into sedimentary rocks and were uplifted and are now exposed in the Rocky Mountains and other ranges of western North America.

23.19 What rock types make up the Sierra Nevada?

Granitic rocks of the Sierra Nevada composite batholith comprise the bulk of the rocks. However, some pieces of the overlying "cover rocks" or "country rocks," into which the batholith intruded, are preserved as roof pendants (remnants of the roof rocks into which the batholith intruded) in the granitic rock. These include both volcanic and sedimentary rocks that are now, of course, metamorphosed.

23.20 The Cordilleran orogeny of western North America is a very complex orogeny with several phases that formed the Rocky Mountains. When and where did it occur?

The mountain building of the Cordilleran orogeny moved eastward like a wave, from Late Jurassic to Eocene time. The Nevadan event formed the first phase of the Sierra Nevada in California and Nevada, lasting from Late Jurassic to Late Cretaceous. The Sevier event was a Late Cretaceous mountain-building episode involving eastward-thrusting in the vicinity of the Rockies. The Laramie event of Latest Cretaceous and Early Cenozoic time involved local uplifts and eastward-thrusting, also in the Rocky Mountains region.

23.21 Does the geologic history of volcanism in the Basin and Range Province have any economic implications?

Yes. For example, more than $16 billion worth of gold reserves had been proven to exist in Nevada by 1988. Some of the porphyry copper deposits of Arizona are also related to this magmatism. Hydrothermal activity, related to the igneous activity, deposited the gold and copper (and molybdenum and other metals, too) in the country rock.

23.22 What age are the abundant coals of western United States?

Paleocene, and the reserves are great. However, these coals are high-sulfur coals, and their use adds to the acid rain problem.

23.23 When was the last epicontinental sea in the interior of North America?

In Paleocene. This was a remnant of the last widespread Cretaceous sea, which was the last major sea on the interior of the continent.

23.24 The Basin and Range Province was once likened to "an army of caterpillars marching northward." Why was this an apt description?

Because the 100 or so small mountain ranges of the Basin and Range Province are all aligned in a north-south direction. The analog is a good one. If you had ever seen army worms (tent caterpillars) on the move, you would agree.

23.25 What is the origin of the spectacular scenery in Bryce Canyon National Park?

This scenery is the continued result of the erosion of fine-grained sediments that had filled an intermontane basin during the Cenozoic.

23.26 How much movement has there been along the San Andreas fault in the last 100 m.y.?

Perhaps 500 km (300 mi).

23.27 On Fig. 23.1, without reading the figure caption, locate each of the following physiographic provinces: Coast Ranges; Rocky Mountains; Western Sierra Madre; Eastern Sierra Madre; Cascade Range; Sierra Nevada; Appalachian Belt (including Plateau, Valley and Ridge, Blue Ridge, Piedmont, Adirondack, and various mountains of New England–Canadian Maritime regions); Canadian and Alaskan Plateau; Columbia Plateau; Colorado Plateau; Mexican Plateau; Great Plains; Interior Lowlands; Coastal Plain; Basin and Range (Sonora Desert to south); Canadian Shield; Interior Highlands (Ozarks and Ouachitas).

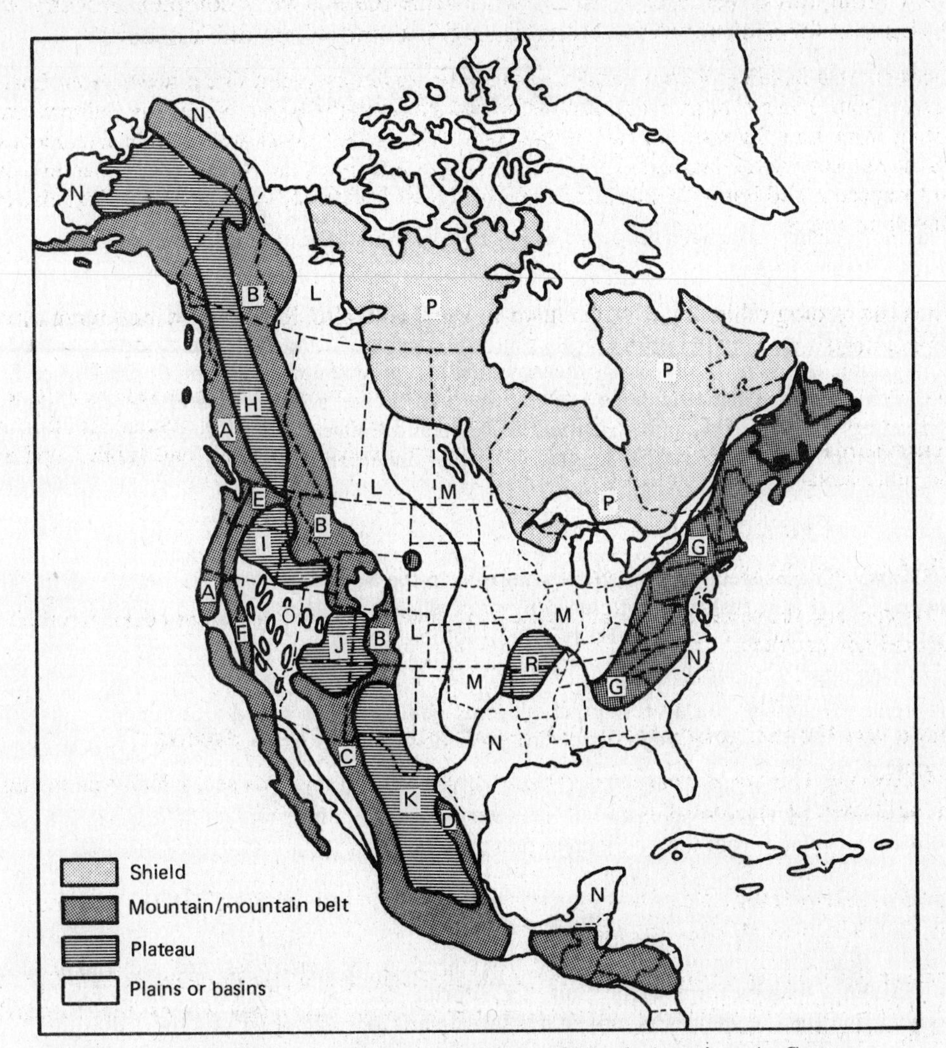

Fig. 23.1 Map of the physiographic provinces of North America. A, Coast Ranges; B, Rocky Mountains; C, Western Sierra Madre; D, Eastern Sierra Madre; E, Cascade Range; F, Sierra Nevada; G, Appalachian Belt (including Plateau, Valley and Ridge, Blue Ridge, Piedmont, Adirondack, and various mountains of New England–Canadian Maritime regions); H, Canadian and Alaskan Plateau; I, Columbia Plateau (lavas); J, Colorado Plateau; K, Mexican Plateau; L, Great Plains; M, Interior Lowlands; N, Coastal Plain; O, Basin and Range (Sonora Desert to south); P, Canadian Shield; R, Interior Highlands (Ozarks and Ouachitas). (*From Ojakangas and Darby, 1976*).

A, Coast Ranges; B, Rocky Mountains; C, Western Sierra Madre; D, Eastern Sierra Madre; E, Cascade Range; F, Sierra Nevada; G, Appalachian Belt (including Plateau, Valley and Ridge, Blue Ridge, Piedmont, Adirondack, and various mountains of New England–Canadian Maritime regions); H, Canadian and Alaskan Plateau; I, Columbia Plateau; J, Colorado Plateau; K, Mexican Plateau; L, Great Plains; M, Interior Lowlands; N, Coastal Plain; O, Basin and Range (Sonora Desert to south); P, Canadian Shield; R, Interior Highlands (Ozarks and Ouachitas).

Supplementary Problems

23.28 What mechanisms have been proposed as possible causes of the extension that resulted in the Basin and Range Province?

It has been suggested that as North America moved westward and collided with the eastward-moving East Pacific Ridge system (spread center) in Mid-Cenozoic time, the San Andreas formed, and a tensional regime developed to the east, thereby kind of "pulling apart" southwestern United States. It has also been suggested that the ridge system was still active beneath western North America, thereby "spreading" it. A third possibility is that the Basin and Range Province is a back-arc system behind the Sierra Nevada magmatic arc, but lagging in time of development. (Does anyone "know"?)

23.29 What was the Tethys Sea?

The seaway that existed between Gondwanaland to the South and Laurasia to the north. The Mediterranean is a remnant that was not closed by the collision of Gondwanaland with Laurasia. The Tethys existed from Late Paleozoic into Early Cenozoic time.

23.30 The Rocky Mountains are a very broad orogenic belt. This has posed a problem for plate tectonocists (unlike the Sierra Nevada, which is easily explained by subduction). So, how is it explained in plate tectonic parlance?

It is explained by some as the product of a very gently dipping subduction zone that has affected the continent far inland from the continental margin.

23.31 Why is there oceanic crust in the Gulf of California?

Because Baja California was torn away from Mexico by northwestward movement along the San Andreas fault, thereby forming the Gulf of California. Baja and Mexico are made of continental crust, and when the Baja segment was moved northwestward, oceanic crust was formed as the "void" developed.

23.32 The "goosenecks of the San Juan River" valley in Colorado are beautiful, sweeping meanders that are cut deeply into the rocks of the area. The river valley has no floodplain. Explain.

The ancestral San Juan River flowed across a peneplain in Cenozoic time, and was undoubtedly meandering on its floodplain in a broad valley. The area was slowly warped upward, and the river continued to erode downward all along its course in an attempt to reach base level. It is a superimposed (superposed) stream, with entrenched meanders.

23.33 What kind of fault is the San Andreas?

It is a strike-slip fault with horizontal movement (as described in Problems 16.20 through 16.23). More technically, in plate tectonic language, it is a transform fault. (See Problems 17.50 through 17.53.)

23.34 What is the age of the Hawaiian Island volcanoes?

The Hawaiian Islands themselves are probably less than 5 m.y. old, and magma is still being extruded on the island of Hawaii nearly every year. However, the individual volcanoes in the chain of volcanoes that continues northwestward from Hawaii are progressively older away from Hawaii. Midway Island is 18 m.y. old, and the oldest of the Emperor Seamounts is about 70 m.y. old. (See Fig. 3.16.)

23.35 When did India collide with Asia?

In Mid-Cenozoic time. However, the collision zone is still active, as indicated by earthquakes.

23.36 How thick is the crust under the Himalaya Mountains?

70 km, the thickest crust known. This is double the general thickness of continental crust.

23.37 Why is the crust so thick under the Himalayas?

Because the continental crust comprising northern India was subducted beneath the continental crust of Asia when the two continents collided.

23.38 What is the origin of the Great Plains or "high plains" physiographic province? (See Fig. 23.1.)

River sediments were eroded from, and deposited east of, the then-young Rocky Mountains in Mid-Cenozoic time.

Chapter 24

The Great Ice Age (The Pleistocene)

INTRODUCTION

The earth has had several ice ages, the most recent being that of the last 1 to 2 m.y. Because these deposits are the most recent sediments on much of northern North America and northern Europe, they are the best studied of all the glacial deposits. Continental glaciers covered 30 percent of the earth's land surface during the zenith of glaciation (Fig. 24.1), whereas only 10 percent is covered today. Nearly all that ice is in Antarctica and Greenland.

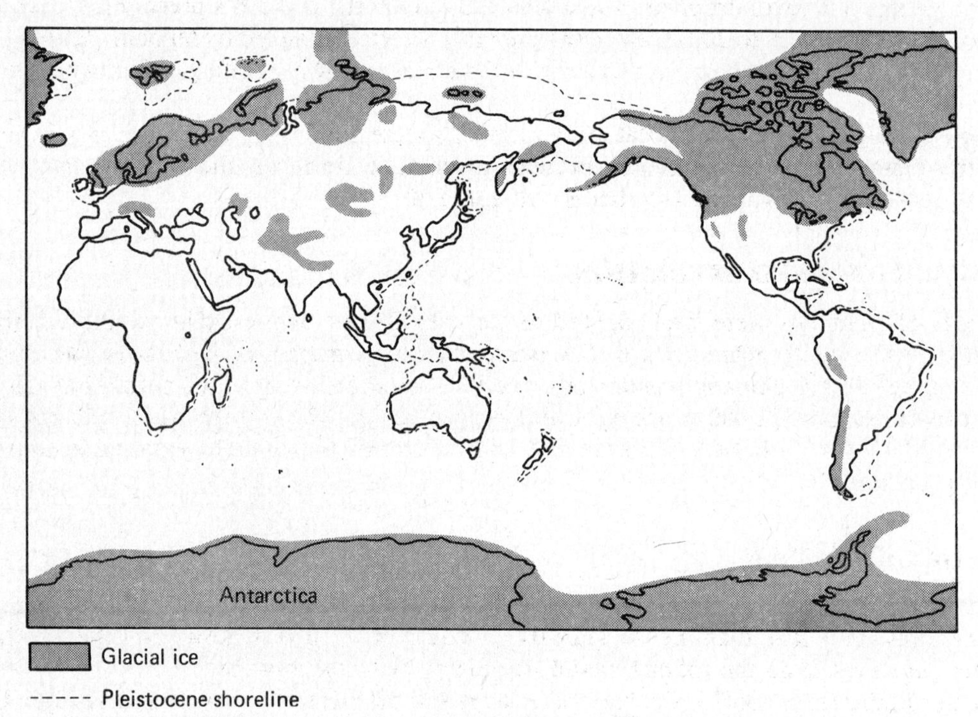

Fig. 24.1 Generalized map of the world showing extent of Pleistocene glaciers.

Earlier ice ages have been documented in rocks about 2300 m.y. old, 800 m.y. old, 700 m.y. old, 440 to 465 m.y. old, and 225 to 325 m.y. old. More ice ages may have occurred, but the deposits may have been eroded away or are as yet undiscovered. Glacial deposits on land have little chance of preservation (i.e., they will be removed by various erosional processes), whereas those deposited on continental margins where subsidence is going on during sedimentation may well be preserved as glaciomarine deposits. Most of the above-mentioned ancient glaciogenic deposits are indeed glaciomarine, having been deposited in the oceans.

GLACIOGENIC SEDIMENTATION

Glacial processes and products, both erosional and depositional, are described in Chapter 10 and should be reviewed before studying this chapter. Glacial deposits include glacial till that is deposited directly from ice and thus is unsorted, meltwater deposits which are deposited by meltwater and are therefore sorted sediments (i.e., river deposits), and ice-rafted detritus dropped from icebergs or floating glaciers into fine-grained, deeper water sediments. The latter are commonly glaciomarine rather than glaciolacustrine (lake) in origin, and many such deposits approach a thickness of 1000 m.

GLACIAL ADVANCES

During the Pleistocene, the continental glaciers advanced southward from cool latitudes and then retreated at least four times. In North America, these advances are named, from oldest to youngest, the Nebraskan, Kansan, Illinoisan, and Wisconsin. In Europe, they are called the Donau, Gunz, Mindel, and Würm.

The glacial deposits of Wisconsin age are on top and are most easily studied. Investigations indicate that there were minor glacial advances and retreats as well and that the glaciers had retreated from most of northern North America and northern Europe by about 10,000 years ago. Therefore, we are still in the Great Ice Age, and a glacial advance will not be unexpected.

Each major advance was accompanied by a drop in world sea level, as so much water was tied up on land as ice. Sea level dropped as much as 120 m (400 ft). If earth's present ice were to melt, sea level would rise by 100 to 200 ft. Even the greenhouse effect, caused by humans putting CO_2 into the atmosphere, may result in enough warming and therefore glacier melting to inundate the world's coastal cities.

The great weight of the continental glaciers, which were commonly thousands of feet thick, depressed the crust. When the ice melted because of climatic warming, the crust began to rebound back up to its original elevation, and this is still going on.

PLEISTOCENE LIFE AND EXTINCTION

Many forms of mammals were well adapted to the cold climate of the Great Ice Age. Perhaps the most famous is the woolly mammoth. It, like many other large mammals, became extinct about 8000 years ago, either due to climatic reasons or "overkill" by Paleo-Indians, or both. Humans entered North America at least 12,000 years ago, and perhaps even 30,000 years ago, crossing over the Bering Strait land bridge between Siberia and Alaska when sea level was lowered because of the continental glaciers.

CAUSES OF GLACIATION

A slight cooling of the earth's climate could cause a major glaciation. The prerequisite is simply that more snow falls each year than melts. Thus there would be a snow buildup, and the weight of the thick snow plus heat from the ground would transform the snow into ice.

Why would the earth cool? A decrease in solar radiation, related to variations in output from the sun or atmospheric properties (e.g., volcanic dust) could be factors. Variations in earth's tilt or a wobble in the earth's rotational axis could be factors. Also, the elliptical orbit of the earth results in its being farther from the sun at certain times, and therefore it receives less solar energy. As the continents move across earth's surface as passive passengers on the lithospheric plates, they may be moved into polar positions where the climate is cooler. All of these factors may be important in initiating and sustaining a continental glaciation.

Solved Problems

24.1 When was the Great Ice Age?

During the Pleistocene, from 1 or 2 m.y. ago nearly to the present, for the last continental glaciers withdrew about 10,000 years ago.

24.2 Did the Great Ice Age end, or are we still living in it?

We are probably still in it. The glaciers have advanced and retreated several times, and since the last retreat was so recent (10,000 years ago), there is no reason to think that the Great Ice Age has ended.

24.3 How much of the earth's land surface was covered by ice?

About 30 percent, as compared to 10 percent today (see Fig. 24.1).

24.4 How were the glacial deposits that are spread across the face of northern North America and northern Europe originally thought to have formed?

They were thought to be sediments deposited during the great flood (the Deluge) as recorded in the Book of Genesis.

24.5 Specifically, how were the large erratic boulders explained?

They were thought to have "drifted" to their present locations as passengers on and within icebergs. As the icebergs melted, they would have dropped their load of debris. (This is the origin of the words *glacial drift,* a general term for glacial sediments; see Problem 10.50.)

24.6 Where did the idea that glacier ice could transport and deposit sediment originate?

In the Swiss Alps where local residents had long observed that valley glaciers deposited stony debris at their lower ends.

24.7 What part did striated and grooved bedrock surfaces play in the development of the theory of continental glaciation?

Striations, grooves, and polished rock were observed in the Swiss Alps in valleys from which valley glaciers had recently melted. When observed at nonmountainous localities, it seemed logical that glacier ice, even far from mountains, had affected those rock outcrops.

24.8 See Fig. 10.3. Name features A, B, C, D, and E.

(A) Striations. (B) Grooves. (C) Chattermarks. (D) Crescentic gouges. (E) Whaleback (roche moutonee).

24.9 How do geologists determine the direction in which a past glacier moved?

By measuring the orientations of glacial striations and grooves; however, these do not give a sense of movement, for two directions are possible. Crescentic gouges on bedrock, also made by ice pressing stones against the bedrock, are much more useful; within each such mark, there is a steep side and a shallow side. The steep face points in the direction from which the ice came (see Fig. 10.3).

24.10 In which direction was the ice that fashioned the features of Fig. 10.3 moving?

Toward the south. Note that the grooves and striations indicate north *or* south. The chattermarks and crescentic gouges both have their steep sides pointing in the directions from which the ice came, which in this diagram is from the north. Thus the ice moved toward the south.

24.11 Does continental glacier ice always move from north to south, as in Fig. 10.3?

No. Many people think that because ice formed in polar latitudes, that it moved from north to south. In general, that is quite true. But at any given spot, the direction can vary, for ice in general moves downhill like water does. Therefore, even the front of a continental glacier has lobes of ice moving in somewhat different directions and at different rates. (Valley glaciers, of course, follow the valleys in which they are situated, always moving downhill.)

24.12 See Fig. 10.5 in Chapter 10. Which portion of the sediment was deposited directly from glacier ice, and which was deposited by meltwater from the melting glacier?

The higher and hilly part was deposited by ice, and the lower and flatter part by meltwater.

24.13 What characterizes sediment deposited directly from glacier ice, as opposed to meltwater deposits? (See Fig. 10.5.)

Poor sorting is the most obvious characteristic of such deposits, which are called *glacial till* (see Problem 10.46). Pieces of sediment range in size from clay to boulders, some as large as houses. Meltwater deposits are sorted into layers of gravel, sand, and silt.

24.14 How does a geologist determine the direction of ice movement from glacial till?

In some tills, the pebbles, cobbles, and boulders show a tendency to be aligned parallel to the direction of ice movement. However, it is usually not easy to determine the sense of ice movement in this manner.

24.15 Which large-scale landforms can be used to reconstruct approximate paleoflow directions of a continental glacier?

Glacial drumlins (see Fig. 10.5), whalebacks (see Fig. 10.3 and Problem 10.58), and relationships of terminal moraines and associated outwash deposits (see Fig. 10.5). The gently sloping lower ends of whalebacks and the steep ends of drumlins point back in the direction from which the ice came.

24.16 What geologic history is revealed by Fig. 24.2?

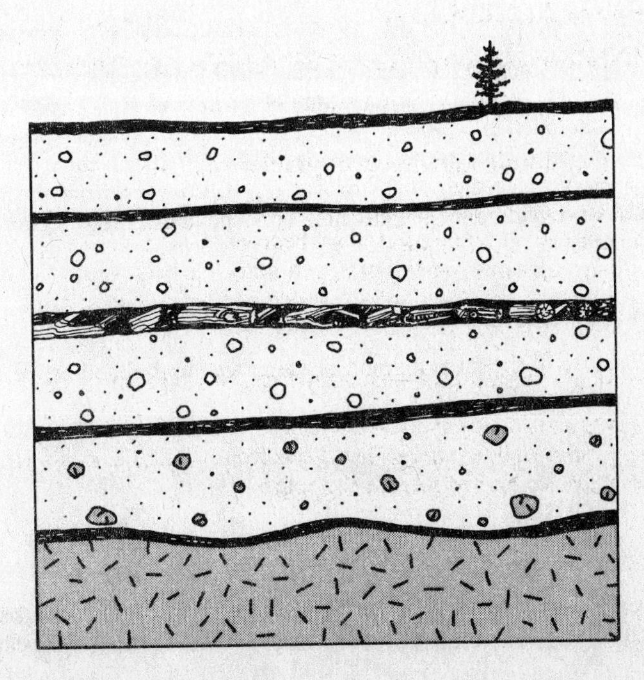

Fig. 24.2 Diagram showing a sequence of glacial tills and soil horizons.

A history of multiple glaciations, with soils developed between ice advances during "interglacial times," which were warmer and more conducive to soil formation.

24.17 How many major ice advances were there in North America and Europe, and what are they called?

Four. In North America, they are, from oldest to youngest, Nebraskan, Kansan, Illinoisan, and Wisconsin. In Europe, the same ice advances are named the Gunz, Mindel, Riss, and Würm (Fig. 24.3).

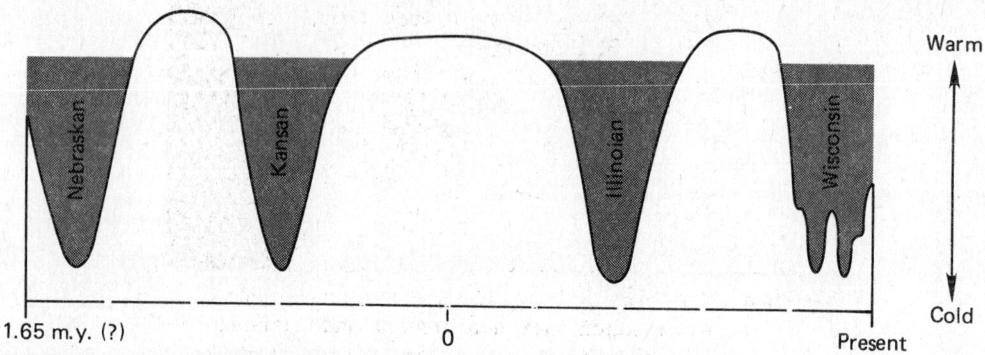

Fig. 24.3 Diagram showing the four major ice advances of the Pleistocene.

24.18 How are sediments deposited in former glacial lakebeds studied in order to ascertain the duration of the lakes?

By coring the lakebeds and counting the glacial varves which are annual layers (see Problem 18–52).

24.19 What does the study of pollen grains preserved in a sediment core from a glacial lake bed tell researchers about geologic history?

It can show climatic changes in the area, based on the types of tree spores and pollen (and other pollen, too) present. These in turn are correlated with, and are evidence of, glacier advances and retreats.

24.20 How can marine fossils from sediment cores be used to provide evidence of past climates?

By measuring the relative abundances of O^{18} and O^{16} isotopes in the shells. Molecules of H_2O^{16} are evaporated, leaving much of the heavier H_2O^{18} behind. During a time of glaciation, H_2O^{16} is locked up in glacier ice, and the ratio of these two isotopes in the ocean water changes, with a higher proportion of O^{18} during such times of glaciation. When the ice melts, the ratio of the O^{18}/O^{16} in the oceans decreases. Marine shelled organisms, when building their shells of $CaCO_3$, use oxygen from the water. Thus the shells reflect the water temperature and hence the climate. To summarize, the O^{18} content of shells decreases as the temperature increases.

24.21 How much did the great continental glaciations affect sea level?

Sea level dropped by 120 m (400 ft) because of the great amount of water tied up on land as ice.

24.22 Where are the major ice masses today?

Antarctica and Greenland.

24.23 See Fig. 24.4. What were the ice conditions and glacial activity in each zone?

In the center, snow was accumulating and being transformed to ice. (See Problem 10.8.) In the zone of maximum erosion, the basal ice was near the melting point of ice and was sliding over the bedrock,

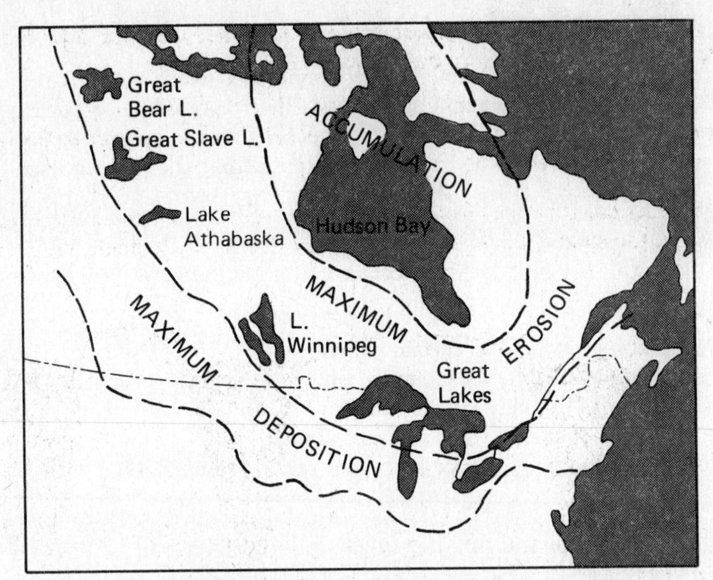

Fig. 24.4 Northern North America showing zones of ice accumulation, maximum erosion, and maximum deposition during the Pleistocene. (*After Matsch, C. L., North America and the Great Ice Age, McGraw-Hill, New York, 1976.*)

abrading it. Also, melting and subsequent refreezing led to plucking of rock pieces from underlying bedrock and incorporation of this material into the basal ice. In the zone of maximum deposition, the glacier was depositing rather than eroding; there, the ice was only melting rather than melting and refreezing as in the zone of erosion.

24.24 How do the glacial processes shown in Fig. 24.4 influence the mineral exploration and mining industries?

In the zone of maximum erosion, the glaciers deposited little sediment. In the zone of maximum deposition, the glacial drift can be as thick as 250 m! Therefore, in the former zone, the mineral deposits may be exposed at the surface, with little if any weathered material on top. In the zone of maximum deposition, they are covered, making exploration and the discovery of mineral deposits very difficult. Canada has a large mining industry with hundreds of exposed mineral occurrences and many mines.

24.25 Based on Fig. 24.4, how are the large lakes related to glaciation?

They are in part the result of glacial erosion of the bedrock.

24.26 What are the dotted lines around the continental coastlines of Fig. 24.1?

The approximate coastlines at the height of glaciation. The continents were larger, as sea level was lower because of the water tied up in ice on the continents.

24.27 How did the glaciers of Fig. 24.1 affect the climate of areas southward from the glaciers, and especially the climate of southwestern United States during the Pleistocene?

The climates became moister and cooler. Southwestern United States was not semiarid as it is today but was much moister with abundant rainfall because of the cooler temperatures on the continent, lowered evaporation rates, and higher precipitation rates. Many lakes formed in the down-faulted portions of the Basin and Range Province (see Problem 23.5).

24.28 What is the history of Great Salt Lake?

It is the remnant of an ice-age lake named Lake Bonneville, which was 10 times larger than Salt Lake, as deep as 1000 ft, and all fresh water. Lake Bonneville evaporated as a consequence of the change in climate that accompanied the demise of the continental glaciers, and it became much saltier. Because irrigation has stopped most fresh water in rivers from reaching the present lake, it has continued to get saltier.

24.29 How did the glaciation of North America and Europe (see Fig. 24.1) affect the vegetation on the land to the south?

Each type of vegetation is adapted to certain climatic zones. Therefore, with the advance of continental glaciers and the accompanying cooler climatic zones, the vegetation pattern was shifted southward by about 1000 km. Tundra, as in today's far north, bordered the southern edge of the glaciers of Fig. 24.1, and southward from that, in order, were zones of spruce, pine, and deciduous forests. The Great Plains of today were largely covered by spruce and pine (see Problem 24.19).

24.30 What caused the retreat of the major glaciers of Fig. 24.1?

A warming of the world's climate resulted in melting of more snow than accumulated, and hence the retreat of the glaciers (see Problem 10.12).

24.31 When did the United States and Canada become essentially ice free? (See Fig. 24.5.)

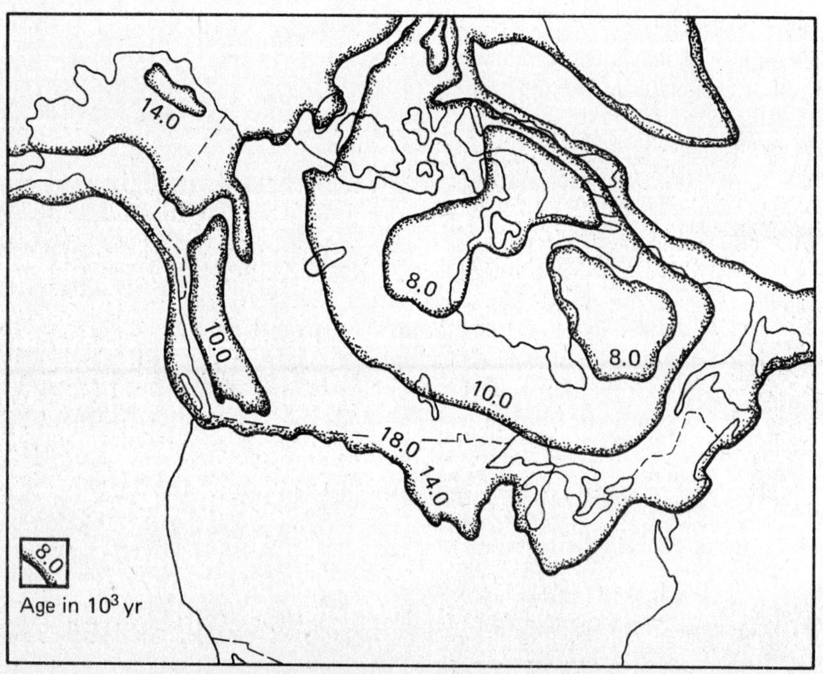

Fig. 24.5 Stages of retreat of the Pleistocene glaciers in North America. Numbers indicate thousands of years. (*After Matsch, C. L., North America and the Great Ice Age, McGraw-Hill, New York, 1976.*)

By about 12,000 years ago, most of the United States was free of ice. By 10,000 years ago, all of the United States was ice-free, except, of course, for small valley glaciers in the mountains. About 8000 years ago, glacier ice remained in Hudson Bay. By 5000 years ago (5000 B.P.), ice on North America was restricted to about its present distribution, mostly on the Arctic islands (e.g., Baffin Island).

24.32 What and where was Glacial Lake Agassiz?

It was a very large postglacial lake, perhaps the largest the world has ever seen, the product of the melting of the Laurentide ice sheet. It probably existed for a few thousand years between 8000 and 12,000 B.P. It extended from northern Saskatchewan (and included Lake Athabasca, the unnamed lake just south of Great Slave Lake on Fig. 24.4) to west-central Minnesota.

24.33 How are the former extents of glacial lakes, such as Lake Agassiz, determined?

By old beaches and by the areal distribution of very flat land that represents the original lake floors. Minor irregularities on the lake floors were eliminated by the settling of silt and clay from the meltwater.

24.34 Who was Lake Agassiz named after?

After Louis Agassiz, a Swiss naturalist, who was largely responsible for the theory of continental glaciation. Later he taught at Harvard.

24.35 What is postglacial rebound?

The weight of the thick ice sheet depressed the crust, especially in Finland, Sweden, and north-central North America (centered on Hudson Bay). Since the ice melted, the crust has been slowly "rebounding," or rising to its former level when it was in isostatic balance (see Problem 16.55).

24.36 From Fig. 24.6, determine the maximum amount of postglacial rebound in North America and Fenno-Scandinavia.

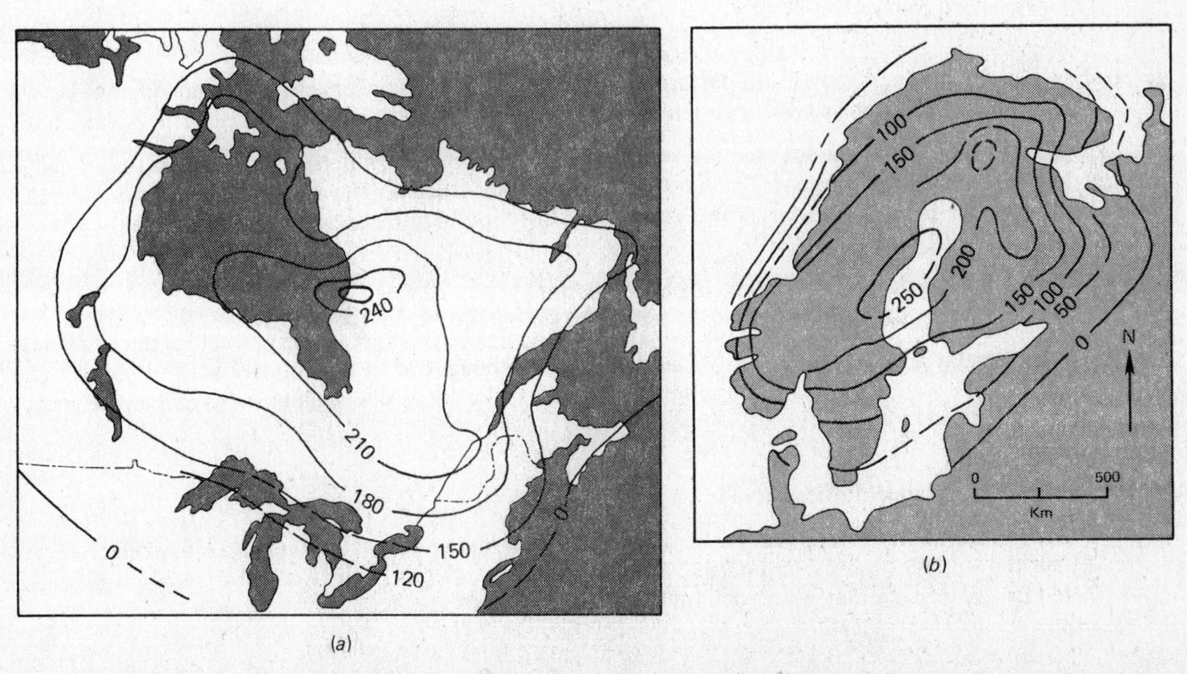

Fig. 24.6 Postglacial rebound in North America and Fenno-Scandinavia. Lines of equal uplift are shown in meters. (*After Matsch, C. L., North America and the Great Ice Age, McGraw-Hill, New York, 1976.*)

The uplift has been more than 270 m in the Hudson Bay area and nearly that amount in Finland and Sweden.

24.37 What is the famous Bering Strait land bridge?

When sea level was lowered during continental glaciation, a bridge of nearly dry land existed between Siberia and Alaska, across the Bering Strait. The water is presently about 60 m (200 ft) deep.

24.38 Why is the Bering Strait land bridge important?

It allowed animals and humans to cross over to North America from Asia. It is thought that human crossings occurred between 12,000 and 30,000 B.P., with the oldest unambiguous evidence of humans in North America about 12,000 years old.

24.39 How was the fauna of South America affected by glaciation in the Northern Hemisphere?

The Isthmus of Panama in Central America was partially under water, but became a dry land bridge in the Pleistocene when sea level was lowered. This allowed animals to cross from north to south and vice versa. Unfortunately for South American mammals who had few natural enemies, many carnivore species migrated from North America. (Also see Problem 19.55.)

Supplementary Problems

24.40 See Fig. 24.2. What is the significance of the variable thickness of the different soil layers?

Thicker soils took longer to form than thinner soils, assuming similar climates. Therefore, thicker soils indicate longer interglacial periods.

24.41 See Fig. 24.3, which shows four major ice advances but with smaller advances and retreats shown within the last (Wisconsin) ice advance. Why is such detail shown only in the Wisconsin advance?

Deposits of the last major advance, the Wisconsin, are the uppermost deposits of the four major glacial advances, and are widespread. Thus it is the most easily studied. The Nebraskan, Kansan, and Illinoisan glacial drifts are largely buried except near their southern extents.

24.42 How has postglacial rebound affected Lake Superior?

The northeast end of Lake Superior is rising at the rate of about 1 m per century, whereas the southwest end is rising at a much slower rate. The result can be thought of as a southwestward tilting of Lake Superior. The southwest end is therefore "drowned" under the water, making an excellent harbor behind a large sand bar.

24.43 What is the so-called driftless area?

It is an area in southwestern Wisconsin and southeastern Minnesota that has no deposits of glacial drift. Either the ice lobes moved around this area, or else the drift has been removed by erosion that also bared the Paleozoic sedimentary rocks of the area. The Mississippi River flows through the western part of the area.

24.44 If you had been viewing Death Valley during the height of Pleistocene glaciation, what would you have seen?

Not glacial ice, but a large lake called Lake Manley. It was as deep as 180 m (330 ft), based on the elevations of old shorelines. This was one of the "pluvial lakes" of the Basin and Range Province (see Problem 24.27).

24.45 What caused the "channeled scablands" of eastern Washington and adjacent areas?

This irregular topography is the result of the catastrophic draining of Glacial Lake Missoula, which had formed as the valley glaciers of the Rocky Mountains were melting. The water was carried away by large streams when the morainal dams were breached; boulders as large as 10 m in diameter were carried by the floodwaters.

24.46 How did the continental glaciation affect the Great Lakes?

The basins of the Great Lakes probably all existed prior to glaciation, but the glaciers further scoured the basins. As the ice lobes of the Wisconsin advance were retreating, water levels in the lakes were much higher than at present because the ice acted as natural dams.

24.47 How can rates of postglacial rebound, as in Fig. 24.6, be determined?

By dating organic remains in old beaches by the C^{14} dating method.

24.48 What is the evidence that the basins of Lake Superior and the other Great Lakes have been tilted by postglacial rebound?

The main evidence is the elevations of individual beaches along the lake shores. A given beach may have an elevation of, say, 10 m above the present lake level at one end of the lake, and 100 m at the other end. Obviously, a beach should be horizontal if it has not been disturbed.

24.49 The term *prehistoric overkill* has been applied to the extinction of many North American Pleistocene species by about 8000 years ago, including the camel, elephant, horse, ground sloth (20 ft tall!), giant bears, saber-toothed cats, and certain wolves, bison, and antelope. What does this imply?

The theory is that they were overhunted by the Paleo-Indians, the first human inhabitants of North America. This does not necessarily mean that they killed all the individuals, for this would have been difficult with primitive weapons. However, elimination of certain species in the food chain may have affected the entire animal population. Alternatively, their extinction may have been related to climatic changes. A big problem with the latter idea is that major extinctions should have occurred after the Nebraskan, Kansan, and Illinoisan ice advances as well as after the Wisconsin, and they did not.

24.50 What is the *Hypsithermal Interval?*

It was a period from 5000 to 7000 B.P. when the earth was warmer than at present by 2 or 3°C. The evidence is in the types of trees present, based upon pollen studies.

24.51 When did the "Little Ice Age" occur?

Between the fourteenth and eighteenth centuries, mountain glaciers all over the world markedly advanced, apparently due to a cooling episode.

24.52 Was the Great Ice Age of the Pleistocene the only major glaciation that the earth has experienced?

No. Ice ages have been documented within the rock record in rocks of several ages: about 2300 m.y., 800 m.y., 700 m.y., 440 to 465 m.y., 225 to 325 m.y., and the last 2 m.y. (i.e., Pleistocene). Others may well have existed, but the evidence is as yet undiscovered or else the evidence was removed by erosion.

24.53 What evidence of ancient glaciations do geologists search for?

There are several. Matrix-supported conglomerate resembling glacial till is one. Another is fine-grained sediment with large stones in the sediment (see Problem 10.22). Striated stones and striated rock beneath the glacial deposits is another. Ideally, several lines of evidence occur together. For example, a matrix-supported conglomerate may have been deposited by a glacier, but it could have formed in other ways, such as by gravity-induced slope processes. But, if the conglomerate contains striated clasts and occurs in association with a fine-grained sediment that contains oversized stones (especially if the stones penetrate the fine-grained laminae), then the evidence is fairly strong.

24.54 What are the causes of ice ages?

Scientists do not know for sure, but suggested causes include variations in the sun's energy output (e.g., the sunspot cycle of 11 years shows a 4 percent energy change); the latitudinal positions of the continents on earth's surface (polar locations are best); the shape of earth's orbit is elliptical, and at certain times the earth is further from the sun; wobble in the earth's rotational axis; changes in the tilt of earth's axis (a 3° variation); and changes in the atmosphere (CO_2 content, volcanic dust, etc.). Perhaps more than one factor is necessary.

24.55 What is the *Milankovitch cycle* or "effect"?

Milutin Milankovitch was a Yugoslav geophysicist who in 1920 suggested that the effect of three factors that alter earth's distance from the sun (see the previous problem) and thus alter the amount of radiation hitting earth are the causes of ice ages (i.e., glacial advances and retreats). The earth's tilt cycle is 41,000 years long, and the wobble of earth's axis (precession) is 22,000 years long. Milankovitch's work has stimulated much additional work, for it alone probably cannot explain glaciations.

24.56 What is the best depositional site for the preservation of ancient glaciogenic deposits?

On continental margins where the crust is subsiding during and after glaciation. Most glacial deposits on land will be eroded away rather than being buried and preserved, whereas glaciomarine deposits on the continental margins may be preserved and uplifted at some distant future time for observation by geologists.

24.57 How widespread is permafrost today? (See Problem 10.59.)

About 20 percent of the earth's land surface is permanently frozen, for the soil temperature remains below freezing except for minor summer melting at the surface. About half of the U.S.S.R. and about half of Canada are included in this periglacial environment.

24.58 What are the LaBrea tar pits of Los Angeles, and why are they important?

Pleistocene seeps from which most of the volatiles are gone left a tarry and sticky residue. When animals came to drink water on the surface of the seep, they became stuck. Predators hearing cries of distress came to eat, and they, too, became trapped. The bodies slowly sank into the tar, and their bones were preserved. The pits have yielded 200 different animal species, including 1000 individual saber-toothed cats and 1600 dire wolves.

Chapter 25

People and the Earth: Environmental Problems

INTRODUCTION

This chapter is the most important chapter in this book. It is too short—it should be expanded to the size of a book. Why? Because people live on this earth and must treat it properly so that it will sustain people and all other forms of life for millennia to come. We are not doing a good job of caring for the earth, and unless we change our outlook and way of life, life itself will cease to exist. Does this sound alarmist? Yes, and it should! We must learn from our mistakes and set about correcting a most dangerous situation.

People use resources, and people create problems of overuse and pollution. Resources, especially food, are not distributed well. The earth's population is already too large, and it is growing larger. We, the people, are the problem. And more and more people mean more and more problems.

Chapter 7 is a review of the origins and distribution of mineral resources and fossil fuels. In this short chapter, the emphasis will be on the environmental problems their use has caused as well as on some of the current and potential shortages that their use (and overuse) has generated and will generate in the future. Americans have a special responsibility, for although they make up only 6 percent of the earth's population, they use more than one-third of earth's resources. Canadians, West Europeans, and Japanese are nearly as guilty as Americans.

FOSSIL FUELS

Coal is abundant, but burning it as a fuel causes pollution. The sulfur in coal combines with water and oxygen to make sulfuric acid and acid rain. Much coal is mined by open-pit methods, and this disturbs the environment. Although laws now require mining companies to restore the mine area to its original state, this is not always easy. In the semiarid West, the scant rainfall makes the reestablishment of vegetation difficult.

Petroleum and natural gas are being rapidly used up. It has been estimated that one-half of the earth's natural gas is already gone, and one-quarter of its oil has been used up. More specifically, the United States has petroleum reserves to last only 11 to 12 years but coal to last 600 years. Higher petroleum prices would stimulate more exploration in the United States, but nevertheless, U.S. production peaked about 1970, and the United States will never again produce as much petroleum per year as it once did.

Therefore, the United States must import more and more petroleum. In 1985, the United States imported 27 percent of its petroleum; in 1990, it imported 48 percent. The number is climbing. Most European countries import most of their petroleum, and Japan imports nearly all its needs. Look back at Problem 7.55 to see how the locations of the world's petroleum reserves will undoubtedly affect geopolitics in the future.

South Africa has plans to become self-sufficient in petroleum by utilizing coal to make synthetic petroleum. Research should be done on the economical utilization of the petroleum in oil shale, which could supply as much petroleum as the United States has ever had. However, mining the oil shale affects the environment in the vicinity of the mines. The oil sands of Alberta will probably be utilized on a larger scale in the near future.

Nevertheless, when petroleum supplies are looked at from a long-range viewpoint, a scenario of shortages in the not-too-distant future emerges.

NUCLEAR FUELS

Nuclear power plants use the fission process, utilizing uranium-235 as the fuel. (See Problems 2.11 through 2.17.) This isotope constitutes only about 1 percent of processed uranium—we don't use the abundant isotope, U-238. If we could perfect the breeder reactor process, U-238 could also be used as fuel. To use U-238, we must use U-235 as an "ignition system," but we are rapidly burning up our ignition system as fuel!

Nuclear power generated by the fission of uranium was once touted as the answer to the world's energy problems. However, this has been reevaluated by most nations, as this energy source poses two main problems: (1) Radioactive wastes must be safely disposed of. Some are dangerous for thousands of years, although 80 percent of the radioactivity in spent nuclear fuel and high-level radioactive waste is dissipated after 1 year of storage. (2) In addition, there is always the danger of a nuclear accident such as the release of radioactive material into the environment, or a dangerous meltdown as at Chernobyl, U.S.S.R., in 1985. Furthermore, nuclear wastes can be used to make nuclear warheads. And, nuclear power plants have proven to be more expensive to build than conventional fossil-fuel plants.

If technological answers can be found for each of these problems, the energy potential from nuclear fission is very large. Whereas high-grade uranium deposits are difficult to find, all rocks contain some uranium, and it is conceivable that low-grade resources could be utilized.

ALTERNATIVE POWER SOURCES

Nuclear fusion (see Problems 2.13, 2.15, and 2.17) may be perfected as a nuclear power source in the future. This is an important objective, as fusion produces no radioactive waste products and the raw material (hydrogen) is abundant. Heavy hydrogen isotopes—hydrogen-2 and hydrogen-3—(see Problem 2.8) will be utilized in the process. The heavy hydrogen from 1 km^3 of seawater could produce more power by nuclear fusion than the power that has been and could be generated by all the fossil fuels the earth has ever had!

Solar power should be utilized much more than it is now (that is, on a larger scale than for heating individual buildings), but that will require great financial investments for research. The earth's surface receives each year 173,000 billion kW of energy from the sun—people generate about 1 billion kW each year from all the processes we use in our civilization! What a fantastic energy reserve to tap into!

Geothermal energy is another possibility. Natural steam at geological hot-spots—New Zealand, Iceland, and northern California—is currently being used for heat and power. The potential is much greater. Recall that the earth's geothermal gradient (see Problem 1.40) increases with depth at the rate of about 1°C per 30 m (100 ft) of depth, or about 33°C per kilometer. Therefore, at a depth of about 7 km, the rocks have a temperature of 230°C, above the boiling point of water. Pressure factors must be considered as well, but if deep wells were drilled anywhere on earth's surface and water were pumped down the wells, the water could be returned to the surface through other wells as steam to turn turbines in power plants. The rock at the bottom of the well would have to be fractured by explosives in order to create more rock surface area from which the water could extract the heat. A major problem is that the rock's heat at the base of the well may be used up too rapidly for it to be an economical process.

In summary, energy sources are a problem. We must use oil and natural gas to supply civilization's short-term needs, but solar and/or nuclear power appear to be the best long-range answers.

POLLUTION

People, especially in the western world, use large quantities of earth's resources and generate large quantities of waste material. The waste products—pollutants—include solid wastes, wastes that go into the surface waters and groundwaters and into the soil, and wastes that pollute the air we breathe and even pollute the upper atmosphere.

Solid wastes are the most obvious of the pollutants, for they are so visible. Barges loaded with solid wastes from New York City, for example, are dumped offshore in the Atlantic Ocean, a shameful albeit economical and easy practice. Such practices must stop, for the oceans are rapidly becoming very polluted from cities, from rivers that bring in wastes from inland, and from ships. Life, both wildlife and human life, are endangered. The food chain is delicate, and if any link in the chain were to vanish, the chain would crumble.

But where are New York City, and other cities, and ships, to put their solid and liquid wastes, which total up to astronomical figures? Should they be buried on land? That takes space, and then the buried wastes can contaminate the groundwaters that people need for drinking and other purposes. Even in Minnesota, not exactly one of the big industrial states, about 1 ton of solid waste per year are generated per person; in the late 1980s, only about 5 percent of this was being recycled and reused. Obviously, recycling of glass, paper, plastic, and metals is a major partial solution to the solid-waste problem.

Dozens of waste chemicals that have been disposed of in the soil and water are carcinogenic (cancer-causing). Other chemicals were used for important reasons without realizing the side-effects they caused. For example, DDT was used as an effective insecticide, until it was realized that not only is it a carcinogen but it also affects the food chain. DDT in insects and fish was consumed by birds and built up in their bodies. This affected their reproduction in that egg shells became thin and weak and therefore few eggs hatched. The bald eagle, for example, was greatly endangered until the use of DDT was banned.

Carbon dioxide generated by the combustion of fossil fuels is building up in the atmosphere. In the United States, nearly one-fourth of the emissions come from motor vehicles (13 percent from autos and 10 percent from trucks and buses). At present rates of increase, the concentration of atmospheric CO_2 will double within the next century. It has already increased about 20 percent since about 1850. This is creating the "greenhouse effect," whereby the sun's heat is able to penetrate the atmosphere but then cannot escape from earth because the CO_2 effectively absorbs the heat. Thus the earth's surface is becoming warmer. In all fairness, this is also a natural process that is necessary to our existence on this planet; without the natural portion of the greenhouse effect (i.e., the original abundance of CO_2 in the atmosphere that traps some of the sun's heat), the earth's surface would be a lot colder and perhaps even freezing. But people may be making it too warm. Was the heat spell of the mid to late 1980s related to the greenhouse effect? We don't know for sure. However, if the effect is indeed real and if it continues, earth's glaciers could melt and sea level could rise 30 to 60 m (100 to 200 ft)! Even a partial melting and a sea level rise of 5 m would be disastrous.

The depletion of the world's rain forests is a factor in pollution, too. As forests are depleted, they use less CO_2 in the process of photosynthesis and thus can't alleviate the excess CO_2 problem. In addition, the burning of the rain forests is contributing an appreciable amount of the CO_2 being added to the atmosphere.

Acid rain is an enormous problem, the result of the burning of fossil fuels. The sulfur in coal (present as sulfides and sulfates) combines with oxygen to form SO_2. This eventually combines with water and oxygen to form H_2SO_4, or sulfuric acid. This can affect the life cycle in lakes, making them sterile of life. Increased acidity also affects vegetation on land. Northern Europe has experienced these problems to a large degree already, and North America is trying to stop acid rain before the lakes and forests are dead. There are other acids, too—incomplete combustion of fossil fuels in automobiles and power plants produces nitrogen oxides that result in nitric acid. Figure 25.1 shows which portions of North America are affected by acid rain.

Chlorofluorocarbons (CFCs) may be destroying the ozone (O_3) layers 12 to 20 mi above earth's surface. There, sunlight changes the molecular structures of the CFCs, creating new molecules that are joined to the ozone molecules, destroying them. Ozone protects us from harmful ultraviolet rays from the sun.

Another pollution problem related to fossil fuels is oil spills by tankers, as most vividly depicted for the world to see by the tanker Valdez spill in Prince Rupert Sound in Alaska in 1989, in which

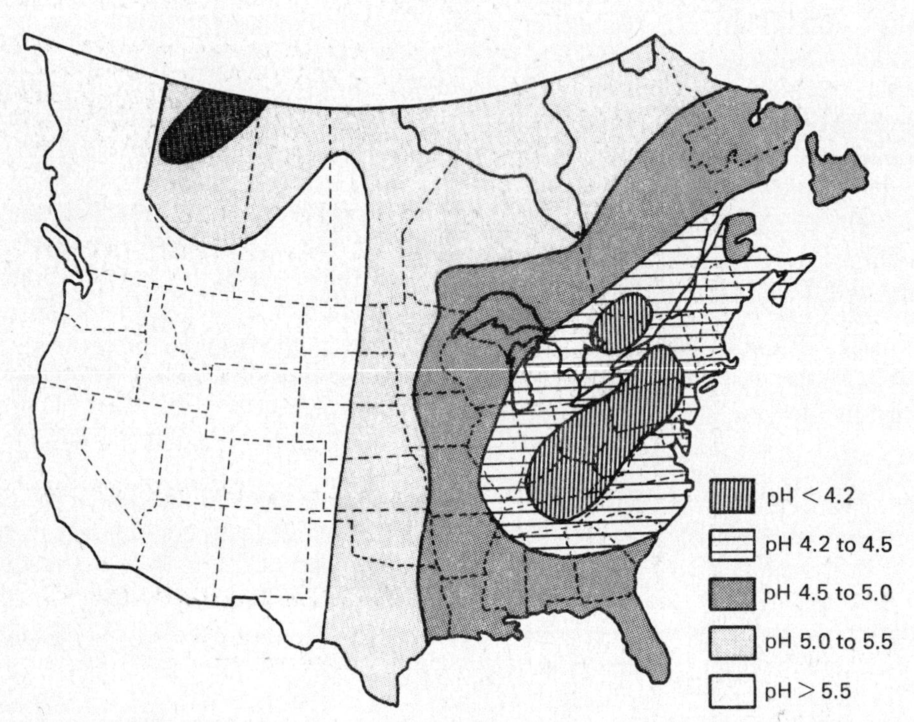

Fig. 25.1 Map showing acid rainfall in North America. (*After U.S. Water News, 1986.*)

millions of gallons of crude oil escaped from the ruptured tanker and contaminated hundreds of miles of shoreline and killed untold quantities of wildlife, including sea otters and bald eagles. Much petroleum is moved by tanker—for example, nearly all Japan's petroleum comes into Japan by tanker. In 1989, 65 percent of the oil consumed in the United States arrived by tanker, and it has been estimated that by the year 2000, this figure will be 90 percent. Therefore, there will be more spills.

Fortunately, there is now worldwide concern about the environmental problems, thanks to a number of environmental organizations that bring the problems to people's attention and to the attention of governments. Concerned citizens activate their governments to do something. Hopefully, worldwide agreements will be forthcoming.

POPULATION

Depletion of earth's resources including fossil fuels, the pollution problem, and the threat of nuclear war are indeed staggeringly big problems! But, *the biggest problem of all is overpopulation!* Earth already has too many people for it to sustain at a comfortable standard of living. Gains made in pollution control or the conservation of energy or improvements in the distribution of food to prevent malnutrition and starvation are quickly negated by the continuing increase in world population.

The increase in population is quite recent, as shown in Fig. 25.2. In 1990, earth had about 5.3 billion people. By the year 2070, it could have 20 billion people. The world's population is increasing at a rate of about 80 to 90 million per year. This is like doubling the population of the United States in 3 years. Could the United States feed these people if they were all born in the United States where food production is unrivaled and where food distribution is not a problem? Yes. What about after the next 3 years? Yes. But what about after 9, or 12, or 15, or 18 years? Even in the United States, it would be a problem. In the next 20 years, the equivalent of 20 countries the size of Bangladesh, and with the same poverty level, will be added to the face of the earth. The greatest population increases are in the poorer countries where the ability to produce and buy food is limited.

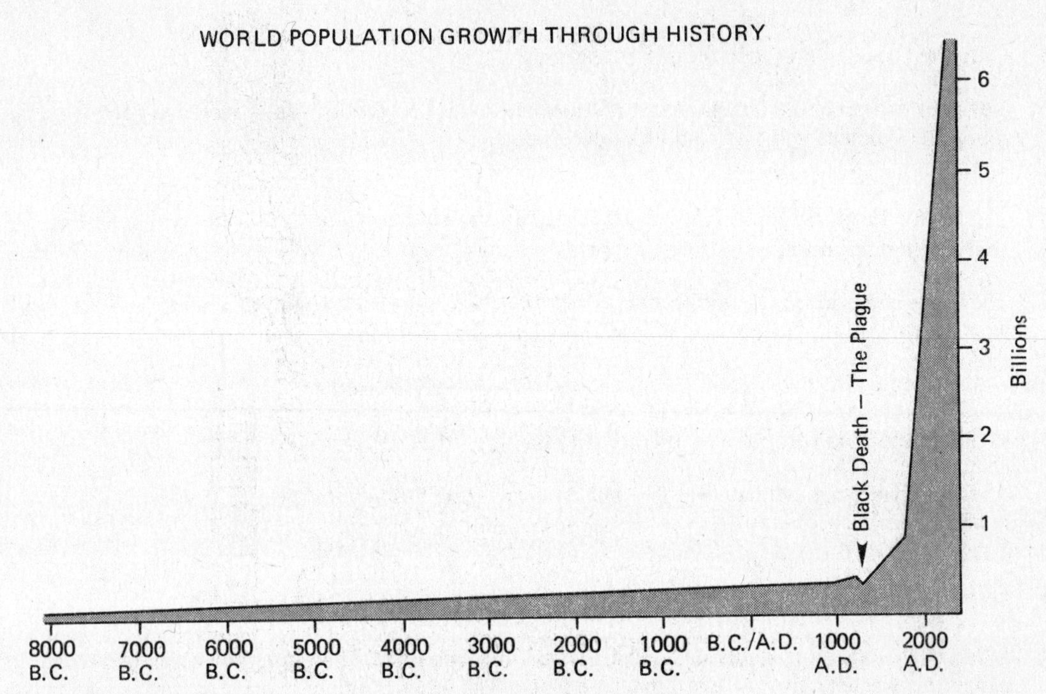

Fig. 25.2 Graph showing the world population growth through history. (*After the Population Reference Bureau and the Population Institute.*)

Solved Problems

25.1 Coal is abundant (see Fig. 7.1), but there are two major problems with its use. What are these?

We can't mine it, and we can't burn it. When coal is burned, sulfur dioxide (SO_2) is given off as a waste product. This eventually combines with water and oxygen to form H_2SO_4 (sulfuric acid), which causes most of the acid rain. Also, much coal is mined by open-pit operations that deface much land. However, regulations now require companies to restore the land to a near-original condition.

25.2 Where does the United States get its petroleum from?

In 1990, 48 percent was imported, and this figure is increasing.

25.3 Could oil companies find and produce more petroleum in the United States?

Yes, if the price of oil went up to, say, $25 per barrel (42 gal). American companies could then afford to explore for more, and would find more, thereby reducing the need for foreign petroleum. This would help both the balance of payments and the nation's security.

25.4 Is the world running out of petroleum?

In spite of periodic gluts of oil on the market, the answer is yes. Oil is a nonrenewable resource, and we are using it up at an ever-increasing rate. We are using it largely as a fuel, rather than saving it for the chemicals it could produce. Our descendants may say, "They *burned* petroleum so they could go from place to place in little vehicles! Can you believe it?"

25.5 Can the rapid use of petroleum be slowed?

Yes. For example, the development of more mass transportation would decrease petroleum use. Also, more fuel-efficient vehicles could be developed.

25.6 There are large oil sand (tar sand) deposits in Alberta (see Problem 7.17 and Fig. 7.7). Are there other such deposits in the world?

There are some in the United States, South America, Africa, Europe, and the U.S.S.R., but they are small compared to Alberta's Athabasca oil sands, and there are no plans to develop these small deposits.

25.7 What is the total amount of bitumen ("tar") in the Alberta deposits?

The total has been estimated at 900 billion barrels, or 5 times the proven crude oil reserves of Saudi Arabia.

25.8 What are the two major problems with using uranium as a nuclear fuel?

Some of the waste products of nuclear fission are long-lived (they last thousands of years), and these dangerous wastes must be stored in a safe place. The other problem is that nuclear accidents can occur.

25.9 What is the difference between high-level and low-level radioactive waste?

High-level nuclear waste is the "spent" fuel from nuclear power plants and the waste from the production of nuclear armaments. *Low-level nuclear waste* results from many medical and industrial processes and involves quite large volumes of low-radioactivity material. High-level radioactive waste is the dangerous one, for it emits highly penetrating radiation and generates much heat.

25.10 How is the United States planning to dispose of nuclear wastes?

By deep burial in casks within "mines" that are 300 m (1000 ft) deep, constructed in rock in a relatively unpopulated area. The designated area at this time is Yucca Mountain, Nevada, where the rock types are volcanic tuffs. (But, plans could change.)

25.11 How is Canada planning to dispose of nuclear wastes?

By storage in sealed rooms in a granitic batholith in Ontario at depths of 500 to 1000 m.

25.12 How does Sweden plan to dispose of its nuclear wastes?

In sealed copper containers stored in solid rock 500 m underground.

25.13 Where are nuclear wastes being stored in the meantime, before the sites of the previous three problems are completed?

Under water in cooled swimming pools.

25.14 What are the most viable short-range and long-range energy sources?

For the near future, oil and gas are essential fuels. For the longer range, the development of solar power and/or the refinement of nuclear power may be the best choices.

25.15 What are some answers to the problem of disposing of solid wastes (i.e., garbage)?

Recycling, conservation, and the development of biodegradable substitute materials.

25.16 Are not the oceans good places to dump our wastes? (Out of sight and out of mind.)

Evidently people used to think so, but environmental awareness has increased. And we now realize that even the vast oceans are badly polluted.

25.17 What would be the effect of a reduced ozone layer, reduced by CFCs put into the atmosphere by modern civilizations?

More of the sun's harmful ultraviolet rays would reach earth's surface, causing more skin cancer, eye damage (cataracts), damage to plants and to animals, global warming, and damaging the ecosystems of the oceans. Without the ozone layer, life may be impossible. Some think that life did not evolve until the ozone layer had developed.

25.18 How long do CFC molecules last, once in the atmosphere?

Depending upon which specific chemical in question, from 8.5 years to 88 years. Therefore, a buildup occurs, and we must stop the problem as soon as possible.

25.19 What are the main sources of CFCs?

Several chemicals are the culprits. Freon-113, CCl_4 (carbontetrachloride), and methyl chloroform are used as solvents. An estimated 40 percent of the CFCs come from refrigeration, air conditioners in automobiles, and aerosol cans.

25.20 What is the ultimate solution to the CFCs problem?

To eliminate totally the use of those chemicals by finding nonharmful substitutes. Chemists are at work on the problem.

25.21 Which countries put the most CFCs into the atmosphere?

The industrialized countries. The United States, Western Europe, and Japan are the main polluters. The United States produces about 40 percent of the world's ozone-depleting chemicals.

25.22 What is the "hole" in the ozone shield over Antarctica?

It is a hole discovered by the British in the mid-1980s. It changes in size with the seasons. On October 5, 1987, it was half as large as Antarctica, or 7 million km^2 (2.5 million mi^2).

25.23 Which countries release the most SO_2 into the atmosphere?

The industrialized nations. Estimates in 1989 of annual emissions in millions of tons are as follows: United States, 22.8; Canada, 4.1; United Kingdom, 4.0; West Germany, 2.9; France, 2.4; Japan, 1.2. (These figures are from the World Resources Institute.)

25.24 What is the "greenhouse effect"?

The sun's heat can penetrate the atmosphere, warming the earth's surface. But the CO_2 in the atmosphere prevents the escape of heat from the earth, as the CO_2 absorbs the heat. Thus, the gases in the atmosphere make the earth comparable to a greenhouse in which the glass keeps in the heat.

25.25 Which countries release the most CO_2 emissions?

The biggest emitters, with the percentage of total indicated, are as follows: United States, 23 percent; Japan, 5 percent; West Germany, 4 percent; United Kingdom, 3 percent; Canada, 2 percent; France, 2 percent; and Italy, 2 percent. (These figures are from the World Resource Institute.)

25.26 Why is the removal of forests a factor in pollution?

Trees (hardwoods) use CO_2 and give off oxygen during photosynthesis. Therefore, fewer forests makes the carbon dioxide buildup even worse. More forests would help to alleviate the "greenhouse effect."

25.27 Would a global reforestation program be useful?

Yes, for the above reason as well as for esthetic reasons. Trees would also have a cooling effect, useful if the earth's surface were to get continually warmer.

25.28 Are there other "greenhouse gases" in addition to CO_2?

Yes. CO_2 accounts for only about one-half of the problem. Others include methane (18 percent), chlorofluorocarbons (13 percent), upper troposphere ozone (8 percent), nitrous oxide (5 percent), and many minor gases.

25.29 How rapidly are rain forests being destroyed?

At the rate of 25 million acres per year, nearly half in the Amazon rain forest.

25.30 Is the agricultural land created by clearing rain forests good agricultural land?

No. Tropical soils are rather poor soils. (See Problems 4.25, 4.26, 4.28, and 4.31.)

25.31 Which rain forests, other than the Amazon, are being depleted?

Most are, although the Amazon is the one everyone hears about and recognizes. For example, the Lacandon Rain Forest in southern Mexico covered 3.2 million acres in 1875, 3.0 million acres in 1960, 1.85 million acres in 1982, and by 1990, there may be less than 1 million acres! That is a rapid destruction.

25.32 Besides the effect on the utilization of CO_2 that the depletion of the rain forest will have, what other side-effects will occur?

The depletion of, and in many cases, the extinction of plants and animals, many as yet undescribed. Rain forests may have 1000 species of plants per hectare (2.47 acres); forests in northern United States, for example, may have only 30 or 40 species per hectare. Many plants of the rain forest could be of medicinal value in the search for the cure of diseases.

25.33 What is the pH ("acidity") of pure rainfall?

About 5.6. This is "acid," too, but acid rain is defined as rain that is more acid than normal rainfall.

25.34 What is the pH of the most acid rain in the United States and Canada?

Less than 4.2 (see Fig. 25.1).

25.35 Over which portion of North America is the rain more acid than normal?

From Fig. 25.1, it can be seen that essentially the eastern half of North America is affected.

25.36 Is there a problem of overpopulation, and if so, why is it a problem? (See Fig. 25.2.)

Yes, there is. The earth is already supporting more people than it should be supporting. Any gains made on environmental problems are overwhelmed by an increase in population. We are losing ground. We must control population. Why put additional stresses on our fragile earth?

Chapter 26

"Geology" of Our Solar System

INTRODUCTION

In the title of this chapter, the word *geology* is in quotes. Why? Because geology means the study of the earth, from the Greek words *geo* (the earth) and *logy* (the science of, or a body of knowledge of), and therefore the word is not technically applicable to other bodies in our solar system. However, because many features on some of these bodies are comparable to geological features on earth, and because many of the same techniques used in geology are used to study the planetary bodies, we have used the word here.

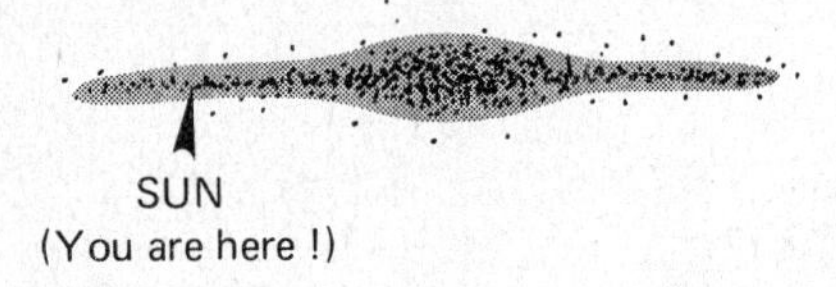

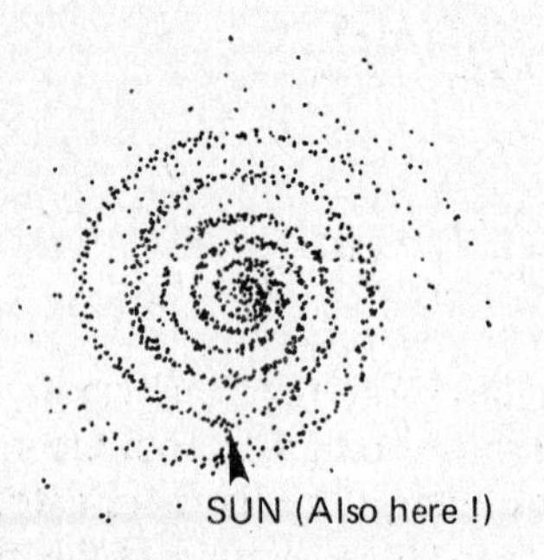

Fig. 26.1 An amateur artist's interpretation of the Milky Way galaxy.

Our solar system, a small part of the Milky Way galaxy (Fig. 26.1), consists of the sun, nine planets, more than 40 satellites or "moons" (at last count), thousands of asteroids, and many comets orbiting the sun (Fig. 26.2). Based on the radiometric ages of meteorites and the moon, the solar system is 4.6 billion years old. Other solar systems could exist, perhaps even billions of them, but the presence of small unlit bodies orbiting stars that are light-years away from earth is extremely difficult to verify. Nevertheless, there is some scant evidence that suggests the presence of planetary bodies around some of the nearer stars.

Several of the bodies in our solar system have been cratered by meteorites, probably mostly about 4.0 billion years ago. Most bear the obvious scars of these impacts, but such scars on earth are uncommon because weathering, erosion, volcanism, sedimentation, and other geological processes on earth have obliterated or covered them. Certainly earth was bombarded as heavily as was the moon.

Comets differ in many ways from the planets and asteroids. They have very elliptical orbits and are composed of ice with rock fragments. They are small, commonly less than 10 km in diameter, and have "tails" of evaporating gases when close to the sun. The tails are caused by the solar wind, and they always point away from the sun.

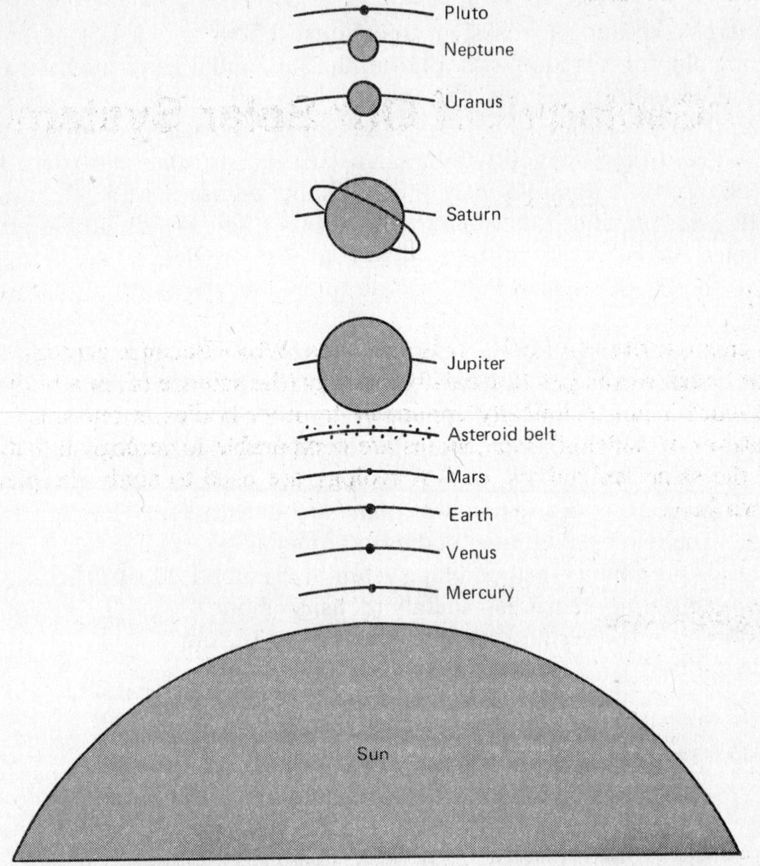

Fig. 26.2 The planets of the solar system, drawn to scale. A portion of the sun is also shown.

THE SUN

The sun is a mass of gases, mostly H and He, but it also contains traces of all the 92 naturally occurring elements found on earth. It is a huge nuclear reactor, although small compared to most stars. Nuclear fusion of hydrogen to helium (see Problem 2.17) produces its energy that is manifested as heat and light. It contains most of the mass of the solar system (more than 99 percent), and this huge mass results in the gravitational attraction that keeps the planets in orbit rather than flying off into space. Its heat and light drive nearly all the processes of earth (e.g., precipitation, winds, oceanic currents, erosion) and support all life on earth by providing the energy for photosynthesis (see Problem 19.61) and hence for the entire food chain. The sun is a class G 2 star in the sequence WOBAFGKMRNS, which classifies stars by intensity and spectral class. The sun is a "yellow dwarf" star.

THE PLANETS IN GENERAL

Planetary arrangement, size, and density are provided in Fig. 26.2 and Table 26.1. The four inner planets (including Earth) are dense, consisting of rock and metal; these are called the *terrestrial planets*. The next four planets are composed of gas and liquid, and are called the *Jovian planets;* Jove is another name for Jupiter. Pluto, at the fringes of the solar system, apparently is another dense planet. However, until 1999, Pluto's orbit is inside the orbit of Neptune.

Mercury, closest to the sun, is too hot to support life as we know it. Venus, Earth, and Mars might be compared to the porridge of the Three Bears—either too hot, too cold, or just right—for the type of life found on earth. Venus is too hot; it is close to the sun, and its CO_2 atmosphere holds in the sun's heat, thereby creating a massive greenhouse effect and a hot surface, roughly 470°C (900°F). Mars is too cold, for it is too far away from the sun, and it has a minimal atmosphere to hold heat on the planet. Fortunately, earth is "just right."

There are more than 40 satellites discovered thus far in the solar system, including the new ones Voyager II found when it flew by Neptune in August 1989. Earth has only one, Jupiter and Saturn have many (15 each), and it's possible there are still some undiscovered.

The planets all orbit the sun in nearly circular orbits (they are slightly elliptical) according to Johannes Kepler's laws of planetary motion. Most rotate counterclockwise if viewed from a spaceship situated above the Earth's North Pole, and all (including the asteroids) orbit in approximately the same plane.

Our knowledge of the composition of the bodies in the solar system is based in part on samples of meteorites and on moon samples collected on six different "field trips" to the moon by Apollo astronauts from 1969 through 1972. Also, an unmanned vehicle placed on Mars, the Viking in 1976, has provided earthlings with the composition of the Martian soil. Other knowledge comes from American and Soviet space probes that have photographed planetary surfaces by either direct photography and radar imagery, measurements of planetary density (see Problem 1.33), and comparisons with the Earth. The two best-situated bodies in our solar system are Earth and Earth's satellite, the moon. In general, the bodies in the solar system are composed of different amounts of these types of materials of differing densities—metals (density about 8 g/cm^3), silicate minerals (density about 2.7 to 3.0 g/cm^3), and water and ice (density about 1.0 g/cm^3).

Table 26.1 Vital Statistics of the Planets

Planet	Diameter (km)	Mass (Earth = 1)	Density (Water = 1)	Gravity (Earth = 1)	Number of Satellites	Time for One Rotation on Axis (Earth Hours or Days)	Time for One Revolution around Sun (Earth Years)	Distance from Sun		Composition of Atmosphere
								(10^6 km)	(10^6 mi)	
Terrestrial planets										
Mercury	4,835	0.055	5.69	0.38	0	59 days	0.241	57.7	36.8	None
Venus	12,194	0.815	5.16	0.89	0	243 days	0.616	107.0	66.9	CO_2
Earth	12,756	1.000	5.52	1.00	1	1.00 days	1.00	149.6	92.6	N_2, O_2
Mars	6,760	0.108	3.89	0.38	2	1.03 days	1.88	226.0	141.0	CO_2, N_2, Ar
Giant planets										
Jupiter	141,600	318.0	1.25	2.64	15	9.83 h	11.99	775	482	H_2, He
Saturn	120,800	95.1	0.62	1.17	18	10.23 h	29.5	1421	883	H_2, He
Uranus	47,100	14.5	1.60	1.03	5	23.00 h	84.0	2861	1777	H_2, He, CH_4
Neptune	44,600	17.0	2.21	1.50	8	22.00 h	165.0	4485	2787	H_2, He, CH_4
Pluto	14,000?	0.8?	4.2?	?	1	6.39 days	248.0	5886	3658	?

(*After Press, P., and Siever, R., Earth, Freeman, 1982.*)

PLANETARY BODIES: SOME SPECIFICS

The *moon* has two types of surface terranes—older (4.6 to 4.3 b.y.) light-colored highlands composed of impact breccias (from meteorite impacts) of broken gabbroic rocks (mostly anorthosite, a rock composed almost entirely of plagioclase), and younger (4.0 to 3.0 b.y.) dark-colored basaltic "seas" or maria. This is a misnomer, of course, since they do not contain water but are actually areas of low relief. Thus, moon rocks resemble earth rocks although there are some differences: They contain no water, contain more iron and titanium, and even contain rare native iron (which could not exist on earth as it would be oxidized in the presence of oxygen). Moon's density indicates it is composed largely of silicate minerals. The highlands are more highly cratered than the maria, with the major cratering about 4.0 b.y. ago. The largest crater is Copernicus. Rill-like features probably were formed by faulting, as they are straight. Other rill-like features are river-like and may be collapsed lava tunnels. The moon has no hydrosphere or atmosphere, and there is no evidence that it ever had them. The same side of the moon faces the earth at all times. A Russian spaceprobe photographed the far side of the moon, and later American astronauts also photographed the "dark far side." Actually, because of the combination of the moon's rotation and the revolution of the Earth around the sun, we see a maximum of 59 percent of the moon's surface and never see 41 percent.

Mercury is highly cratered; one impact crater is 1300 km in diameter. There are long curved cliffs, probably the result of crustal compression. It has a very thin atmosphere, a density similar to earth's, and a magnetic field.

Venus has a dense atmosphere of CO_2 that obscures the surface, but radar has shown craters on a hilly terrain, a large volcano, two large high plateaus, and canyons. The CO_2 atmosphere is 90 times as dense as earth's, and it creates a greenhouse effect so that the surface has a temperature of 470°C (above the melting point of lead!). Its density is similar to earth's, but it has no magnetic field.

Earth is the subject of Chapters 1 through 25, so nothing more need be said here.

Mars has a thin atmosphere and is very cold—as cold as −130°C. A red sandy soil contains hematite; the material has been blown into sand dune fields by strong winds that also create annual dust storms that engulf the red planet. Like the moon, Mars has highlands and maria. Volcanoes are present, and Olympus Mons is the largest known volcano in the solar system; it is 27 km (17 mi) high and 600 km (370 mi) in diameter. Ice caps (mostly frozen water with some carbon dioxide ice as well) are present at the poles, and CO_2 ice or frost annually forms over the northern and southern parts of the Martian surface, to about midlatitudes. Dendritic valleys, now cross-cut by later meteorite craters, suggest the past presence of water on the surface. Straight valleys are interpreted as grabens.

Asteroids, numbered in the thousands, are found in a broad belt between Mars and Jupiter. Two thousand have been named. The largest is Ceres (1020 km in diameter). It is not known for sure whether the asteroids are particles that never came together to form planets, or whether they are remnants of a planet that disintegrated; the former idea is most accepted. Bode's law predicted that a planet should be there.

Jupiter, the largest planet, has a density of only 1.3, showing that it consists of liquids and gases. It is most likely a protostar; that is, a potential star that was just not dense enough to begin nuclear fires burning. Jupiter exudes more heat than it seems it should, probably from nuclear processes. Spectrographic data indicate that most of the gas is hydrogen, with some ammonia and methane. It has planar rings of debris in orbit around the planet at its equator. It is also noted for its Great Red Spot, a gigantic swirling, hurricane-like feature in its atmosphere. The four largest of Jupiter's moons—the Galilean satellites discovered by Galileo with his primitive telescope in 1610—are somewhat like the terrestrial planets. These four moons are visible even with binoculars. *Europa* is covered by thick ice that is broken by multiple series of curved cracks of various wave lengths. *Io* is the most active body in the solar system; at least 10 active sulfur-spewing volcanoes were photographed by the Voyager spacecraft in 1979. This activity is probably caused by tidal forces between the satellite and Jupiter. *Ganymede* and *Callisto* have low densities, indicative of compositions of silicate minerals and ice; they are as large as Mercury. Ganymede has strange furrows that resemble spread centers on Earth, and therefore may have had plate tectonic motions in its history; it is the largest satellite in the solar system.

Saturn, with the lowest density of all planets (0.7—see Table 26.1), is a sphere of gases and liquids. Saturn's famous rings are visible even with binoculars. It has a strong magnetic field, stronger than Earth's. Its many moons, the most famous of which is Titan, the second largest satellite in the solar system, have a variety of compositions and sizes.

Uranus is a low-density planet probably consisting of gases, liquids, and some silicates. It has numerous rings and five moons, one of which (Miranda) shows a structurally complex surface, with canyons and three large areas of ridges and bands that resemble big slabs of crust. The atmosphere of Uranus contains helium and methane, among other gases.

Neptune is a low-density planet like Uranus. A lot was learned about it when Voyager 2 flew by in August 1989. In addition to the two known satellites, Triton and Nereid, Voyager discovered six new ones plus a ring system. Triton was found to contain ice volcanoes.

Pluto, the farthest planet from the sun and the smallest, is quite unknown but appears to be a terrestrial-type planet. Because of a slightly eccentric orbit, it sometimes passes the honor of being the farthest planet from the sun to its neighbor Neptune, which will be the farthest until 1999. Pluto has one known moon, Charon.

ORIGIN OF THE SOLAR SYSTEM

A general scheme for the origin of the solar system is as follows. It probably formed from a swirling disk, or nebula, of gas and dust. It contracted and became hot and dense. At its center, the sun formed as nuclear fusion began. Outward from the center, gravitational attraction between dust particles formed larger bodies which continued to collide and grow, finally resulting in the planets. The planets became hot, largely due to gravitational compression and the decay of radioactive isotopes. This caused melting, and on Earth, differentiation into a core, mantle, and crust. All this probably happened rather quickly 4.6 b.y. ago. The continued escape of heat through the lithosphere (recall that the lithosphere consists of the crust and the upper mantle) is what drives the volcanic and tectonic processes within the earth's crust. Plate movements and accompanying volcanism are a planet's way of getting rid of heat. The moon is volcanically and tectonically inactive because it has built up a thick lithosphere through which heat has not escaped in quantity since 3.2 b.y. ago.

Solved Problems

26.1 What is a solar system?

A *solar system* is a group of planetary bodies in orbit around a star; they are held there by the gravitational attraction of the star.

26.2 What bodies make up our solar system, and how old is it?

The sun (a middle-sized star), nine planets, more than 40 satellites (moons) at last count (but there may be more discovered in the future), and thousands of smaller asteroids and comets. It is approximately 4.6 b.y. old.

26.3 Where in the universe is our solar system located?

Near one edge of the Milky Way galaxy (see Fig. 26.1), in the Orion arm of the galaxy.

26.4 What is the composition of the sun?

More than 90 percent hydrogen, nearly 10 percent helium, and small amounts of all the other 90 naturally occurring elements found on Earth.

26.5 Why is the sun hot?

It has been generally accepted that it is a "nuclear furnace," but the lack of neutrinos casts some doubt on this idea. If indeed it is a nuclear furnace, then hygrogen is being fused or transformed to helium, with energy given off as a by-product. (See Problems 2.13, 2.17, and 2.19.)

26.6 Is the nuclear reaction of the previous problem atomic fission or atomic fusion?

It is fusion, the combining of light nuclei to make heavy nuclei.

26.7 What would happen if the sun's energy output were severely diminished, or if the sun's energy could not reach the Earth due to dust in Earth's atmosphere?

Darkness, the shutdown of photosynthesis and hence plant growth, and starvation for all animals on earth. The extinction of dinosaurs, as mentioned in Problems 19.50, 19.51, and 19.52, may have been related to a dusty atmosphere due to the impact of a large comet or meteorite on Earth.

26.8 Which planets have the most satellites (moons)?

Jupiter with 15 and Saturn with 15 contain most of the more than 40 known moons in our solar system.

26.9 Are all the nine planets generally similar?

No. There are two main types—see Table 26.1. The inner four, the terrestrial planets, have high densities because they consist largely of silicate minerals and metals. The next four, the Jovian planets with low densities, consist largely of gas and liquid. Pluto, the farthest planet, is a kind of terrestrial oddball. (It has been suggested that there may be a tenth planet.)

26.10 What is an AU, or astronomical unit?

One AU is the average distance between the Earth and the sun; this is equal to about 92.9 million mi, or 149.6 million km.

26.11 What is the asteroid belt?

The *asteroid belt* is a broad zone of small planetary bodies (the largest is about 1000 km in diameter) in orbit around the sun between Mars and Jupiter.

26.12 Which is the largest asteroid, and how large is it compared to our (Earth's) moon?

Ceres is the largest, 1020 km in diameter. For contrast, the moon has a diameter of 3476 km.

26.13 Why is the asteroid belt of interest to earthlings?

It is the source of most of the meteorites and therefore provides scientists with samples of the solar system.

26.14 What are comets?

Comets are small and icy (plus included rock fragments) objects as large as 10 km in diameter. When their elliptical orbits take them close to the sun, some ice evaporates, forming "tails" of ionized ions.

26.15 How many solar systems are known?

Only ours is known for sure, but scientists suspect that there are countless others in our galaxy and still countless more in the universe.

26.16 What is the ecliptic?

The *ecliptic* is the plane of the earth's orbit. The other planets and the asteroid belt are approximately in the same plane. Pluto, for example, is a bit off.

26.17 How long does it take the Earth to revolve around (orbit) the sun?

1 year.

26.18 How long does it take the earth to make one rotation on its axis?

24 h.

26.19 Why do the Arctic and the Antarctic experience periods of 24 h of daylight, and periods of 24 h of darkness, at different times of the year?

Because of the tilt of the earth's axis, approximately 23° to the ecliptic, there is a period of several weeks when those areas (one at a time) are so situated that the sun's rays hit or do not hit the Arctic and Antarctic for the entire 24-h period.

26.20 Which body in our solar system, other than Earth, may have experienced plate movement?

Ganymede, one of Jupiter's moons and the largest in the solar system, has strange parallel furrows that may be spread centers comparable to earth's spread centers in the ocean basins.

26.21 Where is the mass of our solar system concentrated?

In the sun (star) where more than 99 percent of the mass is found.

26.22 Which bodies in our solar system, besides the moon, are cratered?

Many, including Mercury, Mars, and several moons such as Uranus's Miranda and Oberon, and Jupiter's Ganymede.

26.23 What kinds of rocks did the astronauts find on the moon?

(1) Basalt and (2) impact breccias consisting of fragments of gabbro and anorthosite (largely plagioclase) in a finer-grained (part glassy) groundmass of what was an impact-generated melt.

26.24 Which is the most famous comet, and why?

Halley's comet returns to view every 76 years in its elliptical orbit around the sun. Its last two visits were 1910 and 1986. It has been known for much longer than that, being first recorded in 240 B.C. It was named for Sir Edmund Halley, who accurately predicted that it would return in 1758.

26.25 Mercury, like Earth, has a strong magnetic field. Based on our knowledge of Earth, what common feature do Earth and Mercury have that could be related to this field?

A molten core made of electrically conductive material (i.e., molten metal) is likely for Mercury as well as Earth. (See Problems 15.21 and 15.29.)

26.26 The moon has no magnetic field. Why not?

The simplest explanation is that it does not have a dense, molten core.

26.27 What is a meteorite? A meteor?

A *meteorite* is a fragment of metal or rock that has collided with a planet or with a moon. A *meteor* is a burning trace of a fragment of asteroid or comet plunging through the atmosphere.

26.28 Why is Earth not as cratered as its satellite, the moon?

It probably had the same density of craters on its surface, but weathering, erosion, sedimentation, mountain building, and volcanism have obliterated most of them.

26.29 How many meteorite craters have been found on Earth?

About 100, the most famous of which is Meteorite Crater, Arizona.

26.30 Earth, Mars, and Venus have oxygen, nitrogen, argon, carbon dioxide, and water vapor in their atmospheres. How does Earth's atmosphere differ from the other two?

Earth has more oxygen—21 percent. Do you recall how it probably formed? See Problem 20.4.

26.31 Of what are the atmospheres of Jupiter and Saturn composed?

Mostly hydrogen, with about 10 percent helium.

26.32 Which planets have rings?

Saturn, Jupiter, and Uranus.

26.33 What are the four main types of meteorites?

Chondritic meteorites: Stony, with small, rounded grains—chondrules—of olivine and pyroxene.

Achondritic meteorites: Stony, without chondrules. They are similar to gabbro and peridotite in composition.

Iron meteorites: Mainly iron and nickel alloy.

Stony-iron meteorites: As the name implies, these contain both iron-nickel and stony (silicate) material.

26.34 Why are carbonaceous chondrites of special interest?

Because they contain some carbon, water, and "organic" molecules.

26.35 How old are most meteorites?

About 4.6 b.y.

26.36 Where do most meteorites come from?

The asteroid belt between Mars and Jupiter. Of the several thousand known, six appear to have come from the moon and eight achondritic meteorites may have come from Mars.

26.37 Why are meteorites of special interest to scientists?

Because if they are 4.6 b.y. old, that is the age of the solar system. They are thought to represent the types of material of which the Earth is made—a metallic core of iron-nickel and a stony mantle of ultramafic and mafic rocks.

26.38 The densities of the different planets vary considerably, as shown in Table 26.1. Why?

The density of a planet depends upon its makeup—is it composed of ice and water (about 1 g/cm^3), silicate minerals (about 2.7 to 3.0 g/cm^3), or metallic (iron-nickel) materials (about 8 g/cm^3)?

26.39 How large is the sun?

It has a diameter of 1,400,000 km, or 866,000 mi.

26.40 Which of the planets has been nicknamed the "water planet"?

Earth. It is the only planet with an abundance of water.

26.41 Which planet, other than Earth, has volcanoes, a "Grand Canyon" larger than the Grand Canyon of Arizona, polar ice caps, and fields of sand dunes?

Mars.

26.42 Where in the solar system is the largest known volcano?

On Mars. It is Olympus Mons. It is 27 km (17 mi) high and is 600 km (370 mi) in diameter at the base. By contrast, the largest volcano on Earth is Mauna Loa on Hawaii; it rises 10 km (6 mi) above the sea floor and has a diameter at its base of 100 km (60 mi).

26.43 The moon's surface can be divided into two major types of terrane. What are they?

The lunar highlands (cratered mountainous uplands) and the maria, or "seas," which are broad dark areas of basalt flows.

Supplementary Problems

26.44 From Table 26.1, what can be concluded about the composition of each of the planets?

The first four—Mercury, Venus, Earth, and Mars—and the last one (Pluto) have high densities and must consist of silicates (rock) and metals. Jupiter and Saturn are the lightest and must include much ice or liquid, plus silicates. Uranus and Neptune should contain silicates and ice/water.

26.45 The moon has a density of about 3.0. What does this suggest?

A composition made up almost totally of silicates (rock).

26.46 Planets orbiting stars other than our own cannot be seen, even with the strongest of telescopes. What, then, is the evidence for the postulated existence of other solar systems?

One possible interpretation of slight variations, or "wobbles," in a star's orbit is that they are caused by the gravitational pull of that star's planets. This has not been substantiated, as there are other possible causes such as the gravitational pull of "brown dwarfs."

26.47 Which planet is farthest from the sun?

Pluto *usually* is, but at certain times Neptune is farthest because its orbit is slightly more elliptical. For example, from 1988 to 1999, Neptune will be the farthest planet.

26.48 When, and how many times, did American astronauts walk on the moon?

Six times, from 1969 to 1972.

26.49 What do straight valleys on the planets or moons (including Earth) commonly indicate?

The presence of faults. Most of the large ones are probably down-faulted blocks (i.e., grabens).

26.50 If you were to fly around the Earth along the equator, how fast would you have to fly to keep the sun about directly overhead?

About 1000 mi/h, for the earth has an equatorial circumference of 24,902 mi (24 h × 1000 mi/h).

26.51 Refer to the previous problem. Should you fly west to east or east to west to stay with the sun?

East to west, the same direction that the sun "moves around the Earth."

26.52 In the answer to the previous problem, why is "moves around the Earth" put in quotes?

Because the sun is "stationary" (although it is circling the galaxy center every 250 m.y.) and it is the Earth that is doing the moving relative to the sun. All the planets orbit the sun, much like our moon orbits the Earth.

26.53 What is the major difference in the way the Earth orbits the sun and the way the moon orbits the Earth?

The Earth is rotating as it revolves around the sun, so that most parts of Earth receive sunlight sometime during a 24-h day. The moon's rotation, however, is synchronized with the Earth's revolution in such a way that only 59 percent of the moon's surface is ever seen from the Earth; only the Apollo astronauts have personally seen the far or dark side of the moon.

26.54 What is a galaxy?

A *galaxy* is a cluster of billions of stars. Our galaxy is called the Milky Way, and it is typical of a type called spiral galaxies (because of their shape). The solar system is located in one of the arms of the galaxy (Fig. 26.1). There are billions of galaxies! Some other types are elliptical and irregular. Our two companion galaxies are the Large and Small Magellanic Clouds, which are visible in the Southern Hemisphere. Another famous galaxy is Andromeda, a beautiful spiral galaxy referred to as M31.

26.55 What is another name for our star, the sun?

Sol, from whence comes the word *solar*.

26.56 Our solar system has been compared to an atom. Why?

Because our solar system is mostly space, it resembles an atom, which is likewise mostly space. However, whereas the electrons of an atom are in 3-D orbits (shells) around the nucleus, the planets are all nearly in a single plane.

26.57 Why are the planets and most moons spherical?

High internal pressures due to the gravity of their large masses cause the total mass to reach stable spherical shapes as they rotate.

26.58 What is Bode's law?

Bode's law (also known as the Titus-Bode law) is a "law" that shows that the planets are located at a regularly increasing distance from the sun. (Each planet's orbit is about 75 percent larger than that of the next inward planet, although the law breaks down somewhat for the placing of Uranus, Neptune, and Pluto.)

26.59 Why was Bode's law useful?

It was used to predict the existence of a planet between Mars and Jupiter. A search of that portion of the sky revealed Ceres, the largest of the asteroids. Further searching revealed the entire asteroid belt.

26.60 Why do large planets give off heat?

Radioactive isotopes produce heat when they decay; large planets have more mass and hence more of these isotopes. Some heat is relict from the planets' origins via gravitational attraction of numerous bodies. The larger planets and moons stay hot longer because they give off heat in proportion to the ratio of their surface areas to their masses. (Large bodies cool more slowly.)

26.61 Why is Io hot?

Io is hot because of expansion and contraction caused by the tidal pull of Jupiter. Its heat is dissipated by volcanism.

26.62 What material is being extruded by Io's volcanoes?

Much of it apparently is liquid sulfur.

26.63 Callisto and Ganymede, two moons of Jupiter, have been extensively cratered by meteorites. On the other hand, Io is not cratered. Why not?

Probably it is related to the age of the cratering. Most craters in our solar system apparently were formed about 4 b.y. ago, whereas Io is so active volcanically that these craters would have been covered by the volcanic deposits.

26.64 Europa is covered by ice. How thick is it?

It has been estimated that it is 100 m thick. It apparently is frozen water rather than some other type of ice such as frozen CO_2 (dry ice).

26.65 In detail, how old are the moon rocks?

The impact breccias are 3.9 to 4.05 b.y. old, the products of meteoric impacts of that age, with the meteorites impacting onto older rocks and hence containing rock fragments 4.3 to 4.6 b.y. old in an impact-generated melt 3.9 to 4.05 b.y. ago. The younger basalts of the maria are about 3.0 to 4.0 b.y. old.

26.66 What do the ages of the previous problem indicate about the moon's history?

The moon formed 4.6 b.y. ago and had a crust of gabbroic and anorthositic rocks. Then it was intensively bombarded by meteorites about 4.0 b.y. ago. The basalts of the maria are younger than that major bombardment. The moon has been a quiet body for about 3.2 b.y.

26.67 What is the origin of the maria, or "seas," that Galileo spotted with his telescope?

One hypothesis is that very large impact craters were formed by large meteorites and that the heat of impact either melted the rocks and caused volcanic activity or caused fractures that reached molten rock at depth. In either situation, basaltic lavas poured out and covered the impact craters.

26.68 How did the moon originate?

There are three traditional hypotheses—capture of an already existent moon by Earth, splitting off from a rapidly spinning molten proto-Earth, and an early origin as part of a moon-Earth binary planet system as the solar system formed. Recently a new idea has been presented—a large-impact hypothesis in which Earth was hit by a large meteorite that threw material from Earth's mantle (after Earth had already been differentiated into a core and mantle) into orbit around Earth. This hypothesis explains the lack of a large metallic core in the moon (it may have a small core) and the similarities of the mantles of Earth and the moon.

26.69 The Earth, like the moon, has been hit by many meteorites, especially in the distant past. Could Earth be hit again?

Yes. We are bombarded continually but most meteors are small and burn up in the Earth's atmosphere. Larger ones do hit the Earth. In March 1989, an asteroid nearly 0.5 mi in diameter missed Earth by only 500,000 mi, about twice the distance from the Earth to the moon. If it had hit land, it would have created an explosion equivalent to about 20,000 hydrogen bombs, and the crater would have been 5 to 10 mi in diameter. If it had hit the ocean, great tidal waves would have resulted, wreaking havoc in coastal regions.

26.70 Why are asteroids of special interest to paleontologists?

Because of the extinctions they may have caused in the geologic past (see Problem 19.51).

26.71 What did Voyager II, an unmanned American space probe launched in 1977, accomplish?

Travelling through space at 40,000 mi/h, it has photographed and recorded scientific data (chemical composition, temperature, atmospheric pressure, and magnetic field strength) as follows: Jupiter in 1979, Saturn in 1981, Uranus in 1986, and Neptune in 1989. Its total journey, as it looped around the sun and then headed toward the outer edge of the solar system, was 4.5 billion miles. Its path will then take it out of the solar system. This mission was made possible by a rare planetary lineup that won't recur until the year 2152.

26.72 What did Voyager II in 1989 find out about Neptune that was previously unknown?

It discovered four new moons; two partial rings or arcs of debris orbiting the planet; clouds; additional data about Neptune's moon Triton, which is the size of Earth's moon and which is the only moon that "travels backward," orbiting Neptune in the opposite direction to Neptune's rotation. At Neptune, about 2.8 billion mi from the sun, the sunlight is 1000 times dimmer than on Earth, so time-exposure photography is necessary. Voyager II came within 4800 km (3000 mi) of Neptune's surface.

Index

Index